Cuisine Foundations: Classic Recipes

THE CHEFS OF LE CORDON BLEU

Australia • Brazil • Japan • Korea • Mexico • Singapore • Spain • United Kingdom • United States

Le Cordon Bleu Cuisine Foundations: Classic Recipes
The Chefs of Le Cordon Bleu

Vice President, Career and Professional Editorial: Dave Garza

Director of Learning Solutions: Sandy Clark

Senior Acquisitions Editor: Jim Gish

Managing Editor: Larry Main

Product Manager: Nicole Calisi

Editorial Assistant: Sarah Timm

Vice President Marketing, Career and Professional: Jennifer Baker

Marketing Director: Wendy Mapstone

Senior Marketing Manager: Kristin McNary

Associate Marketing Manager: Jonathan Sheehan

Production Director: Wendy Troeger

Senior Content Project Manager: Glenn Castle

Senior Art Director: Casey Kirchmayer

Technology Project Manager: Chris Catalina

Production Technology Analyst: Tom Stover

Principle Photography by J. Enrique Chavarria

Supplemental Photography by Lois Siegel, Lois Siegel Productions

© 2011 Le Cordon Bleu International

ALL RIGHTS RESERVED. No part of this work covered by the copyright herein may be reproduced, transmitted, stored, or used in any form or by any means graphic, electronic, or mechanical, including but not limited to photocopying, recording, scanning, digitizing, taping, Web distribution, information networks, or information storage and retrieval systems, except as permitted under Section 107 or 108 of the 1976 United States Copyright Act, without the prior written permission of the publisher.

For product information and technology assistance, contact us at
Cengage Learning Customer & Sales Support, 1-800-354-9706

For permission to use material from this text or product, submit all requests online at **www.cengage.com/permissions.**
Further permissions questions can be e-mailed to
permissionrequest@cengage.com

Library of Congress Control Number: 2010922489

Spiralbound

ISBN-13: 978-1-4354-8138-1

ISBN-10: 1-4354-8138-0

Softbound

ISBN-13: 978-1-1113-0687-8

ISBN-10: 1-1113-0687-7

Delmar
5 Maxwell Drive
Clifton Park, NY 12065-2919
USA

Cengage Learning is a leading provider of customized learning solutions with office locations around the globe, including Singapore, the United Kingdom, Australia, Mexico, Brazil, and Japan. Locate your local office at: **international.cengage.com/region**

Cengage Learning products are represented in Canada by Nelson Education, Ltd.

To learn more about Delmar, visit **www.cengage.com/delmar**

Purchase any of our products at your local college store or at our preferred online store **www.ichapters.com**

Notice to the Reader
Publisher does not warrant or guarantee any of the products described herein or perform any independent analysis in connection with any of the product information contained herein. Publisher does not assume, and expressly disclaims, any obligation to obtain and include information other than that provided to it by the manufacturer. The reader is expressly warned to consider and adopt all safety precautions that might be indicated by the activities described herein and to avoid all potential hazards. By following the instructions contained herein, the reader willingly assumes all risks in connection with such instructions. The publisher makes no representations or warranties of any kind, including but not limited to, the warranties of fitness for particular purpose or merchantability, nor are any such representations implied with respect to the material set forth herein, and the publisher takes no responsibility with respect to such material. The publisher shall not be liable for any special, consequential, or exemplary damages resulting, in whole or part, from the readers' use of, or reliance upon, this material.

Printed in the United States of America
1 2 3 4 5 6 7 14 13 12 11 10

CONTENTS

FOREWORD

I am proud to present *Le Cordon Bleu's Cuisine Foundations*—a project that was two years in the making. We hope that this book will provide a useful reference as you explore the world of cooking and that it will also serve you well as you embark on your own journey, both personally and professionally. At first glance you might think that this is just "another culinary textbook," but on closer examination you will realize that the focus is on technique. To demonstrate those techniques, we have provided visual step-by-step photographs for most of them. We took our cue from the many students and graduates around the world who were looking for a single reference that would explain and show the techniques that have existed and been respected for more than three centuries. With human ingenuity came progress in the kitchen, but the techniques have remained practically unchanged. The tools have changed from wood-burning stoves to induction ovens to the "anti-griddle," and though they certainly influenced the evolution of cooking, they have not replaced the tried-and-true techniques.

What we wanted to do was reset the counter and refresh everyone's history and knowledge of these techniques before they are lost to us. For this reason, we chose to use the recipes that were created throughout the history of French cuisine that best exemplify the application of these techniques, and if you look at the integrity of each recipe, you will recognize the origins of these recipes on today's menus—all around the globe.

We also wanted to pay homage to the generations of chefs who have upheld and passed on their passion for cooking to each succeeding generation, from Taillevent, who as an apprentice probably stood before hot flames, hand-turning a spit, to Ferran Adrià who has used modern technology to redefine gastronomy. These chefs represent the patrimony of *L'art culinaire*—the art of cooking.

Le Cordon Bleu has served its patrimony for more than a century through its chefs, who have chosen a very important calling—teaching. From the moment Le Cordon Bleu opened its kitchens in 1895 on the rue St.-Honoré in Paris, students of all nationalities and all walks of life have come to join us in continuing to respect what French culinary technique represents. It is not about the recipes, but about how you work in a kitchen, whether you are cooking for loved ones or paying customers.

Even with a worldwide network of schools and programs around the world, we can extend our classroom through books and other mediums. I hope you enjoy *Le Cordon Bleu's Cuisine Foundations*, not only as a guide and reference, but as an inspiration.

Amities gourmandes,
André J. Cointreau
President, Le Cordon Bleu International

ACKNOWLEDGMENTS

Le Cordon Bleu would like to thank the chefs and staff of the Le Cordon Bleu schools:

Le Cordon Bleu Paris, Le Cordon Bleu London, Le Cordon Bleu Ottawa, Le Cordon Bleu Madrid, Le Cordon Bleu Amsterdam, Le Cordon Bleu Japan, Le Cordon Bleu Inc., Le Cordon Bleu Australia, Le Cordon Bleu Peru, Le Cordon Bleu Korea, Le Cordon Bleu Lebanon, Le Cordon Bleu Mexico, Le Cordon Bleu Thailand

Le Cordon Bleu College of Culinary Arts in Atlanta, Le Cordon Bleu College of Culinary Arts in Austin, Le Cordon Bleu College of Culinary Arts Inc., A Private, Two-Year College in Boston, Le Cordon Bleu College of Culinary Arts in Chicago, Le Cordon Bleu College of Culinary Arts in Dallas, Le Cordon Bleu College of Culinary Arts in Las Vegas, Le Cordon Bleu College of Culinary Arts in Los Angeles, Le Cordon Bleu College of Culinary Arts in Miami, Le Cordon Bleu College of Culinary Arts in Minneapolis, Le Cordon Bleu College of Culinary Arts in Orlando, Le Cordon Bleu Institute of Culinary Arts in Pittsburgh, Le Cordon Bleu College of Culinary Arts in Portland, Le Cordon Bleu College of Culinary Arts in Sacramento, Le Cordon Bleu College of Culinary Arts in Saint Louis, California Culinary Academy, Le Cordon Bleu College of Culinary Arts in Scottsdale, Le Cordon Bleu College of Culinary Arts in Seattle.

Special acknowledgement to Chef Patrick Martin, Chef Arnaud Guerpillon, Chef Christian Leroy, Chef Jean-Marc Baqué, Chef Christian Faure, M.O.F., Chef Nicolas Rollet, Chef Hervé Chabert, Katharyn Shaw, Carrie Carter, Charles Gregory, Chloe Chandless. Student assistants: Jing Hao Wong, Adam Goodman, Sylvie Dupuis.

Les Potages
(Soups)

Potage (Soups)

Velouté agnès sorel —Smooth chicken and mushroom velouté

Bisque de homard —Lobster bisque

Consommé brunoise—Clear broth with finely diced vegetables

Crème d'argenteuil —Cream of white asparagus

Gazpacho —Gazpacho

Potage julienne d'arblay —Potato and leek soup garnished with julienned vegetables

Potage mulligatawny —Curried chicken soup with apple

Potage clamart —Green pea soup

Potage cultivateur —Garden vegetable soup

Potage parmentier —Potato and leek soup

Soupe à l'oignon gratinée —French onion soup

Soupe de poisson façon bouillabaisse—Bouillabaisse style fish soup

Crème dubarry —Cream of cauliflower

Velouté de volaille —Chicken velouté

VELOUTÉ AGNÈS SOREL

Learning Outcomes

Velouté
Suer
Fouler
Tamponner
Roux blanc
Julienne
Liaison

Equipment

Knives:
Paring knife (*office*), slicing knife (*éminceur*)

Tools:
Bowls, fork, wooden spatula, fine chinois, cutting board, ladle (*louche*), whisk

Pans:
Sauteuse, small russe, marmite, bain marie

Serving

8 persons

VELOUTÉ AGNÈS SOREL

Smooth Chicken and Mushroom Velouté

Quantity		Ingredient
U.S.	*Metric*	*Stock*
2 qt	2 L	Chicken stock
		Liaison
2 oz	60 g	Butter
2 oz	60 g	Flour
		Finishing
5 oz	150 g	Crème fraîche or heavy cream
2 pcs	2 pcs	Egg yolks
3 ½ oz	100 g	Butter, cut into small dice

Method

Stock

1. Heat the chicken stock and bring to a low simmer.

Garniture

1. Clean the mushrooms for the garnish: To obtain clean white caps, first peel and stem the mushrooms, then remove the dark gills. Cut the mushroom caps into ***julienne*** and set aside. Add the skins and stems to the simmering stock.
2. Melt some butter in a small pan over low heat and sweat (***suer***) the ***julienne*** of mushroom until soft. Set aside.
3. Cut the cold cooked chicken breasts and cooked ox tongue into a ***julienne***. Reserve.

Liaison

1. Melt the butter in a medium pan over medium heat. When the butter begins to foam, add the flour and stir well until the mixture begins to bubble. Cook for 1 to 2 minutes to obtain a white roux (*be careful not to let it color*). Transfer the roux to a small plate and set aside to cool.
2. Once the stock is infused, strain through a fine mesh sieve (***chinois***) into a clean pan. Taste the stock and reduce it to concentrate the flavor, if needed.
3. Stirring with a whisk, add the cold roux. Whisking constantly, bring the velouté to a boil over medium heat; then reduce the heat to low and let it cook gently for 15 to 20 minutes.

Finishing

1. Mix the yolks with the crème fraîche or heavy cream. Ladle some velouté into the yolks to temper, off the heat, then mix the tempered yolks back into the pan. Return the pan to the heat and stir with a wooden spoon until thick enough to coat the back of a spoon (***à la nappe***). Strain the velouté through a fine mesh sieve (***chinois***) into a clean pan. Finish by mixing in the butter, then pat (***tamponner***) the surface of the velouté with a knob of butter on the end of a fork to stop the formation of a skin. Set aside and keep warm in a bain marie.

FYI...

Egg yolks are often used as a binding and thickening agent (liaison). In the case of veloutés, they are a secondary liaison. In other recipes (for example, crème Anglaise), egg yolks are the principal liaison.

Quantity		Ingredient
		Garniture
5 oz	150 g	Mushrooms, cut into a ***julienne***
1 pc	1 pc	Cooked chicken breast, cut into a ***julienne***
2 oz	60 g	Cured ox tongue (***langue écarlate***), cut into a ***julienne***
5 brs	5 brs	Chervil

Method

To Serve

1. Divide the tongue, chicken breast, and mushroom ***juliennes*** among heated soup plates, and ladle the hot velouté on top. Decorate with sprigs (***pluches***) of chervil.

Optional: *To make a "Crème" Agnès Sorel: Make a light béchamel sauce instead of the chicken velouté base, and do not add eggs and butter, just more cream.*

À la Agnès Sorel is a French culinary term meaning that the garnish may include mushrooms, truffles, pickled tongue, and, in some cases, a small quantity of white rice. It can also refer to a sauce that is made from cooked meat, Madeira, and demi-glace.

Agnès Sorel was rumored to be the first mistress of King Charles VII of France. She loved food and would often work alongside the chefs to develop dishes, some of which were named after her. Agnès was accomplished in the kitchen and developed some recipes on her own, such as Agnès Sorel soup garnish and Agnès Sorel timbales.

Détailler a lobster
Saisir
Cardinaliser
Mirepoix
Piler
Suer
Pincer
Déglacer
Mouiller
Sucs de cuisson
Flambé

Equipment

Knives:
Cleaver (***couteau à batte***), boning knife (***désosseur***), vegetable peeler (***économe***), paring knife (***office***), slicing knife (***éminceur***), scissors (***ciseaux***)

Tools:
Bowls, fork, wooden spatula, plastic spatula, regular chinois, fine chinois, cutting board, skimmer (***écumoire***), ladle (***louche***)

Pans:
Rondeau, large sauce pan (***russe***), bain marie

Serving

6 persons

Quantity		Ingredient
U.S.	*Metric*	*Base*
3 pcs	3 pcs	Live lobsters (*500 g each*)
		Bisque
2 fl oz	60 mL	Oil
		Mirepoix
1 pc	1 pc	Garlic head, cut in half
2 pcs	2 pcs	Shallots
7 oz	200 g	Carrot
7 oz	200 g	Leek
7 oz	220 g	Onion
5 oz	150 g	Celery
¼ oz	20 g	Tomato paste
3 pcs	3 pcs	Tomatoes, roughly chopped
4 pcs	4 pcs	Garlic cloves
1 pc	1 pc	***Bouquet garni***
1 bunch	1 bunch	Tarragon
1 bunch	1 bunch	Chevril
3 ½ oz	100 g	Short grain rice
1 ¼ oz	50 mL	Cognac
5 fl oz	150 mL	White wine
2 qt	2 L	Water or fish stock
5 fl oz	150 mL	Cream
3 ⅓ oz	100 g	Butter
Pinch	Pinch	Cayenne pepper
		Salt and pepper

BISQUE DE HOMARD

Lobster Bisque

Method

Base

1. Break down the lobsters by separating the tail, claws, and knuckles from the heads. Crack the claws with the cleaver (***demi-batte***). Pierce the center of the lobster tails with a wooden or metal skewer to keep them straight.
2. Heat the oil in a large pan (***rondeau***) over high heat. Add the lobster pieces, color them until their shells turn bright red (***cardinaliser***), and continue to cook for around 5 minutes. Add the ***mirepoix*** and garlic and continue cooking another 5 minutes while stirring. Add the tomato paste and stir 1 to 2 minutes to remove the acidity (***pincer la tomate***). Add the fresh tomato, ***bouquet garni*** and fresh stems of tarragon and chervil. Then add the rice and stir until it becomes translucent. ***Flamber*** with the cognac. Stir with a spatula to dissolve the cooking residues (***sucs***) from the bottom of the pan. Wet (***mouiller***) the contents of the pan with white wine and reduce by half.
3. Remove the lobster tails from the pan and set aside in a medium sauté pan. Add the fish stock to the ***rondeau*** and bring to a boil, and allow the claws to cook for 5 more minutes before removing and reserving with the tails. Continue cooking the liquid for around 25 to 30 minutes until the rice is completely cooked. Season lightly with salt and pepper. During this time, shell the tails and claws, reserving the shells in a separate bowl. Remove the lobster heads and place with the shells and crush (***piler***) together. Add the crushed shells back to the rondeau and stir well before adding the cream. Simmer for 10 to 15 minutes. Allow the bisque to reduce (***reduire***) gently over low heat until it is thick enough to coat the back of a spoon (***à la nappe***). Whisk in three-quarters of the butter (***monter au beurre***).
4. Season to taste and add just enough cayenne pepper to give a hint of spice to the bisque.
5. Strain (***passer***) through a fine mesh sieve. Add the fresh leaves of tarragon and chervil and stir into the bisque. Reserve the bisque in a bain marie. Pat (***tamponner***) the surface with a knob of butter on the end of a fork to stop the formation of a skin.

Garniture

1. Preheat the oven to 350°F (185°C).
2. On a baking sheet lined with a silicone mat or buttered parchment, arrange 12 teaspoonfuls of grated parmesan cheese, leaving at least 2 in.

Quantity		Ingredient
		Garniture
3 ½ oz	100 g	Parmesan cheese, grated
1 oz	30 g	Butter
1 pc	1 pc	Avocado
1 pc	1 pc	Tomato, cut into a ***brunoise***
		Salt and pepper

Method

(5 cm) between each. Spread them out evenly to form circles and top with a small piece of butter. Bake the cheese in the oven until melted and allow it to color to a light brown, no darker.

3. Remove the parmesan chips from the oven and transfer them to a wire rack to cool.
4. Peel the avocado, remove the stone, and coarsely crush with a fork to make a thick purée. Peel and seed the tomato (***émonder, épepiner***) and cut into a small dice (***brunoise***). Remove the skewers and cut 6 medallions from the lobster tails. Finely dice the remaining tail meat and the rest of the lobster meat and mix with the avocado and tomato ***brunoise***. Season to taste with salt and pepper.

Assemble the Garnish

1. Place a medallion on a plate or tray and top with a parmesan disk, then the filling, and cover with a second parmesan disk. Place a second medallion on top. Repeat until there are three layers. Make six garnishes.

To Serve

1. Arrange the garniture in the center of a heated soup plate and ladle the soup around it. Decorate with a sprig of fresh herbs. Serve immediately.

FYI...

Bisque requires a lengthy cooking process that is said to extract layer upon layer of flavor. Traditional recipes call for the chef to use the same cooking pot throughout the entire process to ensure that no flavor is lost. The traditional bisque was not strained through a chinois or blended, but pressed through an *étamine* (a type of thick, finely woven cheese cloth) and the bisque was always served with croutons and sometimes grated cheese. Today the garnish is modern and international because of the use of the avocado and parmesan. Modern versions use olive oil to give a different flavor to the bisque. It is also possible to bind the bisque with rice starch instead of whole rice. Other modern innovations allow the use of Tabasco in place of the cayenne pepper.

CONSOMMÉ BRUNOISE

Learning Outcomes

Clarification
Émonder
Brunoise
Mijoter

Equipment

Knives:
Paring knife (*office*), vegetable peeler (*économe*), chef knife (*couteau chef*)

Tools:
Bowls, wooden spatula, cutting board, skimmer (*écumoire*), ladle (*louche*), fine chinois

Pans:
Russe, small marmite, bain marie

Serving

4 persons

HISTORY

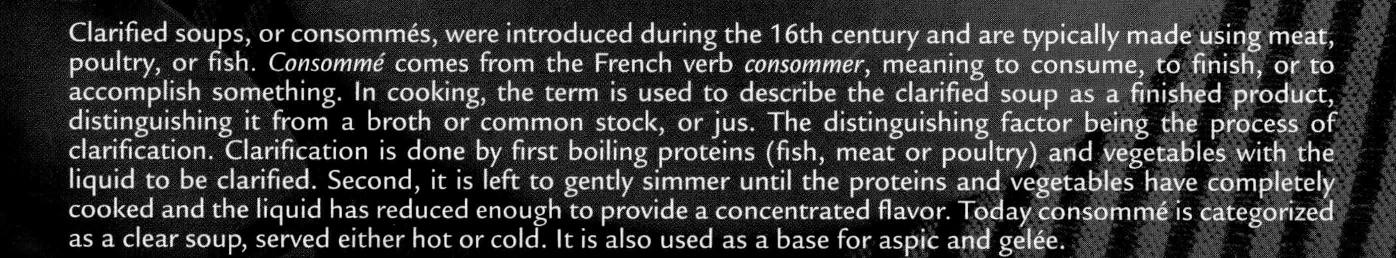

Clarified soups, or consommés, were introduced during the 16th century and are typically made using meat, poultry, or fish. *Consommé* comes from the French verb *consommer*, meaning to consume, to finish, or to accomplish something. In cooking, the term is used to describe the clarified soup as a finished product, distinguishing it from a broth or common stock, or jus. The distinguishing factor being the process of clarification. Clarification is done by first boiling proteins (fish, meat or poultry) and vegetables with the liquid to be clarified. Second, it is left to gently simmer until the proteins and vegetables have completely cooked and the liquid has reduced enough to provide a concentrated flavor. Today consommé is categorized as a clear soup, served either hot or cold. It is also used as a base for aspic and gelée.

CONSOMMÉ BRUNOISE

Clear Broth with Finely Diced Vegetables

Method

Clarification

1. Peel and rinse the carrots and celery. Cut them into a ***brunoise*** along with the leek. Finely chop the trimmings and set aside for the clarification.
2. Skin and seed (***émonder, épépiner***) the tomato.
3. In a pot, assemble the clarification ingredients by combining the ground beef, chopped vegetable trimmings, egg whites, salt, and white pepper. Mix well, then slowly stir in the cold chicken stock.
4. Heat very slowly, stirring often to make sure the proteins do not stick to the bottom of the pot. When the liquid comes to a simmer and the clarification floats to the surface, stop stirring.
5. The cooked proteins (*from the clarification*) will rise to the top of the liquid to form a "raft" (***gâteau***). Do not stir but leave to gently simmer (***mijoter***) for approximately 20 to 30 minutes. During this time, carefully make a hole in the middle of the raft (***gâteau***) with a small ladle. Season the top of the raft (***gâteau***) with salt, coarsely crushed peppercorn, thyme and bay leaf, and gently ladle liquid from the hole onto the raft (***gâteau***) regularly during the cooking. This allows the liquid to be seasoned as it clarifies.
6. Line a fine mesh sieve (***chinois***) with cheesecloth, coffee filter, or damp cloth filter (***étamine***).
7. Once the clarification is completed, carefully ladle the clear liquid from the hole in the raft (***gâteau***) into the lined ***chinois*** and strain into a clean pan. Keep the consommé hot in a bain marie, but do not place it on direct heat nor allow it to boil as this will cloud the consommé.

To Serve

1. Poach the vegetable ***brunoise*** separately in some of the ***consommé***. Drain the ***brunoise*** and divide among preheated serving bowls. Reserve the poaching consommé for another use but do not add it back to the consommé. Pour the hot consommé over the vegetables and decorate with sprigs of chervil.

Quantity		Ingredient
U.S.	*Metric*	
2 ½ qt	2.5 L	White stock
		Clarification
½ pc	½ pc	Vegetable trimmings, finely chopped
1 pc	1 pc	Tomato, small peeled (***émonder***), seeded (***épépiner***), cut into a ***brunoise***
¼ pc	¼ pc	Onion, large, finely chopped (***hacher***)
8 oz	250 g	Ground beef
4 pcs	4 pcs	Egg whites
1 pc	1 pc	Sprig thyme
1 pc	1 pc	Bay leaf
10–12 pcs	10–12 pcs	Black peppercorns
		Salt and peppercorn
		Garniture
2 pcs	2 pcs	Carrots, cut into a ***brunoise***
1 pc	1 pc	Leek white, cut into a ***brunoise***
3 pcs	3 pcs	Celery, peeled and cut into a ***brunoise***
2 oz	60 g	Green peas
1 pt	500 mL	Consommé for poaching the ***brunoise***
5 brs	5 brs	Chervil

Learning Outcomes

Cuisson à l'Anglaise
Roux blanc
Sauter
Fouler

Equipment

Knives:
Paring knife (*office*), slicing knife (*éminceur*)

Tools:
Bowls, wooden spatula, fine chinois, cutting board, skimmer (*écumoire*), ladle (*louche*), food processor, tamis

Pans:
Large sauce pan (*russe*), small stock pot (*marmite*), bain marie

Serving

4 persons

FYI...

Although asparagus has been a food source for over 2,000 years, it was "rediscovered" during the reign of King Louis XIV. The Argenteuil region of France was the first to produce the large speared variety that is so in demand today. The first white asparagus was developed from this variety.

This soup is based on a velouté but can also be made using a béchamel sauce as the base. The white stock would be replaced with milk.

CRÈME D'ARGENTEUIL

Cream of White Asparagus

Method

Base

1. Place the stock over medium-high heat and bring to a simmer. Reduce heat to medium. Bring a large pot of lightly salted water to a boil over high heat.
2. Peel the asparagus spears from top to bottom, turning them to preserve their shape. Be careful not to break the stalks. Line up the spears and trim the ends, then tie them in bundles of 10 using butcher's twine. Add the peelings to the simmering stock.
3. Place the asparagus bundles in the boiling water and poach them until a knife can easily pierce the stems. Drain and place on a towel to cool.
4. Cut the tips to 2 in. (4.5 cm) long and set aside. Thinly slice (***émincer***) the stalks on the bias and set aside to drain.

Optional: *Add lemon juice to the water to help keep the asparagus white.*

Liaison

1. Melt the butter in a medium pan over medium heat. When it begins to foam, add the flour and stir well until it begins to bubble. Cook for 1 or 2 minutes to obtain a white roux (*be careful not to let it color*). Transfer to a small plate and set aside to cool. Strain the asparagus-infused stock into a clean pan. Place over medium heat and whisk in the cold roux. Bring the mixture to a boil over medium heat, stirring constantly. Add the cooked asparagus stems and simmer (***mijoter***) for 20 minutes over low heat, stirring occasionally.

Finishing

1. Mix the soup in a food processor and strain it through a fine mesh sieve (***chinois***). Mix in the cream, then bring the soup to a boil; season and set it aside.
2. Melt the butter in a medium pan over very low heat and roll the asparagus tips in the butter until hot.

To Serve

1. Pour the crème d'Argenteuil into a heated soup bowl and arrange the asparagus tips on top. Finish with chervil or parsley sprigs (***pluches***).

Quantity		*Ingredient*
U.S.	*Metric*	*Base*
2 lb	1 kg	White asparagus
1 pc	1 pc	Lemon, juice of (*optional*)
1 ½ qt	1.5 L	White stock
		Liaison
2 oz	60 g	Butter
		Salt and white pepper
2 oz	60 g	Flour
		Finishing
8 fl oz	250 mL	Crème fraîche or heavy cream (*cold*)
1 oz	30 g	Butter
		To Serve
5 br	5 br	Chervil (or parsley)

GAZPACHO

Learning Outcomes

Cold soup
Émonder
Épépiner
Dégorger
Brunoise
Liaison with bread

Equipment

Knives:
Vegetable peeler (*économe*), paring knife (*office*), slicing knife (*éminceur*)

Tools:
Bowls, rubber spatula, fine chinois, cutting board, ladle (*louche*), whisk, food mill, ring mold

Serving

8 persons

FYI...

Gazpacho is a Spanish soup from the region of Andalusia, made of a purée of raw vegetables. It was originally a white soup, the tomatoes being introduced after the discovery of the New World. It is considered as *soupe étrangere* or foreign soup, and the liaison is stale bread. Before the combination of flour and butter was introduced by La Varenne in the 17th century, bread was used as a liaison in cooking. However, since flour needs cooking, bread is the perfect choice as the thickening agent in this particular soup.

Although the origin of the word *gazpacho* is unknown, some believe the earliest version of this recipe was brought to Spain by the Moors. It is said that horsemen would place the ingredients with bread in a leather pouch and place it under their saddles. The galloping motion would crush the vegetables and the bread would absorb the juices.

GAZPACHO
Gazpacho

Method

Gazpacho

1. Soak the shredded bread in cold water to soften. Wash and trim all the vegetables: Peel the red onion and remove the stem. Remove the stem of the bell peppers. Cut in half and remove the seeds and spongy ribs.
2. Trim the ends of the cucumber and peel. Cut in half lengthwise and, using a spoon, scrape out the seeds and discard. Coarsely chop (***concasser***) the vegetables.
3. *Prepare the tomatoes:* Bring a pot of water to a boil. Cut out the stem end using the tip of a small knife. Add a few tomatoes at a time to the pot and leave in the boiling water for 30 seconds to 1 minute. The skin will start to blister and pull away from the flesh. Remove and immediately place in ice water. Once cooled, drain the tomatoes and peel away the skin (***émonder***). Cut the tomatoes in half and gently squeeze until the seeds come out (***épepiner***). Discard the seeds.
4. Place the chopped vegetables and tomatoes together in a food mill fitted over a large bowl, and purée with the basil, soaked break, garlic, and red onion. Add the tomato paste and stir until smooth. Season with salt and pepper. If it is too thick, add tomato juice as needed. Whisk in the olive oil and sherry vinegar, then season to taste with salt and cayenne. Refrigerate for 3 or 4 hours or preferably overnight to allow the flavors to develop together.

To Serve

1. Serve well chilled with the garniture of the ***brunoise*** of vegetables and fresh basil.

Serving Suggestion (as photographed)

1. Channel (***canneler***) the cucumber, then slice in half lengthwise and cut into thin slices (***émincer***). Line (***chemiser***) a small ring mold with overlapping cucumber slices. Fill the lined ring mold with the ***brunoise*** garnish. Press lightly. Wipe up any excess juice from the vegetables with paper towel. Place the garnish in the center of a chilled soup plate before filling with ***gazpacho***. Once the soup is poured, carefully remove the ring mold. Top with a sprig (***pluche***) of basil. Sprinkle some loose ***brunoise*** into the ***gazpacho*** around the central garnish.

Quantity		*Ingredient*
U.S.	*Metric*	*Gazpacho*
3 pcs	3 pcs	Stale bread slices, shredded
1 pc	1 pc	Red onion
1 pc	1 pc	Green bell pepper, large
1 pc	1 pc	Red bell pepper, large
1 pc	1 pc	English cucumber
4 lb	2 kg	Tomatoes, peeled and seeded (***émonder***)
2 brs	2 brs	Fresh basil
6 pcs	6 pcs	Garlic cloves, finely chopped (***hacher***)
7 oz	200 g	Tomato paste
8 fl oz	250 mL	Tomato juice (optional)
3 ½ fl oz	100 mL	Olive oil
2 fl oz	60 mL	Sherry vinegar
Pinch	Pinch	Cayenne pepper
		Garniture
½ pc	½ pc	English cucumber (*optional*)
5 oz	150 g	Tomatoes, peeled and seeded, cut into a ***brunoise***
1 pc	1 pc	Red pepper, medium, cut into a ***brunoise***
1 pc	1 pc	Green pepper, medium, cut into a ***brunoise***
½ pc	½ pc	English cucumber, seeded, cut into a ***brunoise***
½ pc	½ pc	Red onion, cut into a ***brunoise***
3–4 pcs	3–4 pcs	Garlic cloves, finely diced (***ciseler***)
		Salt and white pepper
		To Serve
2 brs	2 brs	Fresh basil

POTAGE JULIENNE D'ARBLAY

Learning Outcomes

Make a potage parmentier
Julienne cut
Croûtons

Equipment

Knives:
Vegetable peeler (***économe***), paring knife (***office***), slicing knife (***éminceur***)

Tools:
Bowls, cutting board, wooden spatula, food mill or immersion blender, fine chinois, skimmer (***écumoire***), ladle (***louche***)

Pans:
Sauteuse, large sauce pan (***russe***), bain marie, soupière

Serving

4 persons

POTAGE JULIENNE D'ARBLAY

Potato and Leek Soup Garnished with Julienned Vegetables

Method

Parmentier Base

1. Clean the vegetables.
2. Thinly slice (***émincer***) the leeks. Melt the butter in a large pan over low heat and sweat (***suer***) until translucent.
3. Cut the potatoes into a large, thin ***paysanne*** (*do not soak them in water; this will remove the starch that is needed to thicken the potage*). Add them to the pan with the leeks and stir until coated. Add the ***bouquet garni***, then wet (***mouiller***) with the white stock. Season with salt and white pepper and gently simmer (***mijoter***) over low heat for 30 to 40 minutes.
4. Once the potatoes and leeks are cooked, purée them through a food mill into a clean pot or, alternatively, purée them with an immersion blender.
5. Add the cream (***crèmer***) and allow the mixture to gently simmer (***mijoter***). Adjust the seasoning if needed and then strain it (***passer***) through a fine mesh sieve (***chinois***), pressing well with the ladle (***fouler***). Keep the potage warm in a bain marie and pat (***tamponner***) the surface with butter to keep a skin from forming.

Julienne Garnish

1. Melt the butter over medium-low heat in three small pans and sweat the julienned vegetables separately until they begin to soften. Lightly season with salt. Reduce the heat to low, cover, and cook (***étuver***) for 10 minutes or until tender (*avoid coloring*). Remove the julienned vegetables from the pans and set aside.

Croûtons

1. Remove the crust from the bread and cut into cubes. Melt the butter in a sauté pan over medium-high heat and add the bread. Toss the bread in the hot butter until golden, then transfer to a plate lined with paper towels to drain.

To Serve

1. Mix the julienned vegetables together and stir half into the potage.
2. Serve the hot potage in a heated soup plate or a soup tureen (***soupière***). Place a spoonful of ***julienne*** in the middle of the plate. Decorate with sprigs (***pluches***) of chervil and serve croûtons on the side.

Quantity		*Ingredient*
U.S.	*Metric*	*Parmentier*
5 oz	150 g	Leeks, white, sliced (***émincer***)
3 ½ oz	100 g	Butter
1 lb 4 oz	600 g	Potatoes, cut into a ***paysanne***
1 pc	1 pc	***Bouquet garni***
1 ½ pt	750 mL	White stock
1 ¼ pt	600 mL	Cream
1 ¾ oz	50 g	Butter
		Salt and white pepper
		Julienne Garniture
2 oz	60 g	Butter
2 oz	60 g	Carrots, cut into a ***julienne***
2 oz	60 g	Turnips, cut into a ***julienne***
2 oz	60 g	Leeks, cut into a ***julienne***
		Salt
		To Serve
2 pcs	2 pcs	Sliced bread, crusts removed and cut into small cubes
5 oz	150 g	Butter
3 brs	3 brs	Chervil

FYI...

The name of the recipe *Julienne d'Arblay* makes reference to a garnish of julienned vegetables while also paying homage to celebrated English novelist Frances Burney D'Arblay (1752–1840). She was married to a French officer by the name of Alexandre D'Arblay and lived for some time in France where the soup was most likely dedicated to her.

POTAGE MULLIGATAWNY

Learning Outcomes

Soup étranger
Mijoter
Bouquet garni
Ciseler
Singer
Créole rice

Equipment

Knives:
Boning knife (***désosseur***), vegetable peeler (***économe***), paring knife (***office***), slicing knife (***éminceur***), scissors (***ciseaux***)

Tools:
Bowls, cutting board, fork, wooden spatula, plastic spatula, fine chinois, skimmer (***écumoire***), ladle (***louche***), whisk

Pans:
Large sauce pans (***russes***), small stock pot (***marmite***)

Serving

4 persons

FYI...

This soup has its origins in the culinary tradition of the Tamil people of southern India. Literally translated from Tamil, the word *mulligatawny* means "pepper water." However, mulligatawny as we know it today is significantly different from the original soup.

It was likely brought back to England from India by British colonists and then eventually made its way across the channel to France. This is considered a *potage étranger*, or foreign soup. The use of curry echoes its Indian origins and the use of the apples adds a touch of sweetness.

POTAGE MULLIGATAWNY

Curried Chicken Soup with Apple

Quantity		Ingredient
U.S.	*Metric*	*Base*
1 ¾ oz	50 g	Butter
3 pcs	3 pcs	Chicken wings
¼ oz	10 g	Curry powder
1 pc	1 pc	Onion, cut into a large dice (***mirepoix***)
½ pc	½ pc	Celery stalk, cut into a large dice (***mirepoix***)
1 pc	1 pc	Carrots, cut into a large dice (***mirepoix***)
1 ¼ qt	1.25 L	White stock
		Apple trimmings
1 pc	1 pc	***Bouquet garni***
		Salt and pepper
		Liaison
1 ¾ oz	50 g	Butter
½ pc	½ pc	Onion, finely diced (***ciseler***)
1 oz	30 g	Flour
7 fl oz	200 mL	Cream
		Garniture
1 ¾ pts	900 mL	Water
5 oz	150 g	Long grain rice
3 ½ oz	100 g	Smoked slab bacon, cut into lardons
¾ oz	20 g	Butter
5 oz	150 g	Cooked chicken breast, cut into a ***julienne***
2 pcs	2 pcs	Golden Delicious apples, cut into a ***brunoise***
		Salt and pepper

Method

Base

1. Melt the butter in a large saucepan over medium heat and lightly sear the chicken wings until just beginning to color.
2. Add the curry powder and cook for 1 to 2 minutes, then add the onion, celery, and carrot ***mirepoix***. Peel the apple and rub with lemon juice. Add the peelings to the pan along with the apple trimmings. Add the chicken stock and ***bouquet garni*** and bring to a boil, skimming off any foam from the surface (***écumer***). Reduce the heat and leave to simmer for 45 minutes. Place the chicken breast in the simmering stock and allow to poach for 15 minutes. Remove and set aside. Once the stock is finished simmering, strain through a fine mesh sieve (***chinois***) into a clean saucepan and keep warm.

Liaison

1. Melt the butter over medium heat and sweat (***suer***) the diced onion until it begins to release its water. Add the flour and cook for 2 to 3 minutes (***singer***). Whisk in the curried chicken stock and the cream and allow to simmer for 10 minutes.

Garniture

1. Créole rice: Place the water and rice in a pan over medium-high heat and, once the rice comes to a boil, cook for 14 minutes. Remove from the heat, strain, and refresh with cold water. Drain well, then transfer the rice to a colander lined with a cloth. Allow to drain completely.
2. *Blanch the lardons:* Place the lardons in a small saucepan of cold water over high heat and bring the water to a boil. Let the water boil for 1 minute, then drain the lardons and rinse them under cold running water and drain.
3. Melt the butter in a small sauté pan over medium-high heat and sauté the lardons until lightly colored.
4. Cut the cooked chicken breast into a ***julienne***. Set aside.

Finishing

1. Strain the soup and stir in the rice. Allow to simmer for 20 minutes. Season to taste.
2. Just before serving, stir in the apple ***brunoise***.

To Serve

1. Serve in heated bowls. Place a spoonful of lardons and chicken breast ***julienne*** in the middle of the bowl as garnish.

POTAGE CLAMART

Learning Outcomes

Mirepoix
Lardons
Blanchir
Raidir
Suer
Tamponner
Étuver chiffonade
Sauter
Clarifié
Liaison au tapioca
Glacer à blanc

Equipment

Knives:
Paring knife (***économe***), slicer

Tools:
Bowls, cutting board, wooden spatula, plastic spatula, fine chinois, skimmer, ladle, drum sieve (*tamis*)

Pans:
Shallow pan, sauté pan, small saucepan, small stock pot, bain marie

Serving

4 persons

FYI...

The title of *Clamart* indicates that the dish is based on green peas. Clamart, a region located just southwest of Paris, is renowned for its green peas. To give this soup a fresh, bright color, it is preferable to use fresh or frozen peas in lieu of dried.

Liaison au tapioca: The use of tapioca as the liaison in potage clamart demonstrates another way to thicken a soup. Tapioca is a gluten-free thickening agent that is made from the cassava or manioc plant.

POTAGE CLAMART

Green Pea Soup

Quantity		*Ingredient*
U.S.	*Metric*	*Base*
1 lb	500 g	Frozen or fresh green peas
5 oz	150 g	Tapioca
3 ½ oz	100 g	Smoked slab bacon with rind
3 ½ oz	100 g	Butter
3 ½ oz	100 g	Leek, white part, sliced (***émincer***)
4 pcs	4 pcs	Shallots, sliced (***émincer***)
1 ¾ oz	50 g	Onion, sliced (***émincer***)
1 pc	1 pc	***Bouquet garni***
1 ½ qt	1.5 L	White chicken stock
Pinch	Pinch	Sugar
		Salt and white pepper
		Finishing
10 fl oz	300 mL	Cream
3 ½ oz	100 g	Butter
		Garniture à la Française
16 pcs	16 pcs	Pearl onions, peeled
1 ¾ oz	50 g	Butter
Pinch	Pinch	Sugar
		Salt
3 ½ oz	100 g	Salted slab bacon, cut into lardons
3 ½ oz	100 g	Green peas
1 pc	1 pc	Bibb lettuce head, shredded (***chiffonade***)
		To Serve
4 oz	120 g	Pearl tapioca

Method

Base

1. Thaw the green peas or, if using fresh ones, remove them from the pod.
2. Rinse the tapioca in cold water and drain.
3. Remove the rind and reserve and cut the slab bacon into lardons.
4. Melt the butter in a large saucepan over medium heat and add the leek, shallots, and onion to the pan. Sweat (***suer***) in the butter until soft. Add the green peas, ***bouquet garni***, tapioca, and sugar. Stir the vegetables well, add the bouquet garni, then pour in the chicken stock. Add the bacon rind, and reduce the heat and allow to gently simmer (***mijoter***) for 25 minutes or until all the vegetables are tender.

Note: *A pinch of sugar is commonly added to pea dishes to remove bitterness.*

Finishing

1. Once cooked, remove the ***bouquet garni*** and the bacon rind and discard. Strain through a drum sieve (***tamis***), using a plastic scraper (***corne***) to press the solids through. Transfer to a clean pan. Add the cream and bring to a boil over medium heat. Allow to reduce until it coats the back of a spoon (***à la nappe***). Season the potage to taste and keep it warm in a bain marie. Pat (***tamponner***) the surface with cold butter to stop the formation of a skin. Set aside.

Garniture à la Française

1. ***Glacer à blanc*** *the pearl onions*: Place the onions with sugar, salt, and ¾ oz (20 g) of butter in a small pan and fill it with water to just cover the onions. Cover with a buttered parchment lid (***cartouche***) and cook over low heat until the liquid evaporates (*without the onions coloring*). Remove the pan from the heat and set aside.
2. Melt 1 oz (30 g) of butter in a small pan over medium heat. Add the peas and glazed onions. Stir, cover, and reduce the heat to low. Cook for about 5 minutes.
3. Mix in the shredded (***chiffonade***) lettuce, cover, and remove from the heat.

To Serve

1. Poach the tapioca pearls in salted boiling water over high heat until soft. Drain the tapioca.
2. Arrange the *garniture à la française* in the bottom of a heated soup plate and ladle the hot soup around it. Finish with tapioca pearls.

Learning Outcomes

Soupe à légumes taillés
Paysanne
Suer
Mijoter
Blanchir
Croûtons

Equipment

Knives:
Vegetable peeler (*économe*), paring knife (*office*), slicing knife (*éminceur*)

Tools:
Bowls, wooden spatula, cutting board, skimmer (*écumoire*), ladle (*louche*), cheese grater

Pans:
Sauteuse, small sauce pan (*russe*), small stock pot (*marmite*), bain marie

Serving

4 persons

POTAGE CULTIVATEUR

Garden Vegetable Soup

Method

1. Preheat the oven to 350°F (185°C).

Base

1. Clean the vegetables.
2. Cut the leek, cabbage, carrots, turnips, and potatoes into a ***paysanne***.
3. Melt the butter in a large saucepan over medium heat and sweat (***suer***) the carrot and cook until it begins to soften. Then add the leek, the cabbage, the potatoes and finally the turnips, stirring well after each addition.

Tip: *Cabbage contains a gas that can hinder digestion. To remove this, quickly blanch the cabbage leaves in salted boiling water and refresh in an ice bath before cutting for use in the potage.*

4. Add the ***bouquet garni*** and chicken stock to the vegetables and bring to a gentle simmer (***mijoter***) over medium heat for about 20 minutes.
5. Meanwhile, remove the rind from the slab bacon (if needed), and cut into lardons.
6. *Blanch the lardons:* Place them in a small saucepan and cover them with cold water. Bring the water to a boil over high heat and let it boil for 1 minute. Drain the lardons and rinse them under cold water. Drain and set aside.
7. Once the vegetables have been cooking for 10 minutes, add the lardons and continue to simmer (***mijoter***).
8. Cut the green beans the same size as the peas. When the potatoes in the potage are just cooked, add the green beans and peas. Adjust the seasoning.

Croûtons

1. Rub baguette slices with garlic, arrange on a baking sheet, and sprinkle with Gruyère cheese. Toast in the oven until the cheese is melted and begins to brown around the edges.

To Serve

1. Serve the potage in a hot soup plate. Decorate with sprigs (***pluches***) of parsley and serve the croûtons on the side.

Quantity		*Ingredient*
U.S.	*Metric*	*Base*
3 ½ oz	100 g	Leek, white part, cut into a ***paysanne***
3 ½ oz	100 g	Savoy cabbage leaves, central vein removed, cut into a ***paysanne***
5 oz	150 g	Carrots, cut into a ***paysanne***
3 ½ oz	100 g	Turnips, cut into a ***paysanne***
5 oz	150 g	Potatoes, cut into a ***paysanne***
3 ½ oz	100 g	Butter
		Bouquet garni
1 ½ qt	1.5 L	Clear white chicken stock (*or consommé*)
3 ½ oz	100 g	Salted slab bacon, cut into lardons
1 ¾ oz	50 g	Green beans
1 ¾ oz	50 g	Green peas
		Garniture
1 ¾ oz	50 g	Baguette, sliced
1 pc	1 pc	Garlic clove
1 ¾ oz	50 g	Gruyère cheese, grated
2 brs	2 brs	Parsley

FYI...

This soup is characteristically simple and rustic in its preparation. The soup is comprised of vegetables, potatoes, and salted pork. The word *cultivateur* means *farmer*, which is why the vegetables are cut into a paysanne. The name is also meant to indicate that the ingredients used should include those found in the garden patch. The potage cultivateur is a *potage de légumes taillés* because it features cut vegetables.

POTAGE *PARMENTIER*

Learning Outcomes

Potage légumes purée
Make a potage Parmentier
Émincer
Suer
Mouiller
Mijoter
Crèmer
Fouler
Bain marie
Tamponner
Croûtons

Equipment

Knives:
Vegetable peeler (***économe***), paring knife (***office***), slicing knife (***éminceur***)

Tools:
Bowls, cutting board, wooden spatula, food mill or immersion blender, fine chinois, skimmer (***écumoire***), ladle (***louche***)

Pans:
Sauteuse, large sauce pan (***russe***), bain marie, soupière

Serving

4 persons

FYI...

In French cuisine, the use of the name *Parmentier* on a menu indicates that the dish contains potato. The name is from Antoine-Augustin Parmentier (1737–1813), who is responsible for popularizing the potato as a food source in France during the late 18th century. The potato had been introduced to Europe in the 1600s, but in France up until Parmentier's time it was used primarily as pig feed as it was considered to be unsafe to be consumed by humans.

Potage Parmentier is a *potage purée de légumes,* meaning that it is thickened by a purée of vegetables; in this case, the potatoes and leeks are puréed to give this soup its thickness.

POTAGE PARMENTIER
Potato and Leek Soup

Method

Potage

1. Clean the vegetables.
2. Thinly slice (***émincer***) the leeks. Melt the butter in a large pan over low heat and sweat (***suer***) the leeks until soft.
3. Cut the potatoes into a large, thin ***paysanne*** (*do not soak them in water, as this will remove the starch that is needed to thicken the potage*). Add the potatoes to the leeks and stir well. Add the ***bouquet garni*** then wet (***mouiller***) with the stock. Season with salt and white pepper and gently simmer (***mijoter***) over low heat for 30 to 40 minutes.
4. Once the potatoes and leeks are cooked, strain them through a food mill into a clean pot or, alternatively, purée them with an immersion blender.
5. Add the cream (***crèmer***) and allow the mixture to gently simmer (***mijoter***). Adjust the seasoning if needed, then strain (***passer***) through a fine mesh sieve (***chinois***), pressing well with the ladle (***fouler***). Keep the potage warm in a bain marie and pat (***tamponner***) the surface with butter to keep a skin from forming.

Croûton

1. Remove the crust from the bread and cut into cubes. Melt the butter in a sauté pan over medium-high heat and add the bread. Sauté in the hot butter until golden. Transfer to a plate lined with paper towels to drain.

To Serve

1. Serve the hot potage in a heated soup plate or a soup tureen (***soupière***). Decorate with sprigs (***pluches***) of chervil and serve the croûtons on the side.

Quantity		*Ingredient*
U.S.	*Metric*	*Parmentier*
3 ½ oz	100 g	Butter
5 oz	150 g	Leeks, white, thinly sliced (***émincer***)
1 ½ lbs	600 g	Potatoes, cut into a ***paysanne***
1 ½ pt	750 mL	White chicken stock
1 pc	1 pc	***Bouquet garni***
		Salt and white pepper
1 ¼ pt	600 mL	Cream
1 ¾ oz	50 g	Butter
		Finishing
2 pcs	2 pcs	Sliced white bread
1 ¼ oz	40 g	Butter
4 brs	4 brs	Chervil

Learning Outcomes

Suer
Émincer
Singer
Déglacer
Mouiller
Gratiner

Equipment

Knives:
Paring knife (*office*), slicing knife (*éminceur*), serrated knife (*couteau à scie*)

Tools:
Bowls, wooden spatula, cutting board, ladle (*louche*), cheese grater, salamander, heat-proof bowls

Pans:
Large sauce pan (*russe*)

Serving

6 persons

FYI...

Onions are nutrient-rich bulbs that have been valued for different qualities throughout history. They have been used by sailors to stave off scurvy where fresh produce was not available; they have been traded for labor by the Egyptian Pharaohs; and they have provided the less fortunate with nutrients that they would otherwise have been lacking.

There is an old French wedding custom that involves onion soup. Wedding guests stay until the small hours of the morning to "distract" the wedding couple from consummating the marriage by singing and making merry. As the festivities draw to an end, onion soup is served in hopes of rejuvenating the guests before they travel home.

SOUPE À L'OIGNON GRATINÉE

French Onion Soup

Method

1. Preheat oven to 400°F (205°C).

Base

1. Peel the onions, remove the roots, and slice thinly (***émincer***). Melt the butter in a large saucepan over medium heat and cook the onions until soft and lightly colored.
2. Add the flour and cook (***singer***) until it starts to take on some color (*1 to 2 minutes*).
3. Deglaze with white wine and allow the liquid to reduce by half before adding the water. Add the ***bouquet garni***, season to taste, then reduce the heat to a gentle simmer and leave to cook for 45 minutes.

Optional: *Add some* ***glacé de viande*** *for more color.*

Garniture

1. Slice the baguette and lightly toast in the oven, then lightly brush each slice with clarified butter and sprinkle with cheese. Place the bread back in the oven until the cheese is just melted but not colored. Set aside.

Finishing (optional)

1. Remove the ***bouquet garni*** from the soup and discard. ***Optional:*** *Whisk the port wine into the cream and ladle some hot soup into the cream to temper. Off the heat, stir the cream back into the hot soup.*

To Serve

1. Ladle the soup into ovenproof bowls. Top with croûtons to completely cover. Sprinkle with more cheese. Arrange the bowls on a baking sheet and place the tray under a broiler or salamander until the cheese is browned (***gratiner***).
2. Serve immediately.

Quantity		*Ingredient*
U.S.	*Metric*	*Base*
2 lb	1 kg	Onions, sliced (***émincer***)
5 oz	150 g	Unsalted butter
2 oz	60 g	Flour
8 fl oz	250 mL	Dry white wine
1 ½ qt	1.5 L	Water
1 pc	1 pc	***Bouquet garni***
1 ¾ oz	50 g	***Glacé de viande*** (*optional*), for color and flavor
		Salt and white pepper
		Garniture
½ pc	½ pc	French baguette, sliced
3 ½ oz	100 g	Gruyère cheese, grated
3 ½ oz	100 g	Butter, clarified
		Finishing
4 fl oz	120 mL	Heavy cream (*optional*)
1 oz	30 mL	Port wine (*optional*)
3 ½ oz	100 g	Gruyère cheese, grated

Learning Outcomes

Habiller and détailler fish
Raidir
Suer
Mirepoix
Pincer
Déglacer
Mouiller
Mijoter
Piler
Sauté

Equipment

Knives:
Cleaver (***couteau à batte***), vegetable peeler (***économe***), paring knife (***office***), slicing knife (***éminceur***), scissors (***ciseaux***), sole filet knife (***couteau filet de sole***)

Tools:
Bowls, cutting board, wooden spatula, plastic spatula, fine chinois, skimmer (***écumoire***), ladle (***louche***), mortar and pestle whisk

Pans:
Rondeau, marmite, bain marie

Serving

4 persons

SOUPE DE POISSON FAÇON BOUILLABAISSE

Bouillabaisse Style Fish Soup

Method

1. Preheat oven to 425°F (200°C).
2. Prick the potato with a fork and place on a bed of rock salt. Place in the oven and bake for 45 minutes or until easily pierced with a knife. Set aside and allow to cool until warm then peel. Crush with a fork and set aside.

Croûtons à L'ail

1. Cut the baguette into thin slices on the bias. Brush with olive oil and arrange in a single layer on a baking sheet. Place in the oven to lightly toast. Once toasted on both sides, lightly rub with the garlic clove, sprinkle with cheese and put back in the oven until the cheese is melted. Remove and set aside.

Marinade

1. Peel and cut the carrot into a fine julienne. Set aside in a bowl. Clean the leek and cut into a julienne. Add to the carrot. Peel the celery, then cut into a julienne and add to the carrot and leek. Trim the fennel, reserving the trimmings. Cut into a julienne and add to the other vegetables.
2. Trim and clean the whole fish. Cut into thick slices (***tronçons or darnes***). Place in a large bowl. Add the julienned vegetables, the potato, onion, shallot and garlic. Mix well. Add the remaining marinade ingredients and mix until well combined. Set aside to marinate in the refrigerator overnight.

Fonds de Cuisson

1. Finely chop the onion, shallot and garlic, and set aside.
2. Peel the potato and cut into quarters and then thinly slice. Do not rinse and set aside.
3. Heat the olive oil over medium high heat. Add the mirepoix including the fennel root trimmings, shallots, and saffron. Sweat (***suer***) and stir until well coated. Add the tomato paste, stir well, and cook 1–2 minutes (***pincer la tomate***). Deglaze with the pastis and cook until almost dry. Add the dry white wine and reduce by half, scraping the bottom well. Add the tomatoes, garlic head, saffron, and bouquet garni. Stir until combined then wet (***mouiller***) with the fish stock. Lightly season with salt, pepper, and cayenne. Reduce the heat to low, cover, and leave to simmer (***mijoter***) for 45 minutes.

Quantity		Ingredient
U.S.	Metric	
3 lbs	1.5 kg	Fresh, whole fish, such as red snapper, whiting, rockfish, salmon, etc.
		Marinade
4 fl oz	120 mL	Olive oil
1 pc	1 pc	Onion, finely chopped
2 pcs	2 pcs	Shallots, finely chopped
4 pcs	4 pcs	Garlic cloves, finely chopped
2 pcs	2 pcs	Medium tomatoes, peeled, seeded and diced
1 oz	30 g	Tomato paste
Pinch	Pinch	Saffron
1 pc	1 pc	Large carrot, cut into a julienne
1 pc	1 pc	Fennel root, cut into a julienne
2 pcs	2 pcs	Celery stalk, peeled and cut into a julienne
1 pc	1 pc	Leek, white part only, cut into a julienne
1 bq	1 bq	Parsley, finely chopped
1 bq	1 bq	Basil, shredded (optional)
2 pcs	2 pcs	Medium potatoes, peeled, quartered and thinly sliced
1 pc	1 pc	Bouquet garni
		Kosher salt
		Peppercorns
1 fl oz	30 mL	Pastis (optional)
		Fonds de Cuisson
4 fl oz	120 mL	Olive oil
8 oz	250 g	Mirepoix, including trimmings from the fennel
1 ¾ oz	50 g	Tomato paste
1 ¾ fl oz	50 mL	Pastis
8 fl oz	250 mL	Dry white wine
3 pcs	3 pcs	Tomatoes, quartered
1 pc	1 pc	Whole head garlic, cut in two
Pinch	Pinch	Saffron
1 pc	1 pc	Bouquet garni
2 pcs	2 pcs	Shallots
2 ½ qts	2.5 L	Fish stock
		Salt, pepper, cayenne

Quantity		Ingredient
U.S.	Metric	Rouille
2 pcs	2 pcs	Garlic cloves, germ removed
8 oz	250 g	Potato, 1 medium
1 pc	1 pc	Red pepper
2 pcs	2 pcs	Egg yolk
8 oz	250 mL	Olive oil
Pinch	Pinch	Cayenne
Pinch	Pinch	Saffron
		Salt and pepper
		Croûtons à L'ail
½ pc	½ pc	Baguette
2 pcs	2 pcs	Garlic clove
3 ½ oz	100 g	Gruyère cheese, grated
		Olive oil
7 oz	200 g	Grated gruyère cheese

Method

Rouille

1. Clean and trim the red pepper, removing its seeds and spongy ribs. Cut in half and place cut side down in a pan with some olive oil. Place in the hot oven and roast until the skin has blistered and blackened. Remove from the oven, cover with a damp towel, allow to cool, and then remove the skin.
2. Crush the garlic in a mortar with some salt. Add the baked potato and red pepper and work with the pestle into a smooth paste. Add the egg yolks and stir until smooth. If too thick, add some of the fonds to the mixture. Transfer the mixture to a bowl and gradually incorporate the olive oil. Adjust the seasoning, then set aside.

Finishing the Soup

1. Strain the cooking liquid into a rondeau and place over low heat. Transfer the fish and garnish into the hot liquid. Allow to gently poach for 15-20 minutes, or until the potatoes are tender. Adjust seasoning as needed. Transfer to a heated serving bowl and serve accompanied by the croûtons, rouille and additional grated cheese.

FYI...

The etymology of *bouillabaisse* is derived from two French words that describe the method of preparation precisely. This recipe asks the chef to *bouillir* (boil), and then *abaisser* (lower) the temperature with the introduction of each fish ingredient.

Fish steak cuts are called *darnes* and *tronçons*. A *darnes* is a 90° cut that is made across the backbone of a round fish and includes meat from both sides of the fish. A *demi-darnes* is a round fish cut that includes the bone and meat from only one side of the fish. A *tronçons* is a cut that is used on flat fish and also includes the bone.

Learning Outcomes

Cuisson à l'Anglaise
Roux blanc
Cauliflower florettes
Fouler
Panade
Gougères
Velouté
Dessêcher

Equipment

Knives:
Paring knife (***office***), slicing knife (***éminceur***)

Tools:
Bowls, wooden spatula, chinois, étamis, colander, araingée, skimmer (***écumoire***), ladle (***louche***), whisk, cutting board, food processor, piping bag, baking sheet

Pans:
Large sauce pan (***russe***), small stock pot (***marmite***), bain marie, sauteuse

Serving

4 persons

CRÈME DUBARRY

Cream of Cauliflower

Quantity		*Ingredient*
U.S.	*Metric*	*Base*
1 pt	500 mL	Chicken stock (optional)
1 pt	500 mL	Milk
¼ pc	¼ pc	Onion, stuck with a clove (***clouter***)
1 pc	1 pc	***Bouquet garni***
		Coarse salt
1 pt	500 mL	Heavy cream
1 pc	1 pc	Cauliflower
		Gougère Mixture
3 ½ fl oz	100 mL	Water
¾ oz	30 g	Butter
Pinch	Pinch	Salt
1 ¾ oz	50 g	Flour
2 pcs	2 pcs	Eggs
1 ¾ oz	50 g	Gruyère cheese, grated
1 pc	1 pc	Egg for egg wash
		Liaison
1 oz	30 g	Butter
5 pcs	5 pcs	Egg yolks
1 ¾ fl oz	50 mL	Heavy cream
3 ½ oz	100 g	Butter
1 oz	30 g	Flour

Method

1. Melt the butter in a medium pan over medium heat. When the butter begins to foam, add the flour and stir well until the mixture begins to bubble. Cook for 1 or 2 minutes to obtain a white roux (*be careful not to let it color*). Set aside to cool.
2. Preheat the oven to 400°F (205°C).

Pâte à Choux

1. *Make a pâte à choux:* Combine the water, butter, and salt in a medium pan and bring the mixture to a boil over high heat. Once the butter has completely melted, remove from the heat and add the flour all at once. Stir with a wooden spatula until combined; then, over medium heat, stir until the mixture forms a clean ball coming away from the sides of the pan (***dessêcher***). Transfer the hot dough to a clean bowl and spread out to cool slightly. Using a spatula, mix the eggs into the ***dough*** one by one, making sure the last one is completely incorporated before adding the next. The dough should be elastic and slightly sticky. Mix in half the cheese and transfer the mixture to a piping bag fitted with a plain 10 mm tip.
2. Pipe out (***coucher***) small balls (***choux***) of the gougère mixture onto a lightly greased baking sheet. Brush with egg wash and sprinkle with the remaining cheese. Bake until golden (*15 to 20 minutes, depending on the size of the balls*), being careful not to open the oven door during the cooking process. Once cooked, transfer to a wire rack to cool. Reserve.

Garniture

1. *Prepare the cauliflower:* Bring a large pan of salted water to a boil. Remove the green leaves, trim away any brown spots, and cut away the core. Cut the cauliflower into large pieces. Select a few pieces, and cut small florettes (***inflorescences***) and set aside.
2. Blanch the cauliflower in the boiling salted water (***à l'anglaise***) until tender. Refresh, then drain and set aside.

Base

1. Heat the milk with the onion stuck with a clove (***clouté***), and the ***bouquet garni***. Season with salt, white pepper, and nutmeg. Bring to a low simmer and allow to infuse. Strain the hot milk into the cold ***roux*** all at once, then stir with a whisk. Bring to a boil over medium heat, stirring continuously until thick. Add the cauliflower and cook until

Method

tender, 20 to 30 minutes. Purée through a food mill into a clean pan. Add the cream, season to taste, and strain through a fine mesh sieve (***chinois***).

Finishing

1. Combine the egg yolks and cream and cut the butter into a small dice. Strain the hot liquid through a fine mesh sieve (***chinois***) into a clean pan, pressing well with a small ladle (***fouler***). Add 1 to 2 ladles of the hot liquid to the egg yolk mixture and stir well. Remove the pan from heat and whisk in the tempered yolks; then stir in the butter until melted. If the crème is too thick, fix the consistency with a little hot milk or chicken stock.
2. Pat (***tamponner***) the surface with a knob of butter on the end of a fork to stop the formation of a skin. Cover and reserve in a bain marie.
3. Bring a small pan of salted water to a boil. Add the florettes (***inflorescences***) and cook until tender. Drain and reserve.

To Serve

1. Pour the crème into a heated bowl or tureen (***soupière***) and serve the gougères on the side.
2. If needed, reheat the cauliflower florettes (***inflorescences***) in simmering water (***chauffante***). If serving in a bowl, garnish each bowl with florettes (***inflorescences***) before filling. Finish with a sprig of chervil.

Quantity		Ingredient
		Finishing
2 pcs	2 pcs	Egg yolks
3 ½ oz	100 g	Heavy cream
3 ½ oz	100 g	Butter
		Salt and white pepper
		Gougères
		To Serve
3 ½ oz	100 g	Cauliflower, cut into florettes
4 brs	4 brs	Chervil (***pluches***)

FYI...

This preparation was named after the last mistress of King Louis XV of France, the Comtesse du Barry, who was said to have had a penchant for cauliflower.

Crème vs Velouté? This recipe can easily be adapted as a velouté by replacing the milk with the same amount of white stock (as pictured above). Both the béchamel and the velouté are thickened by a roux blanc.

VELOUTÉ DE VOLAILLE

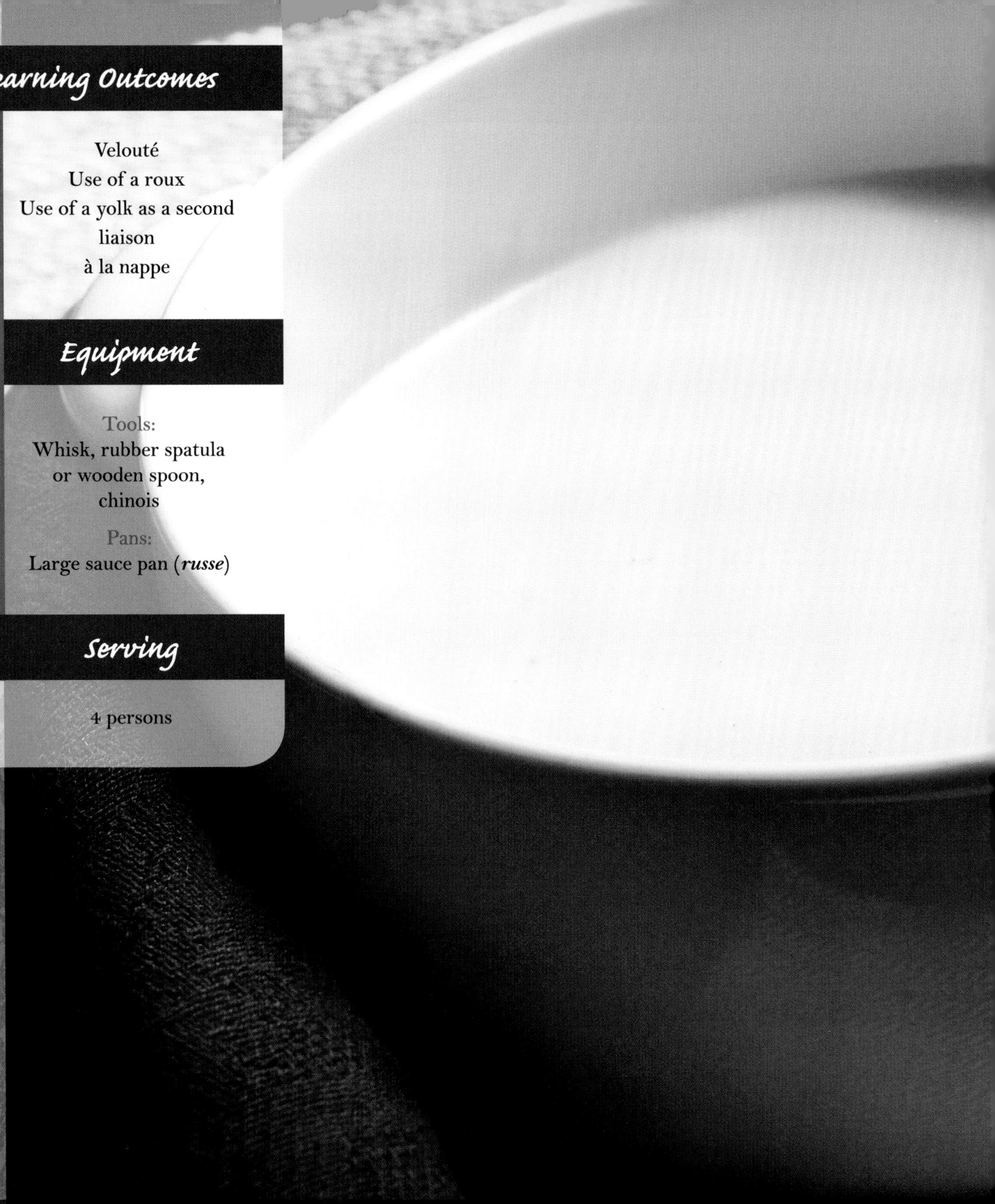

Learning Outcomes

Velouté
Use of a roux
Use of a yolk as a second liaison
à la nappe

Equipment

Tools:
Whisk, rubber spatula or wooden spoon, chinois

Pans:
Large sauce pan (*russe*)

Serving

4 persons

Tamponner is a technique that is used in French Cuisine to eliminate a skin from forming on a sauce (or, in this case, a velouté) while it rests on the hot stove top. Dotting the surface of a hot preparation with butter creates a layer of fat that acts as a preventative against the formation of a skin.

VELOUTÉ DE VOLAILLE
Cream of Chicken

Method

Velouté

1. Melt the butter in a medium pan over medium heat. When the butter begins to foam, add the flour and stir well until the mixture begins to bubble. Cook for 5 to 10 minutes over low heat to obtain a light golden-colored roux (*this is called a* ***roux blond***).
2. Add the cold consommé all at once to the hot roux and whisk. Stirring continuously with the whisk, raise the heat to medium high, and bring the velouté to a boil. Add the ***bouquet garni*** and peppercorns and season lightly. Reduce the heat to low and simmer for 30 minutes, skimming the surface (***écumer***) regularly.

Finishing

1. Add half the cream to the velouté and bring it to a boil over high heat.
2. Mix the egg yolks with the remaining cream. Stir in 1 to 2 ladles of the velouté to temper the yolks.
3. Remove the velouté from the heat and whisk in the tempered yolks. Strain through a fine mesh sieve (***chinois***) into a clean pan. ***Tip:*** *Once the yolks have been incorporated, the mixture should never be brought up to a temperature above 185°F (85°C) or it will curdle.*
4. Place back over low heat and stir the velouté in a figure 8 motion with a wooden spoon over low heat until thick enough to coat the back of the spoon (***à la nappe***). Taste and adjust seasoning.
5. Pat (***tamponner***) the surface with a knob of butter on the end of a fork to stop the formation of a skin and keep warm in a bain-marie.

To Serve

1. Strain the velouté through a fine mesh sieve (***chinois***).
2. Cut the cold butter into a small dice and stir into the velouté until melted.
3. Serve in a hot soup bowl or tureen (***soupière***).

Quantity		*Ingredient*
U.S.	*Metric*	*Base*
1 qt	1 L	Chicken consommé
¾ tsp	2 g	Black peppercorns, wrapped and tied in gauze
1 pc	1 pc	***Bouquet garni***
		Liaison (Roux Blond)
1 ¾ oz	50 g	Butter
1 ¾ oz	50 g	Flour
		Finishing
8 fl oz	250 mL	Cream
2 pcs	2 pcs	Egg yolks
¾ oz	25 g	Butter, small dice
		Salt to taste

Les Entrées-Froid

(Cold Appetizers)

Les Entrées—Froid (Cold Appetizers)

Asperges froides, sauce hollandaise —Asparagus with hollandaise sauce

Céleri rémoulade —Celery root remoulade

Champignons à la turque —Stewed mushrooms with currants

Les crudités et leurs sauces —Raw vegetables and sauces

Macédoine de légumes —Diced vegetable salad

Poireaux vinaigrette —Leeks with vinaigrette

Salade de lentilles —Lentil salad

Salade bretonne —Seafood and artichoke salad

Salade des nonnes —Molded rice salad with truffles and chicken

Salade d'épinards —Spinach salad with bacon and poached egg

Salade niçoise —Salad with tuna confit and provençal vegetables

Salade de riz au crabe —Rice salad with crab

Salade de pommes de terre —Potato salad

Tomates garnies à la bretonne —Tomatoes filled with couscous salad

ASPERGES FROIDES, SAUCE HOLLANDAISE

Learning Outcomes

Hollandaise sauce
Cuisson à l'Anglaise
Beurre clarifié
Émulsion
Sabayon

Equipment

Knives:
Vegetable peeler (*économe*), paring knife (*office*), slicing knife (*éminceur*)

Tools:
Bowls, cutting board, araignée, ladle (*louche*), whisk

Pans:
Small and large sauce pans (*russe*), bain marie

Serving

4 persons

FYI...

In this recipe, the asparagus is boiled à *l'Anglaise*, which translates loosely as "English style." When the words *English* and *cooking* are used in the same sentence, it is often in the context of a culinary joke. Irish playwright George Bernard Shaw once said, "If the British can survive their meals they can survive anything!" However, putting to rest stereotypes surrounding English cuisine, the *Grand Dictionnaire du Gastronome* devotes two pages in praise of traditional English fare. In any event, *cuisson à l'anglaise* is only one aspect of this preparation. For the history of sauce hollandaise see page 465 in *Cuisine Foundations*.

ASPERGES FROIDES, SAUCE HOLLANDAISE

Asparagus with Hollandaise Sauce

Quantity		Ingredient
U.S.	Metric	Base
½ pc	½ pc	Lemon, juice of
2 lb	1 kg	White asparagus
		Salted water
		Hollandaise Sauce
3 pcs	3 pcs	Egg yolks
1 ½ fl oz	40 mL	Water
½ pc	½ pc	Lemon, juice of
8 oz	250 g	Butter, clarified
		Salt and white ground pepper
		To Serve
4 brs	4 brs	Chervil (or parsley)

Method

Prepare the Asparagus

1. Bring a large pot of lightly salted water to a boil over high heat. Add the juice of half a lemon to the water.
2. Peel the asparagus spears from top to bottom, turning them to preserve their shape. Be careful not to break the stalks. Line up the spears and trim the ends to equal lengths, then tie them in bundles of 10 using butcher's twine. Place the asparagus bundles in the boiling water and cook them (***cuisson à l'anglaise***) until a knife can easily pierce the stems. Transfer to an ice bath and, once cooled, remove and place on a towel to drain.

Note: *The asparagus spears should be tender but still hold their shape. When cooked to be served immediately, white asparagus should be cooled on a towel and served at room temperature. When prepared in advance (**mise en place**), the spears should be refreshed quickly in an ice bath to stop the cooking process, to maintain flavor, and to prevent spoilage.*

Sauce Hollandaise

1. Using a whisk, beat the egg yolks, water, and lemon juice together with a little salt in a large mixing bowl. Place the bowl over a bain marie and whisk the mixture vigorously for several minutes. At first it will become light and frothy, then gradually it will increase in density. When it becomes a thick, foam that creates ribbons when the whisk is lifted from the bowl, take the mixture off the bain marie. At this point, incorporate the butter (in a thin, steady stream), whisking continually. Season to taste and reserve in a warm bain marie until ready to use.

Tip: *If the sauce splits, transfer a little sauce to another bowl, add a little cold water, and stir until it is homogeneous. Then, slowly whisk the broken sauce into the repaired sauce until re-emulsified.*

To Serve

1. Serve the asparagus on a folded napkin on a serving dish with the hollandaise sauce on the side. Decorate with sprigs (***pluches***) of chervil.

CÉLERI RÉMOULADE

Learning Outcomes

Sauce mayonnaise
Ciseler
Hacher
Julienne

Equipment

Knives:
Paring knife (*office*), chef knife (*couteau chef*)

Tools:
Bowls, cutting board, wooden spatula, plastic spatula, whisk

Serving

4 persons

FYI...

A rémoulade is a mayonnaise-based sauce with the addition of mustard, chopped herbs, capers, and gherkins. This particular preparation traditionally accompanies a ***julienne*** of celeriac. Likely stemming from the Picardie word for radish (*remola*), the word *rémoulade* first shows up in print in *La Cuisinière Bourgeoise* in 1786.

CÉLERI RÉMOULADE

Celery Root Remoulade

Method

Mayonnaise

1. Place the egg yolks and mustard in a bowl, then season and mix until homogeneous.
2. Incorporate the vegetable oil in a thin stream, whisking continuously until the mixture is dense and glossy.
3. Stir in the lemon juice.

Garniture Rémoulade

1. Into the mayonnaise stir the mustard, chopped capers, gherkins, chervil, and parsley (*optional*).

Base

1. Clean and peel the celery root and immediately rub it with the cut side of the lemon to avoid discoloration.
2. Cut the celery root into a ***julienne***, toss with the lemon juice, and season to taste.

To Serve

1. Combine the ***julienne*** of celery root and the mayonnaise. Mix well and serve.

Quantity		*Ingredient*
U.S.	*Metric*	*Base*
2 ½ lb	1.2 kg	Celery root, cut into ***julienne***
½ pc	½ pc	Lemon, juice of
		Salt and pepper
		Mayonnaise
2 pcs	2 pcs	Egg yolks
¼ oz	10 g	Dijon mustard
8 fl oz	250 mL	Vegetable oil
2 pcs	2 pcs	Lemon, juice of
		Garniture Rémoulade
¾ oz	20 g	Dijon mustard
		Salt and pepper
1 ¾ oz	50 g	Capers, finely chopped (***hacher***)
1 ¾ oz	50 g	Gherkins, finely chopped (***hacher***)
¾ oz	20 g	Chervil, finely chopped (***hacher***)
¾ oz	20 g	Parsley, finely chopped (***hacher***) (*optional*)
Pinch	Pinch	Cayenne pepper
		Salt and pepper

CHAMPIGNONS À LA TURQUE

Learning Outcomes

Court mouillement
Émonder and concasser tomatoes
Torréfier
Cuisson à la Turque

Equipment

Knives:
Paring knife (*office*), slicing knife (*éminceur*)

Tools:
Bowls, wooden spatula, cutting board, skimmer (*écumoire*), ladle (*louche*)

Pans:
Large sauce pan (*russe*)

Serving

6 persons

French cuisine has been both influenced and complimented by other cultures. This recipe is a variation of "légumes à la grecque," which is known for the tang from the white wine and lemon juice and the flavor of the whole coriander seeds. Preparations "À la turque" find their roots in Turkish cuisine which is itself a culmination of Asian, Middle Eastern, and European influences. The Ottoman Empire (ruled over by what we now call Turkey) spanned three continents and integrated the culinary traditions of its provinces.

CHAMPIGNONS À LA TURQUE

Stewed Mushrooms with Currants

Method

Cuisson à la Turque

1. Soak the currants in white wine.
2. Toast (***torréfier***) the coriander seeds and black peppercorns in a small pan over low heat.

Note: *To bring out their fullest flavor, aromatic seeds should be lightly toasted before use.*

1. Heat the olive oil in a medium sauté pan over medium heat and add the pearl onions, diced onions, and shallot. Sweat (***suer***) until soft, then add the toasted coriander seeds and peppercorns. Add the ***bouquet garni***, then cover and cook slowly (***étuver***) for about 10 minutes (*being careful not to let it color*).
2. Strain the currants (*reserve the wine*) and add the currants to the pan. Add the tomato paste and cook for 2 to 3 minutes (***pincer la tomate***), then add the mushrooms and lemon juice and season lightly. Wet (***mouiller***) with the white wine and reduce by half to decrease the acidity. Add the tomato ***concassée*** and let the mixture cook until thick (about 15 minutes). Pour in the chicken stock and leave the mixture to cook gently for 5 to 10 minutes more.Remove from the heat.

Finishing

1. Taste and adjust seasoning. Add fresh cilantro and cayenne to taste.

To Serve

1. This dish is best served cold, but it can also be served lukewarm. Serve in the cooking liquid.

Tip: *Let rest overnight in the refrigerator for best flavor.*

Quantity		*Ingredient*
U.S.	*Metric*	*Base*
1 ¼ lb	600 g	Button mushrooms, cleaned (*not peeled*)
1 pc	1 pc	Lemon, juice of
		Salt and pepper
		Cuisson à la Turque
1 ¾ oz	50 g	Currants
8 fl oz	250 mL	White wine
¾ oz	20 g	Coriander seeds
1 tsp	3 g	Black peppercorns
5 fl oz	150 mL	Olive oil
4 oz	120 g	Pearl onions, peeled
½ pc	½ pc	Onion, finely diced (***ciseler***)
1 pc	1 pc	Shallot, finely diced (***ciseler***)
1 pc	1 pc	***Bouquet garni***
1 ¼ oz	40 g	Tomato paste
3 pcs	3 pcs	Tomatoes, peeled (***émonder***), seeded (***épépiner***), and diced (***concasser***) (*see page 115 in Cuisine Foundations*)
8 fl oz	250 mL	Chicken stock
		Salt
		Finishing
1 bq	1 bq	Cilantro, roughly chopped (***ciseler***)
Pinch	Pinch	Cayenne pepper

LES CRUDITÉS ET LEURS SAUCES

Learning Outcomes

Vegetable cuts
Sauce rémoulade
Horseradish sauce
Emulsion

Equipment

Knives:
Vegetable peeler (*économe*), paring knife (*office*), slicing knife (*éminceur*)

Tools:
Rubber spatula, mixing bowl, whisk

Serving

6–8 persons

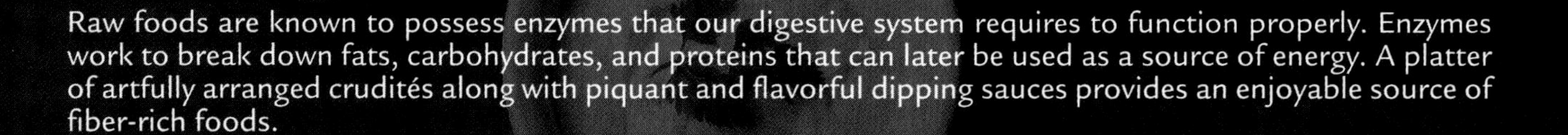

FYI...

Raw foods are known to possess enzymes that our digestive system requires to function properly. Enzymes work to break down fats, carbohydrates, and proteins that can later be used as a source of energy. A platter of artfully arranged crudités along with piquant and flavorful dipping sauces provides an enjoyable source of fiber-rich foods.

LES CRUDITÉS ET LEURS SAUCES

Raw Vegetables and Sauces

Method

Base

1. Prepare the vegetables: Trim the peppers of their stems, seeds, and spongy ribs. Cut into a ***julienne***.
2. Peel the carrots and cut into a ***julienne***.
3. Peel the tomatoes (***émonder***) by plunging them into boiling water for about 30 seconds and refreshing them in an ice water bath.
4. Slice (***émincer***) the radishes, cucumber, and peeled tomatoes.
5. Peel the celery stalk and slice on the bias (***sifflet***).

Horseradish Dip

1. Whisk the cream and sugar to soft peaks. Fold in the grated horseradish, lemon juice, breadcrumbs, and season to taste with salt and pepper.

Sauce Rémoulade

1. Make a mayonnaise: Place the egg yolks and mustard in a bowl, then season and mix until homogeneous.
2. Incorporate the vegetable oil in a thin stream, whisking continuously until the mixture is dense and glossy.
3. Stir in the vinegar.
4. Mix in the chopped capers, gherkins, parsley, chervil, tarragon, anchovy essence, and the remaining mustard.

To Serve

1. Arrange the vegetables on a platter and garnish with mixed greens. Serve the sauces and remaining mixed greens on the side.

Note: *If making the sauces in advance, reserve in the refrigerator in a covered container. Because they contain raw egg yolk, these sauces are fragile and should be made the day of serving.*

Quantity		Ingredient
U.S.	*Metric*	*Base*
1 pc	1 pc	Red pepper, cut into ***julienne***
1 pc	1 pc	Yellow pepper, cut into ***julienne***
1 pc	1 pc	Green pepper, cut into a ***julienne***
3 pcs	3 pcs	Carrots, cut into ***julienne***
4 pcs	4 pcs	Tomatoes, peeled (***émonder***), thinly sliced (***émincer)***
8 pcs	8 pcs	Radishes, thinly sliced (***émincer***)
1 pc	1 pc	Cucumber, thinly sliced (***émincer***)
3 pcs	3 pcs	Celery stalk, cut into ***sifflet***
		Mixed greens

Quantity		Ingredient
U.S.	*Metric*	*Horseradish Dip*
4 oz	125 mL	Heavy cream
1 tsp	4 g	Sugar
¾ oz	25 mL	Grated horseradish
1 tsp	5 mL	Lemon, juice of
2 tbsp	7 g	Fresh breadcrumbs
		Salt and freshly ground pepper
		Sauce Rémoulade
2 pcs	2 pcs	Egg yolks
½ oz	10 g	Dijon mustard
8 oz	250 mL	Vegetable oil
½ oz	10 mL	White wine vinegar
¾ oz	25 g	Capers, finely chopped (***hacher***)
¾ oz	25 g	Gherkins, finely chopped (***hacher***)
2 brs	2 brs	Parsley, finely chopped (***hacher***)
2 brs	2 brs	Chervil, finely chopped (***hacher***)
2 brs	2 brs	Tarragon, finely chopped (***hacher***)
1 tsp	5 mL	Anchovy essence
		Salt and pepper

MACÉDOINE DE LÉGUMES

Learning Outcomes

Macédoine cut
Cuisson à l'Anglaise
Mayonnaise
Stable cold emulsion

Equipment

Knives:
Vegetable peeler (*économe*), slicing knife (*éminceur*)

Tools:
Mixing bowl, whisk, strainer, savarin mold

Pans:
4 small sauce pans (*russes*)

Serving

4 persons

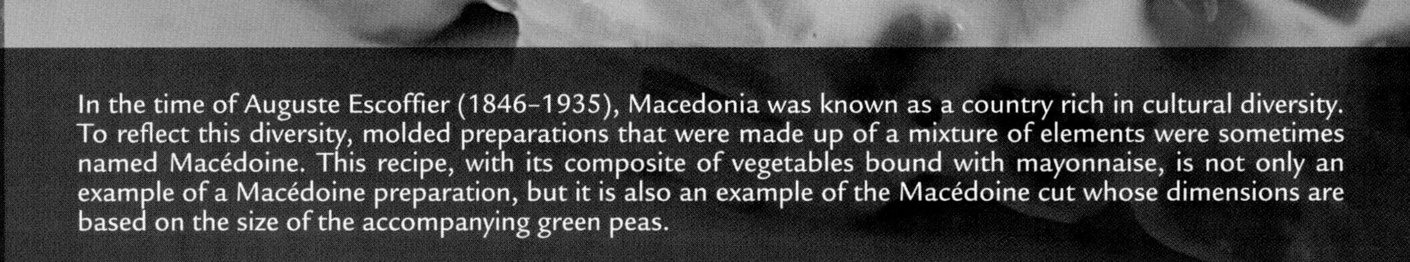

FYI...

In the time of Auguste Escoffier (1846–1935), Macedonia was known as a country rich in cultural diversity. To reflect this diversity, molded preparations that were made up of a mixture of elements were sometimes named Macédoine. This recipe, with its composite of vegetables bound with mayonnaise, is not only an example of a Macédoine preparation, but it is also an example of the Macédoine cut whose dimensions are based on the size of the accompanying green peas.

MACÉDOINE DE LÉGUMES

Diced Vegetable Salad

Quantity		*Ingredient*
U.S.	*Metric*	*Base*
8 oz	250 g	Carrot, cut into ***macédoine***
8 oz	250 g	Turnip, cut into ***macédoine***
8 oz	250 g	Green beans
8 oz	250 g	Peas
		Coarse salt
		Mayonnaise
2 pcs	2 pcs	Egg yolks
½ oz	10 g	Dijon mustard
1 pt	500 mL	Vegetable oil
½ fl oz	10 mL	White wine vinegar
		Salt and white pepper
		To Serve
		Fresh herbs or micro greens

Method

Mayonnaise

1. Mix the egg yolks and mustard in a bowl until homogeneous, then season to taste. Pour in the vegetable oil in a thin stream, whisking continuously until the mixture is dense and glossy. Stir in the vinegar. Reserve the mayonnaise in a covered container in the refrigerator until needed.

Macédoine

1. Cut the carrot and turnip into a ***macédoine*** [¼ in. (5 mm) cubes, 1.4 in. (5 mm) per side]. Cut the green beans to the same size as the ***macédoine***.

Note: *The size of the* ***macédoine*** *may need to be adjusted to match the dimensions of the peas.*

2. Cook the vegetables separately in boiling salted water. Cook each vegetable until tender, then refresh immediately in ice water, strain, and place on a clean towel to drain.

Finishing

1. Line the savarin mold with plastic wrap and reserve in the refrigerator.
2. Mix the vegetables together and bind them with the mayonnaise. Taste and adjust the seasoning.
3. Fill the plastic-lined mold, pressing well, and tapping the pan lightly on the work surface to help the vegetables settle. Refrigerate for at least 1 hour or overnight.

To Serve

1. Place a chilled plate upside down on the mold. Then, holding it securely, turn it over. Remove the mold and gently peel away the plastic wrap from the ***macédoine***. Decorate with fresh herbs or micro greens.

POIREAUX VINAIGRETTE

Learning Outcomes

Stable emulsion
(*émulsion stable*)
Cuisson à l'Anglaise
Vinaigrette sauce
Hard cooked eggs

Equipment

Knives:
Paring knife (*office*),
slicing knife (*éminceur*)

Tools:
Bowls, cutting board,
araignée, ladle (*louche*),
whisk, kitchen twine

Pans:
Marmite

Serving

4 persons

FYI...

Leeks were originally known as "poor man's asparagus" and did not find their way onto the tables of the wealthy until the 19th century. Prior to this time, the leek was reserved for applications like pot-au-feu and stock flavoring; however, in poireaux vinaigrette, the leek is not hidden away but is rather featured and dressed liberally with vinaigrette and crumbled egg.

A classic vinaigrette is a stable emulsion; the presence of mustard acts as a stabilizer preventing the oil and vinegar from separating.

POIREAUX VINAIGRETTE

Leeks with Vinaigrette

Method

Cuisson

1. Cut the dark green leaves off the leeks and make a 1 ½ in. (3 to 4 cm) long incision into the middle of the green end (*to help remove dirt*). Soak in a large bowl of cold water to remove the dirt. If the leeks are especially sandy, agitate in the water to remove the sand from the inner leaves.
2. Tie the leeks in a bunch and cook them ***à l'Anglaise:***
3. Bring a large pot of salted water to a boil over high heat and cook the leeks until they can be pierced with a knife without resistance. Refresh them quickly in an ice bath. Drain the leeks thoroughly and squeeze firmly to extract the maximum amount of water. Remove the string and allow to continue to drain on kitchen towels.

Vinaigrette

1. Dissolve the salt and pepper in the vinegar and whisk in the Dijon mustard. Gradually incorporate the oil in a steady stream, whisking continuously until thick. Mix in the chopped garlic, chervil, and shallots.

To Serve

1. Place the leeks on a chilled platter, coat with some vinaigrette on top, and serve the rest on the side. Crumble one hard boiled egg over the leeks and serve the other halved.
2. Decorate with a sprig (***pluche***) of chervil.

Quantity		*Ingredient*
U.S.	*Metric*	*Base*
2 lbs	1 kg	Leeks
		Coarse salt
		Vinaigrette
1 fl oz	30 mL	Red wine vinegar
½ oz	15 g	Dijon mustard
5 fl oz	150 mL	Oil
2 pcs	2 pcs	Garlic cloves, finely chopped (***hacher***)
5 brs	5 brs	Chervil (or parsley), finely chopped (***hacher***)
2 pcs	2 pcs	Shallots, finely diced (***ciseler***)
		Salt and white ground pepper
		To Serve
2 pcs	2 pcs	Eggs, hard cooked (***durs***), peeled (***écaler***)
2 brs	2 brs	Chervil

Learning Outcomes

SALADE DE LENTILLES

Cuisson des légumes secs
Émulsion instable
Lardons
Blanchir
Rafraîchir
Hacher
Ciseler
Suer
Mijoter

Equipment

Knives:
Vegetable peeler (*économe*), paring knife (*office*), slicing knife (*éminceur*)

Tools:
Bowls, cutting board, wooden spatula, colander, araignée, skimmer (*écumoire*), ladle (*louche*), whisk

Pans:
Sauteuse, sautoir, small sauce pan (*russe*), small stock pot (*marmite*)

Serving

4 persons

FYI...

High in protein and minerals, lentils have been used as a food source for over 8,000 years. Native to southwestern China and northern Asia, this legume was being cultivated even as wild lentils were still growing in abundance. In this preparation, the lentils' natural nutty flavor is contrasted and complemented by the sharpness of the vinaigrette.

SALADE DE LENTILLES
Lentil Salad

Method

Cuisson

1. Blanch (***blanchir***) the lentils by placing them in a large pan of cold water and bring to a boil over high heat. Strain and return the lentils to the pan. Cover the lentils with fresh water and add the ***bouquet garni***, garlic head, carrot, and onions ***clouté***. Bring to a simmer over high heat, then reduce the heat to low and let the lentils cook until tender but not mushy (*about 35 minutes*). Season the lentils 20 minutes into the cooking process.

Note: *Adding salt too early in the cooking process can prevent the lentils from soaking up water during cooking, resulting in a hard texture.*

2. While the lentils are cooking, blanch (***blanchir***) the salted slab bacon: Place in a pan of cold water, and bring to a boil over high heat. Allow to boil for 5 minutes, then strain and add to the lentils.

Vinaigrette

1. Dissolve the salt and pepper in the vinegar and whisk in the Dijon mustard. Gradually incorporate the oil in a steady stream, whisking continuously until thick. Mix in the chopped garlic and shallots.

Garniture

1. *Blanch the cabbage:* Bring a large saucepan of salted water to a boil over high heat. Remove and discard the outer leaves and cut out the core. Separate the leaves, then place a few at a time and boil for 1 minute. Refresh in an ice bath and drain well.
2. Finely mince the cabbage, finely chop the garlic and parsley, and finely dice the shallot and onion. Set aside.
3. Once cooked, strain the lentils and pork. Remove and discard the rind from the bacon and cut four slices. Set aside. Cut the remaing bacon into lardons.
4. Heat the olive oil in a medium sauté pan over low heat and sweat (***suer***) the shallots, garlic, onion, and cabbage. Add the lardons and wet (***mouiller***) with the chicken stock. Simmer for 10 minutes and adjust the seasoning. Stir in the lentils, remove from the heat, and allow to cool until warm.

To Serve

1. Dress the lentils with the vinaigrette and mix in the chopped parsley. Transfer to a serving bowl and top with the cooked slices of bacon. Serve lukewarm.

Tip: *Some of the lentils' cooking liquid can be added to the finished salad if it is too dry. If the lentils are cooked in advance, cool them down and reserve in the cooking liquid to prevent them from drying out.*

Quantity		*Ingredient*
U.S.	*Metric*	*Base*
10 oz	300 g	Lentils, soaked overnight
1 pc	1 pc	***Bouquet garni***
1 pc	1 pc	Garlic head, whole
1 pc	1 pc	Carrot, peeled, whole
2 pcs	2 pcs	Onions, studded (***clouté***) with a clove
4 oz	125 g	Salted slab bacon
		Salt and pepper
		Vinaigrette
1 ¾ fl oz	50 mL	Red wine vinegar
1 oz	30 g	Dijon mustard
5 fl oz	150 mL	Olive oil
2 pcs	2 pcs	Garlic cloves, finely chopped (***hacher***)
2 pcs	2 pcs	Shallots, finely diced (***ciseler***)
		Salt and pepper
		Garniture
½ pc	½ pc	Cabbage, blanched, finely minced (***ciseler***)
2 pcs	2 pcs	Garlic cloves, finely chopped (***hacher***)
2 pcs	2 pcs	Shallots, finely diced (***ciseler***)
1 pc	1 pc	Onion, finely diced (***ciseler***)
2 tsp	10 mL	Olive oil
7 fl oz	200 mL	Chicken stock
		Salt and pepper
		To Serve
5 brs	5 brs	Italian parsley, finely chopped (***hacher***)

Learning Outcomes

Cuisson des légumes à sec
Rafraîchir
Hacher
Émulsionner
Concasser
Infuser
Cuisson à l'Anglaise and marinière
Ébarber
Tourner
Brunoise
Blanc de cuisson

Equipment

Knives:
paring knife (*office*), slicing knife (*éminceur*)

Tools:
Bowls, fork, wooden spatula, fine chinois, colander, cutting board, skimmer (*écumoire*), ladle (*louche*), whisk, tamis

Pans:
Small sauce pan (*russe*), small stock pot (*marmite*)

Serving

4 persons

FYI...

Characteristic of the cuisine of Brittany, this recipe includes many ingredients that are either grown in, or fished off the coast of this region of France. Artichokes and cauliflower, for instance, are widely cultivated in the region. Salade bretonne also makes use of haricots (white beans) as well as mussels—both of which are quintessential Breton ingredients.

SALADE BRETONNE

Seafood and Artichoke Salad

Quantity		Ingredient
U.S.	**Metric**	***Base***
7 oz	200 g	White beans
6 pcs	6 pcs	Artichokes, turned
3 ½ oz	100 g	Flour
½ pc	½ pc	Lemon, juice of
		Salt
		Garniture
14 oz	400 g	Mussels, cooked marinière
2 pcs	2 pcs	Shallots, finely diced (***ciseler***)
3 ½ fl oz	100 mL	White wine
1 pc	1 pc	***Bouquet garni***
8 pcs	8 pcs	Medium shrimp tails, shelled, deveined
1 ¾ oz	50 g	Leek whites, cut into ***brunoise***
1 ¾ oz	50 g	Celery branch, cut into ***brunoise***
10 brs	10 brs	Italian parsley, roughly chopped (***concasser***)
½ pc	½ pc	Cauliflower (florettes)
		Salt and pepper
		Vinaigrette
3 ½ fl oz	100 mL	Cream
1 tsp	5 g	Saffron
½ fl oz	15 mL	Mussel cooking juice
1 ¾ oz	50 g	White beans (cooked)
4 pcs	4 pcs	Garlic cloves, finely chopped (***hacher***)
1 ¾ fl oz	50 mL	Red wine vinegar
3 ½ fl oz	100 mL	Oil
		Salt and pepper
		Finishing
4 pcs	4 pcs	Tomatoes, peeled, seeded, and diced (***concasser***)
		To Serve
1 ¾ oz	50 g	Parsley, finely chopped (***hacher***)
		Fresh herbs

Method

Base

1. Soak (***tremper***) the white beans in cold water for 12 hours, then drain them and place them in a stock pot. Cover the beans with cold water and bring to a boil over high heat. Reduce the heat to a simmer and cook the beans until tender (*about 40 minutes*); add salt 20 minutes into the cooking process and periodically skim (***écumer***) the surface of the pot. Once cooked, drain the beans and set aside to cool.
2. Prepare a ***blanc de cuisson***: In a large pan, dissolve some flour in cold water, add the lemon juice, and season with salt.
3. Peel and turn the artichokes (see page 160). Rub them with a lemon half to prevent discoloration. Place the artichokes in the ***blanc de cuisson*** over low heat, cover with a parchment paper lid, and cook them until they are easily pierced with the tip of a knife. Cool the artichokes in their cooking liquid, and reserve in the refrigerator until ready to use.

Garniture

1. *Cook the mussels **à la marinière***: Clean the mussels and rinse them in several changes of water. Heat some butter in a large pan over low heat and sweat (***suer***) the shallots until they are soft. Add the white wine and bouquet garni, and reduce by two-thirds. Increase the heat to high and bring the liquid to a boil. Add the mussels, then cover and shake the pot as a way of stirring throughout the cooking process. Cook until the mussels have opened, about 5 minutes. Let the mussels cool in their cooking liquid before decanting them. Discard any closed mussels. Remove the mussels from their shells and tear off the mantles (***ébarber***). Strain the cooking liquid through a fine mesh sieve (***chinois***) and set aside.
2. Peel and devein (***dénerver***) the shrimp and place them in a medium pan with the leek whites, shallots, celery, and parsley. Pour in the cooking liquid from the mussels and cook over low heat for 5 minutes or until pink. Strain and reduce the cooking liquid by half. Allow the shrimp to cool, then reserve in the refrigerator.
3. Cut the cauliflower into florettes (***inflorescences***) and cook them ***à l'Anglaise***: Bring a large pot of salted water to a boil over high heat and cook the florettes until tender. Refresh the florettes in an ice bath, then strain and reserve in the refrigerator.

Vinaigrette

1. Infuse the cream with saffron over low heat for 5 minutes. Add a tablespoon of the seafood cooking liquid and strain through a fine mesh sieve (***chinois***).
2. Press 1 ¾ oz (50 g) of the cooked beans through a drum sieve (***tamis***) using a plastic scraper and stir this purée into the cream. Then add the garlic, vinegar, and seasoning.
3. Whisking continuously and incorporate the oil in a thin stream until emulsified.

Finishing

1. Drain the artichokes and remove the choke. Reserve the four nicest bottoms. Cut the remaining two in small pieces and combine with the beans, cauliflower, tomatoes, mussels, and shrimp. Season to taste with the vinaigrette, then stir in the parsley.

To Serve

1. Place the artichoke hearts on a chilled plate, and fill with the seafood salad. Decorate with fresh herbs and drizzle with the remaining vinaigrette.

SALADE DES NONNES

Learning Outcomes

Pocher
Mouler
Beurrer
Chemiser
Clouter

Equipment

Knives:
Paring knife (*office*), slicing knife (*éminceur*)

Tools:
Bowls, fork, wooden spatula, plastic spatula, colander, cutting board, skimmer (*écumoire*), whisk, round cutter

Pans:
Medium sauce pan (*russe*)

Serving

4 persons

FYI...

Truffles are like hidden gems beneath the soil. These elusive fungi grow with only the roots of select tree species and in select environments. Unlike mushrooms (which rely on air currents to transfer spores), truffles rely on animals to consume them and then to spread the spores through defecation. To promote this process (called mycophagy), truffles have evolved to emit a scent to attract animals to ensure their species' survival.

The aromatic chicken stock can be used to make an aspic with which the salade de nonnes can be coated to keep it from drying out.

SALADE DES NONNES

Molded Rice Salad with Truffles and Chicken

Method

Base

1. Heat the chicken stock, whole black truffle, truffle juice, studded onion, and ***bouquet garni*** in a large pan over low heat. Season, add the chicken breast and gently poach the chicken for 20 minutes without letting the liquid boil. Remove the chicken and truffle from the saucepan and set aside to cool.
2. Strain the poaching liquid into a clean pan, bring to a boil, and add the rice. Reduce the heat to a simmer and cook the rice until tender (*20 to 25 minutes*). Strain the rice but do not rinse it as this will remove the starch and flavor. Reserve the cooking liquid for another use. Spread the rice out on a tray (*to cool faster*). Once the rice is at room temperature, reserve it in the refrigerator. Thinly slice the chicken and truffle (***émincer***) and cut small circles out of the slices using a round cutter. Finely chop (***hacher***) the truffle trimmings and reserve them separately from the chicken trimmings.

Montage

1. Whip the cream to soft peaks, fold in the lemon juice, and season to taste.
2. Mix the finely chopped truffle into the rice, adding enough of the lemon cream to bind the mixture.
3. Keeping in mind that the design will be visible when the salad is unmolded, arrange the chicken and truffle circles in a decorative pattern around the inside of the bowl. Cover the chicken and truffle rounds with a thin layer of rice (***chemiser***) and fill the rest of the cavity with alternating layers of rice and chicken trimmings. Press well, cover with plastic wrap and chill in the refrigerator.

To Serve

1. To unmold the salad, loosen the edges with the tip of a small knife. Flip the bowl onto a plate. Holding the bowl securely with the plate, give it a few good, strong shakes. Remove the bowl and serve the remaining cream on the side.

Tip: *If desired, the bowl can be lined with a sheet of plastic wrap to make unmolding easier.*

Quantity		*Ingredient*
U.S.	*Metric*	*Base*
1 pt	500 mL	Chicken stock
1 pc	1 pc	Black truffle
3 ½ fl oz	100 mL	Truffle juice
¼ pc	¼ pc	Onion, studded with a clove (***clouter***)
1 pc	1 pc	***Bouquet garni***
5 oz	150 g	Chicken breast
6 oz	175 g	Rice
		Salt and pepper
		Montage
7 fl oz	200 mL	Cream
½ pc	½ pc	Lemon, juice of
		Salt and pepper
		To Serve
As needed	As needed	Clarified butter, to add shine (lustrer)

SALADE D'ÉPINARDS

Learning Outcomes

Dégorger
Escaloper
Lardons
Émulsion
Croûtons
Oeuf poché hors coquille
Chauffante

Equipment

Knives:
Paring knife (*office*), slicing knife (*éminceur*)

Tools:
Mixing bowls, skimmer (*écumoire*), whisk

Pans:
Frying pan, large pan

Serving

4 persons

FYI...

This particular *salade d'épinards* is a variation on both a traditional *salade de gésier* (gizzard salad) and a *salade frisée* (curly endive salad). In this recipe, spinach accompanies the curly endive and the sautéed chicken livers stand in for the gizzards. Additionally, the presence of poached eggs and lardons are aspects of the original recipes.

SALADE D'ÉPINARDS

Spinach Salad with Bacon and Poached Egg

Method

Base

1. *Clean and trim the livers:* Remove and discard any green bile spots. Place in cold milk to degorge.
2. Tear the spinach and curly endive into bite-size pieces and reserve in the refrigerator, covered by a wet kitchen towel.
3. *Prepare the poached eggs*: Fill a large saucepan three-quarters full of water and generously season with white vinegar and coarse salt. Bring the water to a boil over high heat, then reduce to a simmer. Break the eggs one at a time into a ramekin then slip the eggs into the simmering water. ***Note:*** *The rising bubbles will prevent the eggs from sticking to the bottom of the pan and will also help the whites envelop the yolks.*
4. When the whites begin to coagulate, use a skimmer to help them wrap firmly around the yolks. Poach the eggs for about 3 minutes or until the whites are set but the yolks are still very soft. Gently remove them from the hot liquid with a skimmer and transfer to an ice bath to cool. Once cooled, trim off any loose whites (***ébarber***) and transfer the eggs to a towel-lined tray.

Vinaigrette

1. Dissolve the salt in the vinegar, add the pepper, and whisk in the Dijon mustard. Gradually incorporate the oil in a steady stream, whisking continuously until emulsified. Stir in the garlic and herbs.

Garniture

1. *Blanch the lardons:* Place the lardons in a pan of cold water and bring to a boil over high heat. Blanch the lardons for 1 minute, then refresh under cold running water and drain. Heat the oil in a sauté pan over high heat and sauté the lardons until golden. Drain on a paper towel and reserve the cooking pan (*along with the fat from the lardons*) for the livers.
2. *Prepare the croutons:* Heat the clarified butter in a large, shallow pan over medium-high heat and toss the diced bread into the butter until golden on all sides. Drain on a paper towel.
3. Bring a pot of salted water to a boil to reheat the poached eggs (***chauffante***).
4. *Cook the livers:* Drain and pat the livers dry, denerve them, slice them on the bias (***escaloper***), and season. Reheat the pan used for cooking the lardons over medium-high heat and sauté the livers until they are pink in the middle (***rosé***). Set aside the livers on a wire rack near the stove to keep them warm. Deglaze the pan with a little vinaigrette.

To Serve

1. Toss the spinach and curly endive in the vinaigrette and place in the bottom of a salad bowl. Arrange some lardons, croûtons, and chicken livers on top.
2. Using a skimmer (***écumoire***), dip the eggs in the boiling water for 20 seconds to reheat. Drain the eggs and arrange them on top of the salad.
3. To finish, drizzle a little vinaigrette over the salad.

Quantity		*Ingredient*
U.S.	*Metric*	***Base***
10 oz	300 g	Spinach, washed, destemmed
5 oz	150 g	Curly endive
4 pcs	4 pcs	Eggs
3 ½ fl oz	100 mL	White vinegar
		Coarse salt
		Vinaigrette
1 fl oz	30 mL	Red wine vinegar
1 oz	30 g	Dijon mustard
3 ½ fl oz	100 mL	Vegetable oil
2 pcs	2 pcs	Garlic cloves, finely chopped (***hacher***)
4 brs	4 brs	Parsley, finely chopped (***hacher***)
2 brs	2 brs	Chives, finely chopped, ***ciseler***
		Salt
		Pepper
		Garniture
3 ½ oz	100 g	Salted slab bacon, cut into lardons
1 fl oz	30 mL	Oil
8 fl oz	250 mL	Butter, clarified
7 oz	200 g	Chicken livers, degorged (***dégorger)***
8 pcs	8 pcs	White bread, crust removed, small dice
		Salt and pepper

SALADE NIÇOISE

Learning Outcomes

Confire
Cuire à l'Anglaise
Canneler
Dégorger
Émincer
Rafraîchir
Oeufs durs
Évider
Unstable emulsion

Equipment

Knives:
Paring knife (*office*), slicing knife (*émınceur*)

Tools:
Bowls, plastic spatula, colander, cutting board, araignée, skimmer (*écumoire*), ladle (*louche*), whisk

Pans:
Sauteuse, sautoir, small sauce pan (*russe*), small stock pot (*marmite*)

Serving

4 persons

FYI... *Niçoise* in culinary preparations indicates a culinary style that is characteristic of the region of France that surrounds the city of Nice. Local produce such as olives, garlic, and tomatoes, as well as local Mediterranean fish such as anchovies, are all well represented in salade niçoise.

SALADE NIÇOISE

Salad with Tuna Confit and Provençal Vegetables

Method

1. Degorge (***dégorger***) the anchovies in a bowl of cold milk.

Tuna Confit

1. Cut the tuna into 2 in. (5 cm) thick slices. Heat the olive oil in a medium, heavy-bottomed pan over low heat. Add the thyme, bay leaves, garlic, shallot, basil, black peppercorns, rosemary, and sea salt. Gently cook the tuna for about 30 minutes, then remove the pan from the heat and allow to cool in the oil.

Note: *During cooking, the oil should stay within a temperature range of 150 °F to 160 °F (65 °C to 70 °C). The tuna is ready when it crumbles easily.*

Salade

1. Wash and trim the lettuce, radishes, cucumber, celery, tomatoes, green beans, potatoes, and peppers in cold water.
2. *Cook the potatoes:* Place the potatoes in a saucepan with enough cold water to cover them. Bring the water to a boil over high heat and cook the potatoes until a knife can be inserted without resistance (*about 20 minutes*). Drain the potatoes, then peel them while they are still hot. Set aside.
3. Remove the stem and peel the tomatoes (see page 114). Set aside.
4. *Cook the green beans **à l'Anglaise:*** Bring a saucepan of salted water to a boil over high heat, then add the green beans. Once the beans are tender, refresh them in an ice bath, then split them in two along the seam. Set aside.
5. *Hard cook the eggs:* Place the eggs in a small pan with cold water and bring to a boil over high heat. Reduce the heat to a simmer and leave the eggs to cook for 10 minutes. Refresh them under cold running water, peel (***écaler***), and set them aside.
6. Trim and thinly slice (***émincer***) the radishes.
7. Channel (***canneler***) the cucumber, then cut it in half lengthwise and scoop out the seeds with a spoon (***évider***). Thinly slice (***émincer***) and set aside.
8. Peel the celery and slice it thinly (***émincer***).
9. Cut the peppers into a ***julienne***.
10. Cut the hard cooked eggs and tomatoes into quarters and thickly slice the potatoes.

Vinaigrette

1. Dissolve the salt in the vinegar, then add the pepper and the chopped garlic. Gradually incorporate the oil in a steady stream, whisking continuously until emulsified.

To Serve

1. Season the potato slices by tossing them in some of the vinaigrette, then strain the tuna and break it into pieces.
2. Line a serving platter with the lettuce leaves and arrange the vegetables on top along with the eggs, anchovies, olives, and tuna. Drizzle the salad with some vinaigrette and serve the rest on the side.

Quantity		*Ingredient*
U.S.	*Metric*	*Tuna Confit*
14 oz	400 g	Fresh tuna
7 fl oz	200 mL	Olive oil
2 brs	2 brs	Thyme
2 pcs	2 pcs	Bay leaves
1 pc	1 pc	Garlic clove, finely chopped (***hacher***)
1 pc	1 pc	Shallot
1 br	1 br	Basil
1 ½ tsp	5 g	Black peppercorns
1 br	1 br	Rosemary
		Sea salt
		Salade
1 pc	1 pc	Boston lettuce
16 pcs	16 pcs	Radishes, sliced (***émincer***)
1 pc	1 pc	Cucumber, channeled (***canneler***), cored (***évider***), and sliced (***émincer***)
2 pcs	2 pcs	Celery stalks, peeled, sliced (***émincer***)
1 lb	500 g	Tomatoes, peeled and quartered
5 oz	150 g	Green beans
10 oz	300 g	Potatoes (*fingerling*)
½ pc	½ pc	Red pepper, cut into ***julienne***
½ pc	½ pc	Green pepper, cut into ***julienne***
16 pcs	16 pcs	Black olives (*small*), pitted
4 pcs	4 pcs	Eggs, hard cooked, quartered
1 oz	30 g	Anchovies, degorged in milk
		Vinaigrette
2 ½ fl oz	75 mL	White wine vinegar
4 pcs	4 pcs	Garlic cloves, finely chopped (***hacher***)
7 fl oz	200 mL	Olive oil
		Salt and pepper

SALADE DE RIZ AU CRABE

Learning Outcomes

Riz créole
Rafraîchir
Brunoise
Émulsion

Equipment

Knives:
Paring knife (*office*), slicing knife (*éminceur*), channeling knife (*canneleur*)

Tools:
Bowls, cutting board, fork, wooden spatula, plastic spatula, colander, skimmer (*écumoire*), ladle (*louche*), whisk, round-bottomed bowl

Pans:
Small stock pot (*marmite*)

Serving

4 persons

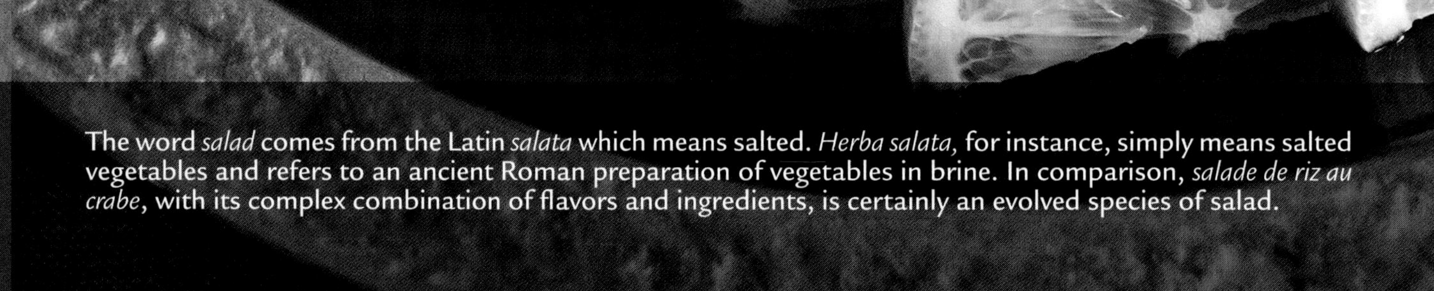

FYI...

The word *salad* comes from the Latin *salata* which means salted. *Herba salata,* for instance, simply means salted vegetables and refers to an ancient Roman preparation of vegetables in brine. In comparison, *salade de riz au crabe*, with its complex combination of flavors and ingredients, is certainly an evolved species of salad.

SALADE DE RIZ AU CRABE

Rice Salad with Crab

Method

Base

1. *Cook the créole rice:* Rinse the rice in a large amount of cold water and drain. Place the rice in a large pan and add the water and coarse salt. Place the pan over medium-high heat and, once the rice comes to a boil, cook the rice for 14 minutes. Remove from the heat, strain, and refresh with cold water. Drain well, then transfer the rice to a colander lined with a kitchen towel. Allow to drain completely.
2. Crumble the crab meat and cut all the vegetables into a ***brunoise***. In a small saucepan over low-to-medium heat, reduce the crab ***bisque*** until it coats the back of a spoon (***à la nappe***).
3. Once the bisque is reduced, combine it with the brunoise of vegetables and the crab meat. Season to taste and mix into the rice.

Vinaigrette

1. Line a round-bottomed bowl with plastic wrap. Dissolve the salt in the vinegar, add the white pepper, and stir in the mustard. Gradually incorporate the oil in a steady stream, whisking continuously until emulsified. Mix in the diced shallots and chopped garlic and adjust the seasoning with salt, white pepper, and a pinch of cayenne.
2. Season the rice mixture to taste with the vinaigrette, then transfer the salade de riz au crabe to the plastic-lined bowl and press well. Cover and chill until ready to serve.
3. Peel tomatoes and cut in 4 to 6 wedges. With a paring knife, cut out the hearts of the tomatoes and discard. Set aside.

To Serve

1. Unmold the salad onto a chilled serving plate and decorate with tomato petals, sliced lemon (*optional*), and fresh herbs.

Quantity		*Ingredient*
U.S.	*Metric*	*Base*
7 oz	200 g	Rice
1 ½ qt	1 ½ L	Water
		Coarse salt
5 oz	150 g	Crab meat, cooked
1 pc	1 pc	Red bell pepper, cut into ***brunoise***
1 pc	1 pc	Green bell pepper, cut into ***brunoise***
3 pcs	3 pcs	Celery stalk, cut into ***brunoise***
3 ½ fl oz	100 mL	Crab ***bisque***
		Salt and pepper
		Vinaigrette
1 ¾ fl oz	50 mL	Red wine vinegar
1 oz	30 g	Dijon mustard
3 ½ fl oz	100 mL	Oil
1 pc	1 pc	Shallot, finely diced (***ciseler***)
1 pc	1 pc	Garlic clove, finely chopped (***hacher***)
Pinch	Pinch	Cayenne pepper
		To Serve
2 pcs	2 pcs	Tomatoes, peeled (***émonder***)
3 brs	3 brs	Parsley, coriander, or chervil
		Optional
1 pc	1 pc	Lemon, channeled (***canneler***), sliced (***rondelles***)

SALADE DE POMMES DE TERRE

Learning Outcomes

Cuisson à l'anglaise
Making an unstable vinaigrette

Equipment

Knives:
Vegetable peeler (*économe*), chef knife (*couteau chef*)

Tools:
Whisk

Pans:
Large sauce pan (*russe*)

Serving

4–6 persons

SALADE DE POMMES DE TERRE
Potato Salad

Method

Cuisson

1. Wash the potatoes well, peel them into regular cyclinders and cut them into thin slices (***émincer***) 1.8 in. (3 mm).
2. Place the slices in a large saucepan of cold salted water. Bring the water to a boil over high heat and cook the potatoes until they can be easily pierced with a knife. Refresh the potato slices under cold running water and drain well.

Vinaigrette

1. Season the vinegar with salt and white pepper. Pour in the oil in a thin stream, whisking continuously to create an emulsion. Set aside.

To Serve

1. Mix the chopped herbs into the vinaigrette. Season the potatoes by delicately tossing them in the vinaigrette until completely coated. Arrange the potatoes on a chilled platter. Serve the remaining vinaigrette on the side.

Quantity		*Ingredient*
U.S.	*Metric*	*Base*
1 lb	500 g	Potatoes, peeled and sliced (***émincer***)
		Vinaigrette
1 fl oz	30 mL	Vinegar
3 fl oz	90 mL	Olive oil
		Salt and pepper
		To Serve
1 oz	30 g	Parsley, finely chopped (***hacher***)
1 oz	30 g	Chervil, finely chopped (***hacher***)
1 oz	30 g	Tarragon, finely chopped (***hacher***)
1 oz	30 g	Chives, finely chopped (***hacher***)

FYI...

English botanist John Gerard, in the 1597 book *Great Herball,* was the first to write, and then to publish, a description of the potato and its possible culinary uses.

TOMATES GARNIES À LA BRETONNE

Learning Outcomes

Évider
Dégorger
Gonfler
Brunoise
Revenir
Chiffonade
Pincer la tomate
Déglacer
Mouiller
Réduire

Equipment

Knives:
Vegetable peeler (*économe*), paring knife (*office*), slicing knife (*éminceur*)

Tools:
Bowls, cutting board, wooden spatula, plastic spatula, fine chinois, skimmer (*écumoire*), ladle (*louche*), melon baller, fork, whisk

Pans:
Sauteuse, baking or shallow tray

Serving

4 persons

FYI...

This recipe makes liberal use of the North African ingredient known as couscous. Couscous is made by first moistening semolina flour, then rolling it out by hand until it forms miniature grains. One theory of the origin of the word *couscous* is that it is an evolution of the old Arabic word for the grainy food passed from the parent bird to its young.

TOMATES GARNIES À LA BRETONNE

Tomatoes Filled with Couscous Salad

Method

Base

1. Cut the top off the tomatoes, then cut a thick slice from the bottom and reserve. The tomatoes will be served upside down. Using a melon baller (***cuillère parisienne***), remove the pulp and ribs from the bottom. Sprinkle the cavity with coarse salt and set aside to degorge upside down on a paper towel-lined rack in a cool place.
2. *Prepare the couscous*: Spread out the couscous on a baking pan in an even layer. Ladle enough hot water or stock evenly over the couscous until just dampened. Cover the pan with a kitchen towel and leave the couscous to absorb the liquid (***gonfler***). Remove the towel when the liquid has been absorbed and gently stir the couscous to separate. Sprinkle with more hot liquid, cover, and allow to absorb. Continue until all the liquid has been absorbed by the couscous, stirring before each addition. Separate the grains using a fork and allow to cool. Optional: the couscous can be rubbed with olive oil before the hot liquid is added as insurance against clumping.)

Sauce

1. Heat the oil in a large sauté pan over medium-high heat and sauté the carrot, onion, celery, and leeks until lightly colored. Add the tomato paste and cook for 1 to 2 minutes (***pincer la tomate***), stirring well. Deglaze with white wine and allow to reduce until almost dry. Add the fish ***fumet*** and the mint and bring the contents of the pan to a boil. Turn down the heat to medium and simmer until the sauce has reduced by half [skimming (***écumer***) off any impurities that might rise to the surface]. Strain the sauce through a fine mesh sieve (***chinois***). Taste and adjust the seasoning, then let the sauce cool to room temperature.

Farce

1. Chop the lobster tails (***hacher***). Peel, seed, and dice (***concasser***) the tomato. Peel and seed the cucumber and cut it into ***brunoise*** and shred the basil (***chiffonade***).
2. Transfer the cooled couscous to a bowl and mix in the lobster, crab meat, tomato, cucumber, and basil.

Finishing

1. Whisk together the sauce, lemon juice, chopped shallot, and cumin. Gradually whisk in the olive oil in a steady stream. Pour some of the finished sauce into the couscous and mix thoroughly. Adjust the seasoning, adding more sauce if needed.

To Serve

1. Thickly slice the lobster tails into medallions.
2. Turn over the tomatoes and fill them with the couscous mixture. On top of each tomato, place a lobster medallion and a sprig of mint, then cover with the reserved bottoms.

Quantity		*Ingredient*
U.S.	*Metric*	*Base*
8 pcs	8 pcs	Medium tomatoes
		Coarse salt
8 oz	250 g	Couscous
¾ fl oz	20 mL	Olive oil (optional)
1 pt	500 mL	Hot water or white stock
		Sauce
1 fl oz	30 mL	Olive oil (optional)
5 oz	150 g	Carrot, cut into ***mirepoix***
5 oz	150 g	Onion, cut into ***mirepoix***
3½ oz	100 g	Celery, cut into ***mirepoix***
3 ½ oz	100 g	Leeks, cut into ***mirepoix***
2 oz	60 g	Tomato paste
10 fl oz	300 mL	White wine
1 qt	1 L	Fish ***fumet***
5 brs	5 brs	Mint
		Farce
2 pc	2 pc	Lobster tails, cooked and shelled
1 pc	1 pc	Tomato, peeled (***émonder***), seeded (***épépiner***), and diced (***concasser***)
3 ½ oz	100 g	Cucumber, cut into ***brunoise***
5 brs	5 brs	Basil, shredded (***chiffonade***)
7 oz	200 g	Crab meat
		Finishing
1 pc	1 pc	Lemon, juice of
1 pc	1 pc	Shallot, finely diced (***ciseler***)
¼ oz	10 g	Cumin
12 fl oz	350 mL	Olive oil
		To Serve
2 pcs	2 pcs	Lobster tails, cooked and shelled
5 brs	5 brs	Mint

Les Entrées-Chaud
(Hot Appetizers)

Les Entrées—Chaud (Hot Appetizers)

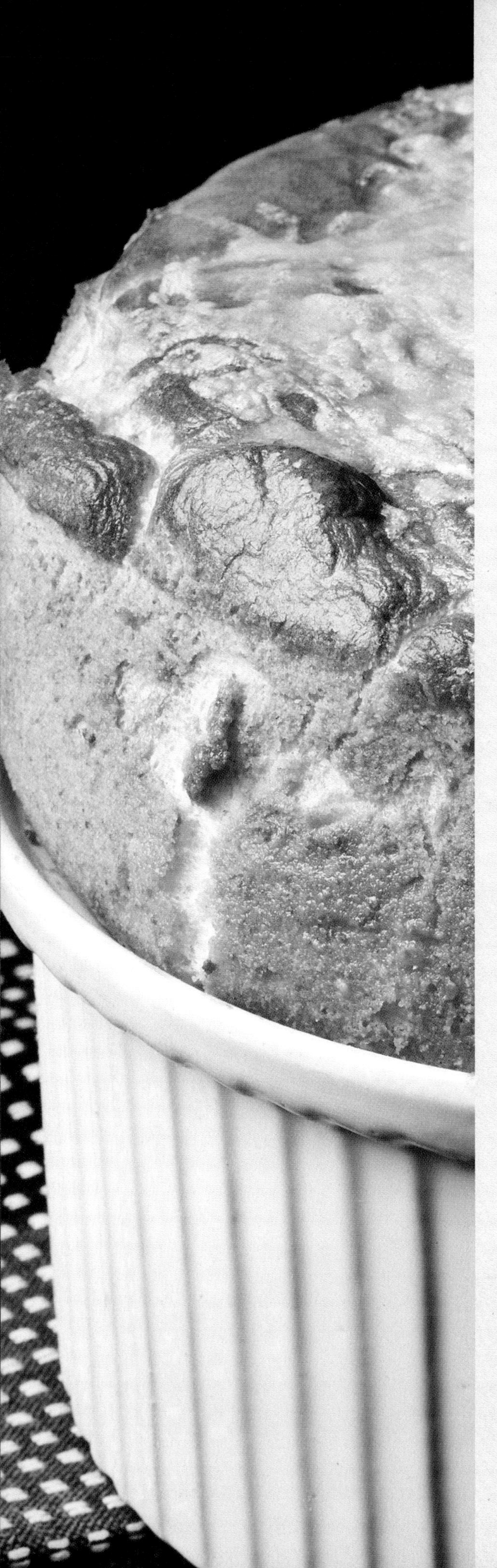

Aubergines imam bayaldi
—Eggplant stuffed with tomatoes and onions

Cromesquis de langoustines
—Deep fried langoustine parcels

Croquettes de poulet et jambon
—Chicken and ham croquettes

Gnocchi à la parisienne
—Tartlets filled with gnocchi in Mornay sauce

Gratin dauphinois
—Scalloped potatoes with cheese

Fritots de ris de veau, sauce tomate, persil frit
—Sweetbread fritters with tomato sauce and fried parsley

Laitue braisée
—Braised lettuce

Petits farcis niçois
—Veal and pork stuffed vegetables

Petits gâteaux de foies blonds
—Chicken liver creams

Pissaladière
—Provençal onion pie

Quiche lorraine
—Savory custard and bacon tart

Ratatouille niçoise
—Provençal stewed vegetables

Rissoles de veau, sauce ravigote
—Veal filled fried pastries

Soufflé au fromage
—Cheese soufflé

AUBERGINES IMAM BAYALDI

Learning Outcomes

Légumes farcis
Tomates concassée
Gratiner
Sauce tomate
Dégorger un légume

Equipment

Knives:
Paring knife (*office*), slicing knife (*éminceur*)

Tools:
Wooden spatula, plastic spatula, food mill, fine chinois, colander, cutting board, skimmer (*écumoire*), cheese grater, bowls

Pans:
Sauteuse, gratin dish, bain marie

Serving

4 persons

FYI...

This preparation is also known as *Imam Bayaldi,* which translates from the Turkish as "the Imam (holy man) has fainted." The legend surrounding the preparation tells the story of a young woman whose veil slips off her face as she leans forward to serve a holy man. Either due to his surprise at the young girl's beauty or because of the intoxicating aroma of the dish, the holy man is said to have fainted. Another version of the story is that he collapsed when he realized how much olive oil was used at a time when the ingredient was very costly.

AUBERGINES IMAM BAYALDI

Eggplant Stuffed with Tomatoes and Onions

Method

1. Preheat the oven to 400°F (205°C).

Prepare the Eggplants

1. Wash the eggplants, remove the tops, and slice in half lengthwise. Scoop the flesh out of the center of each half, leaving a ¼ in. (½ cm) of flesh on the inside of the skins (*this ensures that the eggplants will keep their shape during cooking*).
2. Cut the extracted flesh into a medium dice, place in a colander, and sprinkle with coarse salt to draw out the water (***dégorger***). Let the diced eggplant rest over a bowl for 20 to 25 minutes. After the degorging process is complete, rinse the diced eggplant under running water and pat them dry.

Prepare the Sauce Tomate

1. Heat the olive oil in a small sauté pan over low heat and sweat (***suer***) the chopped shallots until translucent. Add the chopped garlic and sweat for 2 minutes, then add the tomato paste and cook for another 1 to 2 minutes (***pincer la tomate)***. Add the chopped tomatoes (with tomato trimmings and skins), fresh basil, and parsley. Season with salt and pepper, cover, and cook over low heat for 25 to 30 minutes.
2. Purée the mixture through a food mill, then strain it through a fine mesh sieve (***chinois***) into a clean pan. Place back over medium heat and cool until the sauce thickly coats the back of a spoon (***à la nappe***). Reserve in a bain marie until needed.

Farce

1. Heat the olive oil in a large sauté pan over medium heat and sweat the onions until translucent. Add the garlic and stir. Then add the tomato paste and cook for 1 to 2 minutes (***pincer la tomate***), stirring well. Add the diced eggplant and cook until soft. Stir in the diced tomatoes and season with salt, white pepper, and a pinch of cayenne. Then simmer over low heat for 30 minutes until the tomatoes and eggplant have a compote-like texture (***compoter***). Stir from time to time, to prevent the bottom from burning. Taste and add lemon juice if desired.
2. Remove from the heat and add the grated parmesan, thyme, parsley, and freshly shredded basil (***chiffonnade***) and stir gently.

Finishing

1. Fill the eggplant shells with the eggplant and tomato farce. Arrange the filled eggplants in a greased ***gratin*** dish. Top with sliced tomatoes and sprinkle with the grated Gruyère cheese. Place in the oven and bake for 10 to 15 minutes or until heated through.

To Serve

1. Transfer the filled eggplants directly from the oven to a hot serving dish and pour the *sauce tomate* into the bottom of the dish. Top the eggplants with sprigs of fresh thyme.

Quantity		*Ingredient*
U.S.	*Metric*	*Base*
2 pcs	2 pcs	Italian eggplants
		Coarse salt
		Sauce Tomate
1 fl oz	30 mL	Olive oil
3 pcs	3 pcs	Shallots, finely diced (***ciseler***)
2 pcs	2 pcs	Garlic cloves, finely chopped (***hacher***)
1 oz	30 g	Tomato paste
4 pcs	4 pcs	Tomatoes, peeled (***émonder***), seeded (***épépiner***), and diced (***concasser***)
		Tomato parings
2 brs	2 brs	Fresh basil
¼ bq	¼ bq	Parsley, finely chopped
		Salt and pepper
		Farce
1 ¾ fl oz	50 mL	Olive oil
3 pcs	3 pcs	Onions, finely diced (***ciseler***)
2	2	Garlic cloves, finely chopped (***hacher***)
8 pcs	8 pcs	Tomatoes, peeled (***émonder***), seeded (***épépiner***), and diced (***concasser***)
Pinch	Pinch	Cayenne pepper
1 pc	1 pc	Lemon, juice of (*optional*)
3 oz	90 g	Parmesan cheese, grated
3 brs	3 brs	Thyme, rubbed (***émietter***)
2 brs	2 brs	Parsley, finely chopped (***hacher***)
2 brs	2 brs	Basil, shredded (***chiffonade***)
		Salt and white pepper
		Finishing
4 pcs	4 pcs	Tomatoes, cut in half and thinly sliced
3 ½ oz	100 g	Gruyère cheese, grated
		To Serve
4 brs	4 brs	Fresh thyme

CROMESQUIS DE LANGOUSTINES

Learning Outcomes

Mousseline
Pâte à crêpes
Tamiser
Frire

Equipment

Knives:
Chefs knife (*couteau chef*)
scissors (*ciseaux*)

Tools:
Bowls, wooden spatula, plastic spatula, food processor, fine chinois, cutting board, araignée, ladle (*louche*), tamis, corne, pastry bag, large round tip

Pans:
Poêle noir or non-stick crêpe pan, deep-fryer

Serving

8 persons

FYI...

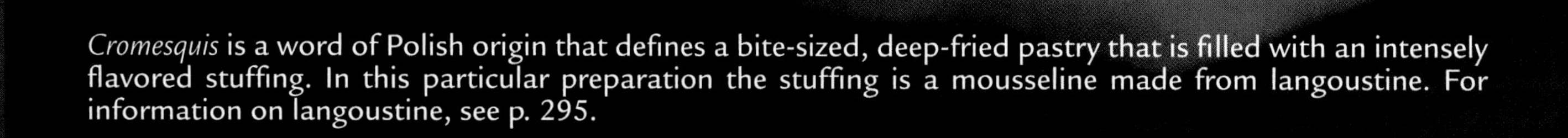

Cromesquis is a word of Polish origin that defines a bite-sized, deep-fried pastry that is filled with an intensely flavored stuffing. In this particular preparation the stuffing is a mousseline made from langoustine. For information on langoustine, see p. 295.

CROMESQUIS DE LANGOUSTINES

Deep Fried Langoustine Parcels

Quantity		Ingredient
U.S.	*Metric*	*Pâte à Crêpes*
3 ½ oz	100 g	Flour
3 pcs	3 pcs	Eggs, lightly beaten
8 fl oz	250 mL	Milk
1 ¼ oz	40 g	Butter, clarified
¼ oz	10 g	Tarragon, blanched, finely chopped (***hacher***)
¼ oz	10 g	Parsley, blanched, finely chopped (***hacher***)
¼ oz	10 g	Chives, blanched, finely chopped (***hacher***)
		Oil or clarified butter for cooking
		Salt
		Mousseline de Langoustine
12 pcs	12 pcs	Langoustine tails, shelled
7 oz	200 g	Cream
¼ oz	10 g	Saffron
4 pcs	4 pcs	Egg whites
1 pc	1 pc	Garlic clove, finely chopped (***hacher***)
1 pc	1 pc	Shallot, finely diced (***ciseler***)
Pinch	Pinch	Cayenne pepper
		Salt and white pepper

Method

1. Heat the deep fryer to 320°F (160°C).

Crêpes

1. Sift (***tamiser***) the flour into a bowl and make a well in the center. Pour the beaten eggs and salt into the well. Slowly incorporate the eggs, then gradually mix in the milk to avoid lumps from forming. Mix in the clarified butter, then strain the batter through a fine mesh sieve (***chinois***), into a clean bowl.
2. Let the mixture rest for a minimum of 30 minutes in a cool place.
3. Remove the herb leaves from their stems and blanch (***blanchir***) in boiling water for 10 seconds, then refresh immediately. Drain the herbs well, squeezing out any excess water, chop them finely (***hacher***), and mix them into the batter.
4. *Cook the crêpes*: Heat an 8-in. nonstick pan over medium heat with a small amount of clarified butter or oil. Pour a small ladle of the batter into the pan while tilting and turning it to make sure the batter covers the bottom of the pan evenly. Return to the heat and allow to cook for 1 minute, or until the edges of the crêpe have browned. Flip the crêpe, cook for 30 seconds, then transfer it to a plate and repeat until all the batter has been used. Cover the crêpes with a damp cloth and reserve.

Mousseline de Langoustine

1. In a small saucepan over low heat, combine 2 or 3 tablespoons of cream with the saffron and bring to a gentle simmer. Stir this back into the rest of the cream and refrigerate until needed.
2. Place the langoustine tails in the bowl of a food processor and process until finely minced. Press the pulp through a drum sieve (***passer au tamis***) using a plastic scraper (***corne***) and transfer to a large bowl set over an ice bath. Work the egg whites one at a time into the pulp using a rubber spatula, mixing well after each addition. When the egg whites are completely incorporated, mix in the cream in a steady stream (***monter la farce***), beating well until completely incorporated. Stir in the garlic and shallot and season with salt, white pepper, and cayenne. Reserve in the refrigerator until needed.

Montage

1. Fill a large piping bag fitted with a large round tip with the *mousseline de langoustine*. Leaving enough room at the edges for folding, pipe a thick line of the mousseline just below the middle of the crêpe, leaving about 1 ½ in. (4 cm) on each side. Fold the bottom up over the mousseline and press to flatten. Fold the sides in and finish by folding the crêpe into an enclosed rectangle. Arrange on a tray, seam side down. Repeat until all the crêpes are filled.

Cuisson

1. Fry the cromesquis in the deep fryer until golden. Drain them well on a paper towel and season immediately.

To Serve

1. Serve on a folded napkin set on a platter.

CROQUETTES DE POULET ET JAMBON

Learning Outcomes

Croquette
Dénerver
Passer au tamis
Mousseline de viande
Beurre en pommade
Paner à l'Anglaise
Frire

Equipment

Knives:
Paring knife (*office*), slicing knife (*éminceur*)

Tools:
Piping bag with a large plain tip, bowls, cutting board, plastic spatula, araignée, whisk, food processor, tamis, corne

Pans:
Large sauce pan (*russe*), deep fryer

Serving

8 persons

FYI...

Derived from the French verb *croquer* which means "to take a bite," croquettes are breaded, deep-fried, and stuffed with a number of ingredient combinations. Aside from the poultry and ham filling (*farce*) that is used in this recipe, other croquette fillings include a variety of meats, fish, and vegetables. Croquettes are always bite-sized.

CROQUETTES DE POULET ET JAMBON

Chicken and Ham Croquettes

Method

1. Heat the deep fryer to 320°F (160°C).

Mousseline de Viande

1. *Make a mousseline de viande*: Trim and denerve the chicken breasts and place them in the bowl of a food processor, and process until finely minced. Press the meat through a drum sieve (***passer au tamis***) using a plastic scraper (***corne***) and transfer to a medium bowl set over an ice bath. Work the egg whites one at a time into the purée using a rubber spatula, mixing well after each addition. When the egg whites are completely incorporated, mix in the cream in a steady stream (***monter la farce***), beating well until completely incorporated. Season with salt and white pepper.
2. Cook a spoonful of the mousseline in simmering water and taste to verify the seasoning. Adjust the seasoning if needed.
3. Mix the ***brunoisette*** of truffle, ***brunoise*** of cooked chicken, and ***brunoise*** of ham into the mousseline. Leave to rest in the refrigerator until it stiffens (*about 30 minutes*).
4. Line a tray with parchment paper. Transfer the mousseline into a piping bag fitted with a plain tip. Pipe long logs of the mousseline onto the prepared tray. Using a wet paring knife, cut into cork shapes (***bouchon***) 2 in. (5 cm) long, around 1 ¾ to 2 ½ oz (50 to 70 g) each. Reserve in the freezer until they harden.

Truffle Butter

1. Combine the butter, truffle juice, cognac, and seasonings and mix until homogeneous.

Tip: *The compound butter can be piped with a pastry bag and refrigerated, or served soft (**pommade**).*

Paner à l'Anglaise

1. Lay out 3 shallow containers side by side. Starting from the left, pour the flour into the first container. In the middle container, beat the egg yolks, oil, salt, and white pepper (***anglaise***). In the last container, pour in the breadcrumbs. Prepare a tray lined with parchment paper to the right of the breadcrumbs.
2. Remove the ***croquettes*** from the freezer and roll them one by one in the flour and gently tap off any excess. Transfer them to the ***anglaise*** mixture and coat completely. Finally, roll the ***croquettes*** in the breadcrumbs and transfer them to the prepared tray.

Cuisson

1. Gently immerse the ***croquettes*** in the hot deep fryer until crisp and golden brown. Drain them on a paper towel and season immediately.
2. Serve hot in a folded napkin with truffle butter on the side.

Quantity		*Ingredient*
U.S.	*Metric*	
10 oz	300 g	Chicken cooked, cut into ***brunoise***
10 oz	300 g	Ham cooked, cut into ***brunoise***
1 pc	1 pc	Truffle, cut into ***brunoisette***
		Mousseline de Viande (Farce Fine)
10 oz	300 g	Chicken breast, boneless, skinless
2 pcs	2 pcs	Egg whites
7 oz	200 g	Cream
		Salt and white pepper
		Truffle Butter
5 oz	150 g	Butter (***pommade***)
½ fl oz	10 mL	Truffle juice
1 fl oz	30 mL	Cognac
		Salt and white pepper
		Paner à l'Anglaise
10 oz	300 g	Flour
3 pcs	3 pcs	Egg yolks
½ fl oz	10 mL	Oil
10 oz	300 g	Breadcrumbs
		Salt and white pepper

Learning Outcomes

Pâte à choux
Pocher
Béchamel sauce
Mornay sauce
Gratiner
Pâte brisée
Abaisser
Fond de tarte cuit à blanc
Tamiser
Fontaine
Foncer

Equipment

Knives:
Paring knife (*office*), scissors (*ciseaux*)

Tools:
Wooden spatula, plastic spatula, fine chinois, cutting board, skimmer (*écumoire*), ladle, cheese grater, pastry bag, plain tip [3/8 in. (10 mm)], baking beans or beads, rolling pin, docker, round cutter, whisk

Pans:
Bowls, small sauce pan (*russe*), small marmite, bain marie, 3 in. (8 cm) tartlet molds

Serving

8 persons

FYI...

The word *gnocchi* stems from the earlier Italian usage *nocca*, which translates as "knuckle." This is certainly an apt visual description for this versatile and rustic fare. However, as a tartlet topped with Mornay sauce, Gnocchi à la Parisienne does not fall very comfortably under the moniker of rustic!

GNOCCHI À LA PARISIENNE
Tartlets Filled with Gnocchi in Mornay Sauce

Quantity		*Ingredient*
U.S.	*Metric*	Pâte Brisée
3 ½ oz	100 g	Butter
7 oz	200 g	Flour
1 tsp	5 g	Salt
1 pc	1 pc	Egg yolk
½ fl oz	10 mL	Water
		Sauce Mornay
2 ½ oz	75 g	Butter
2 ½ oz	75 g	Flour
1 ½ pt	750 mL	Milk
0.5 pc	0.5 pc	Onion, studded (***clouté***)
1 pc	1 pc	***Bouquet garni***
Pinch	Pinch	Nutmeg, grated or ground
Pinch	Pinch	Cayenne pepper
3 ½ oz	100 g	Gruyère cheese, grated
2 pcs	2 pcs	Egg yolks
½ oz	15 g	Butter
		Salt and pepper
		Pâte à Choux
8 fl oz	250 mL	Water
2 oz	60 g	Butter
4 oz	125 g	Flour
4 pcs	4 pcs	Eggs
Pinch	Pinch	Salt
		To Finish
5 oz	150 g	Gruyère cheese

Method

1. Preheat the oven to 350°F (180°C).

Pâte Brisée

1. *Make a pâte brisée:* Cut the cold butter into small dice. Sift (***tamiser***) the flour onto the work surface. Using your fingertips, work the butter into the flour until the mixture resembles fine yellow sand (***sabler***). Make a well in the flour (***fontaine***) and add the egg yolk, salt, and water. Using a plastic scraper (***corne***), incorporate the liquids into the dry ingredients with a cutting motion. If the dough seems dry, sprinkle it with some cold water. Using the heel of the hand, smear the dough away from yourself to ensure no lumps remain (***fraiser***). Once the dough forms, gather it into a smooth ball, flatten it into a disc, and wrap it in plastic. Let the dough rest in the refrigerator for at least 20 minutes (see page 368 in *Cuisine Foundations*).

Fonçage

1. Lightly dust the marble with flour (***fleurer***) and place the dough in the center. Roll out the dough (***abaisser***), giving it quarter turns as you roll. Continue rolling and turning until the dough is ⅛ in. (3 mm) thick; then prick it with the docker (***pique pâte***) or a fork. Cut the dough using a round cutter that is slightly larger than the tartlet molds. Line the molds with the dough rounds (***foncer***), leaving a slight border extending above the edges. Let the tartlets rest for 20 minutes in the refrigerator.

Cuisson à Blanc

1. Arrange the tartlets on a baking sheet without touching. Line each one with a circle of parchment paper slightly larger than the tart. Fill the cavities with baking beads. Place the baking sheet in the oven and bake the tartlets until lightly colored (about 15 minutes).
2. Remove the baking sheet from the oven and place it on a rack to cool. Remove the baking beads from the tartlets.

Tip: *If the bottoms of the tarts look too pale, return the tarts to the oven without the lining for 2 to 3 minutes.*

Sauce Mornay

1. *To make a Mornay sauce, start by making a béchamel sauce:* Melt the butter in a medium pan over medium heat. When the butter begins to foam, add the flour and stir well until the mixture begins to bubble. Reduce the heat to low and cook for 1 to 2 minutes to obtain a white roux (*be careful not to let it color*). Transfer to a small plate and set aside to cool.
2. Combine the milk, studded onion, and ***bouquet garni*** in a medium saucepan over medium heat. Season with salt, pepper, and nutmeg. Once the milk comes to a boil, whisk the cold roux, piece by piece, into the hot milk until it is completely incorporated and begins to thicken. Bring to a boil over medium heat, then reduce the heat to low and leave to simmer gently for 10 minutes. Remove the sauce from the heat, remove the ***bouquet garni*** and studded onion, and strain the béchamel through a fine mesh sieve (***chinois***) into a large bowl. Mix in a pinch of cayenne.
3. *To finish the Mornay:* Gradually stir the cheese into the béchamel and allow the sauce to cool until warm. Beat the egg yolks in a small bowl, then stir some of the sauce into the egg yolks, mixing well to temper them. Stir the warmed egg yolks back into the sauce until completely incorporated. Pat (***tamponner***) the surface of the sauce with a knob of butter on the end of a fork to stop the formation of a skin.
4. Cover the sauce with plastic wrap and reserve in a bain marie until needed.

Pâte à Choux

1. Combine the water, butter, and salt in a medium pan and bring the mixture to a boil over high heat. Once the butter has completely melted, remove from the heat and add the flour all at once. Stir with a wooden spatula until combined; then, over medium heat, stir until the mixture forms a clean ball coming away from the sides of the pan (***dessêcher***). Transfer the hot dough to a clean bowl and spread out to cool slightly. Using a wooden spatula or spoon, mix the eggs into the dough one by one, making sure the last one is completely incorporated before adding the next. The dough should be elastic and slightly sticky. Transfer the pâte à choux to a piping bag fitted with an 10 mm plain tip.
2. Set aside.

Cuisson des Gnocchi

1. Bring a large pan of salted water to a boil over high heat. Applying an even pressure, pipe out the pâte à choux above the boiling water and cut the dough into regularly sized pieces with the back of a paring knife.
2. Skim off the gnocchi once they float to the surface (*floating is an indication that they are cooked*). Roll the hot gnocchi in a little melted butter to cool them and to prevent them from sticking to each other.

To Finish

1. Mix some of the Mornay with the gnocchi and fill the tartlet shells. Top the gnocchi with more Mornay and sprinkle with grated Gruyère cheese. Arrange on a baking sheet and place in the oven until the filling puffs up and the tops are nicely browned (***gratiner***).

GRATIN DAUPHINOIS

Learning Outcomes

Gratiner
Infusing milk

Equipment

Knives:
Vegetable peeler (***économe***), chef knife (***couteau chef***)

Tools:
Fine chinois, cheese grater

Pans:
9-by-13-in. gratin dish, medium sauce pan (***russe***)

Serving

8–10 persons

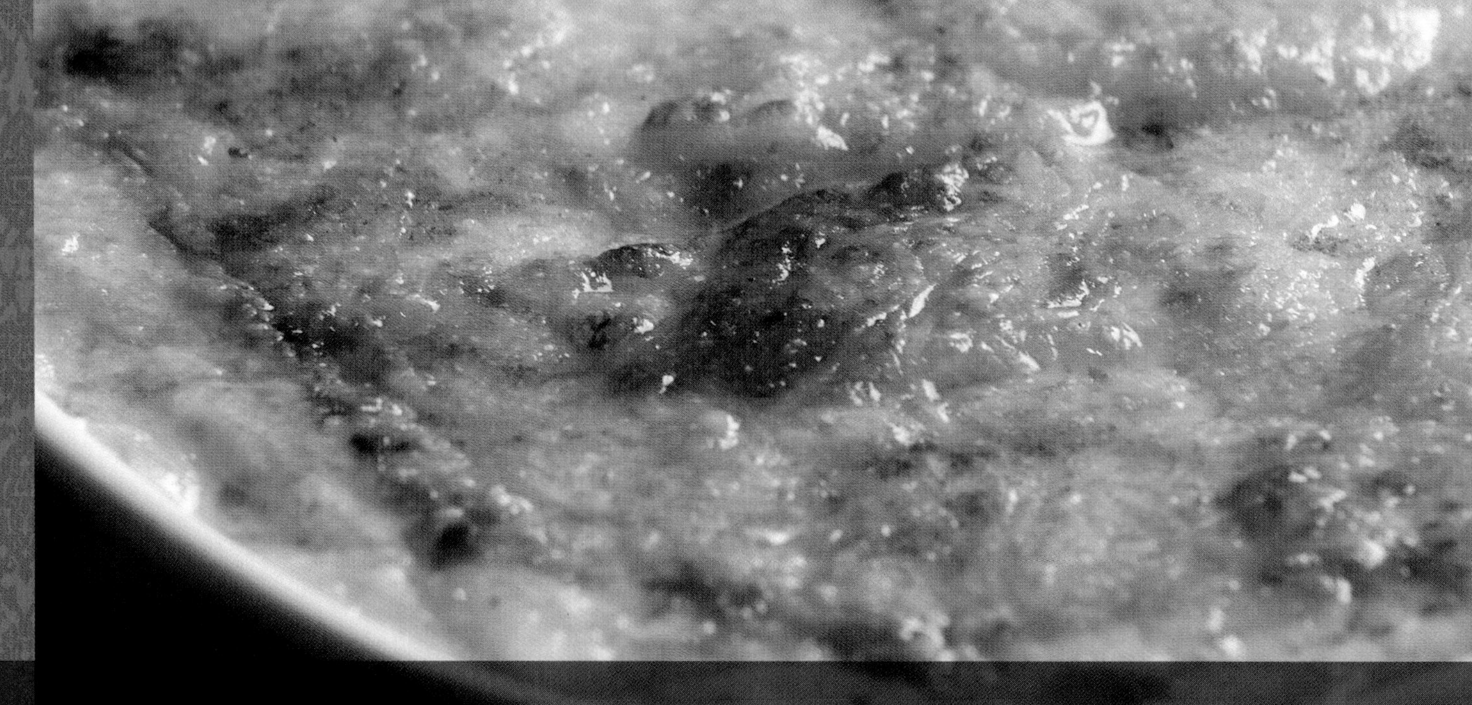

FYI...

In the late 1500s the French word *gratin* referred to the skin on the bottom of a pan that was scraped off (*en grattant*). By the 1800s, the term was being used more or less as we use it now: as a culinary technique that involves covering a dish in cheese or breadcrumbs and browning it in the oven. True to its original meaning, the process of *gratiner* produces a dish with a palatable crust. Gratin dauphinois, as its name would suggest, is a specialty of the old Dauphiné province of southeastern France.

GRATIN DAUPHINOIS

Scalloped Potatoes with Cheese

Method

1. Preheat the oven to 380°F (190°C).

Cream Mixture

1. Coat a deep, ovenproof dish with softened butter and reserve in the refrigerator.
2. Combine the milk, freshly grated nutmeg, ***bouquet garni***, and garlic in a medium saucepan over medium heat. Season well with salt and pepper. Bring the liquid to a boil, remove it from the heat, and leave it to infuse for 5 minutes.

Cuisson

1. Meanwhile, wash and peel the potatoes; slice (***émincer***) them thinly, about 1.8 in. (3 mm). Do not rinse the potatoes as the starch is needed. Layer the potato slices tightly in the buttered dish and season each new layer with salt and pepper.
2. Stir the cream into the hot milk, then pour through a fine mesh sieve (***chinois***) onto the potatoes until the dish is full. Let rest for 1 to 2 minutes. The potatoes will absorb the liquid and the level will decrease. Top with the remaining liquid and sprinkle generously with grated cheese. Cover the dish with buttered parchment paper and foil. Place it in the oven to cook for 1 hour or until the potatoes are easily pierced with a knife.

Finishing

1. Remove the foil and parchment paper and return the dish to the oven to brown (***gratiner***) until the surface is golden.
2. Remove from the oven and let rest for 5 to 10 minutes before serving.

Quantity		*Ingredient*
U.S.	*Metric*	*Cream Mixture*
1 ½ qt	1.5 L	Milk
Pinch	Pinch	Freshly grated nutmeg
1 pc	1 pc	***Bouquet garni***
2 pcs	2 pcs	Garlic cloves
5 ½ fl oz	165 mL	Heavy cream
		Cuisson
2 ½ lbs	1.2 kg	Potatoes, firm-fleshed
7 oz	200 g	Gruyère cheese, grated
4 oz	120 g	Unsalted butter, softened, for gratin dish
		Salt and pepper

FRITOTS DE RIS DE VEAU, SAUCE TOMATE, PERSIL FRIT

Learning Outcomes

Working with white offal
Pâte à frire
Sauce tomate
Degorger
Presser
Blanchir

Equipment

Knives:
Paring knife (*office*), slicing knife (*éminceur*)

Tools:
Bowls, cutting board, wooden spatula, fine chinois, araignée, ladle (*louche*), whisk, food mill

Pans:
medium sauce pan (*russe*)

Serving

4 persons

FYI...

Fritots, pronounced "freeto," are savory fritters that can be made with sweetbreads, offal (usually brains), meat, fish trimmings, and shellfish. In this application, veal sweetbreads are degorged and battered, whereas in other fritot recipes the stuffing is marinated. Fritots are traditionally served with fried parsley and a sharp tomato sauce.

FRITOTS DE RIS DE VEAU, SAUCE TOMATE, PERSIL FRIT

Sweetbread Fritters with Tomato Sauce and Fried Parsley

Method

1. Heat the deep fryer to 340°F (170°C).

Prepare the Sweetbreads

1. Soak the sweetbreads for 6 to 24 hours in cold water in the refrigerator, changing the water every few hours (***dégorger***).
2. *Blanch the sweetbreads*: Place the degorged sweetbreads in a large pan and cover with cold water. Season generously with coarse salt and bring to a boil. Allow the sweetbreads to boil for 1 to 2 minutes, then turn off the heat and let cool in the water before removing them. With a small paring knife, remove the surrounding membranes and any fat.
3. Place the sweetbreads between two wire racks lined with towels and place a weight on top to press out the excess water. Leave for at least 1 hour or preferably overnight.

Sauce Tomate

1. Heat the olive oil in a small sauté pan over low heat and sweat (***suer***) the diced shallots until translucent, stir in the crushed garlic and cook for 1 to 2 minutes more, then add the tomato paste and cook for 1 to 2 minutes (***pincer la tomate***). Add the diced tomatoes, the stems from the basil and parsley, and season to taste. Cover and cook over low heat for 25 to 30 minutes.
2. Purée the sauce in a food mill and strain it through a fine mesh sieve (***chinois***). The consistency of the sauce should be thick enough to coat the back of a spoon (***à la nappe***). Finely chop the basil and parsley leaves and stir in. Season to taste and reserve in a bain marie until needed.

Pâte à Frire

1. Mix the sifted flour and salt together in a bowl and make a well (***fontaine***) in the center. Combine the egg yolks, water, and oil. Pour this mixture into the well, whisk it into the flour, then let the batter rest until needed.

Finishing

1. Roll the parsley in flour; shake off the excess and deep fry it for 10 seconds. Drain the parsley on a paper towel and season immediately.
2. Pat the sweetbreads dry and cut them into small, bite-sized pieces.
3. With a pinch of salt, whip the egg whites in a large bowl until soft peaks form. Delicately fold one-third of the egg whites into the pâte à frire to lighten it (*this will make it easier to incorporate the remaining egg whites*). Fold in the rest of the egg whites.
4. Roll the sweetbreads in the flour (*making sure to tap off the excess*), then dip them in the batter and deep fry until golden.
5. Transfer the fritots to a paper towel and season immediately.

To Serve

1. Arrange the fritots in a folded napkin and serve the tomato sauce on the side, hot or cold.

Note: *Always degorge sweetbreads for a minimum of 6 to 24 hours in several changes of cold water before using them.*

Quantity		*Ingredient*
U.S.	*Metric*	*Base*
14 oz	400 g	Veal sweetbreads
		Coarse salt
		Sauce Tomate
1 fl oz	30 mL	Olive oil
3 pcs	3 pcs	Shallots, finely diced (***ciseler***)
1 pc	1 pc	Garlic head, crushed
1 oz	30 g	Tomato paste
4 pcs	4 pcs	Tomatoes, peeled (***émonder***), seeded (***épépiner***), and diced (***concasser***)
2 brs	2 brs	Basil
2 brs	2 brs	Parsley
		Salt and pepper
		Pâte à Frire
8 oz	250 g	Flour, sifted
Pinch	Pinch	Salt
2 pcs	2 pcs	Egg yolk
7 fl oz	200 mL	Water
1 oz	30 mL	Oil
2 pcs	2 pcs	Egg whites
Pinch	Pinch	Salt
		To Serve
2 brs	2 brs	Parsley
		Flour, for dredging

LAITUE BRAISÉE

Learning Outcomes

Stuffing vegetables
Blanching a leafy vegetable
Making a glaze from a jus de braisage
Braiser

Equipment

Knives:
Vegetable peeler (*économe*), paring knife (*office*), chefs knife (*couteau chef*)

Tools:
Kitchen twine, mixing bowls, ice bath, trays, wooden spatula, wire rack, chinois, pastry brush

Pans:
Large pot, large sauté pan

Servings

4 persons

FYI...

White or brown stock? Laitue braisée is a classic garnish served with roasted meat or fowl. The color of the stock should match the meat it is accompanying: white stock for poultry and brown stock for red meat such as beef or lamb. It can be prepared with or without a filling (farce), as well.

LAITUE BRAISÉE

Braised Lettuce

Quantity		Ingredient
U.S.	Metric	
4 pcs	4 pcs	Boston lettuce
		Vinegar
		Coarse salt
		Salt and pepper
		Farce
1 ¾ oz	50 g	Butter
5 oz	150 g	Smoked slab bacon, cut into very small lardons
2 pcs	2 pcs	Garlic cloves, finely chopped (***hacher***)
2 pcs	2 pcs	Shallots, finely chopped (***hacher***)
¼ oz	10 g	Parsley, finely chopped (***hacher***)
5 oz	150 g	Fresh breadcrumbs
1 pc	1 pc	Egg
		Salt and pepper
		Garniture de Braisage
3 ½ oz	100 g	Butter
1 pc	1 pc	Carrot, cut into large ***paysanne***
1 pc	1 pc	Onion, cut into large ***paysanne***
1 pc	1 pc	Garlic clove, finely chopped (***hacher***)
½ oz	15 g	Parsley, finely chopped (***hacher***)
3 ½ oz	100 g	Smoked slab bacon, cut into lardons and blanched
1 pc	1 pc	***Bouquet garni***
1 pt	500 mL	White or brown stock
1 pc	1 pc	Bacon rind
1 ¾ oz	50 g	Butter

Method

1. Preheat the oven to 350°F (180°C).
2. Wash the whole lettuce heads thoroughly in several changes of cold water. Remove the outer leaves and reserve. ***Note***: *The outer leaves will be used to cover the stuffed hearts to protect them from dehydration during braising.*
3. Pour the vinegar into a large pot of cold water and soak the lettuce hearts for 10 minutes.
4. *Blanch the lettuce hearts:* Bring a large pot of salted water to a boil over high heat. Add the lettuce hearts and blanch them until soft to the core (about 5 minutes). Refresh the hearts in an ice bath and squeeze out all the excess water; holding the lettuce by its stem, squeeze downward toward the top, and repeat this several times. Place the drained lettuce hearts on a tray lined with a kitchen towel and reserve in the refrigerator.

Farce

1. Melt the butter in a large sauté pan over medium-high heat and sauté the lardons bacon until it begins to color. Reduce the heat to medium, add the garlic and shallots, and cook tender, being careful not to allow them to color. Stir in the parsley and mix well, then stir in the fresh breadcrumbs. Taste and adjust the seasoning. Remove from the heat and mix in the egg. Transfer to a dish and spread out the farce to cool.

Montage

1. Carefully remove one of the outer leaves from each lettuce heart and set aside. Lay out the blanched lettuce heart on a cutting board and spread out the leaves. Remove the stem, cutting at a shallow angle. Season with salt and pepper and brush with clarified butter. Divide the farce into 4 equal portions and place one portion in the center of each lettuce heart. Neatly wrap the leaves around the farce until it is completely enclosed. Spread the outer leaves and wrap the filled lettuce heart. Turn over with the seam on the bottom, and trim off the stem from the end. Brush with more clarified butter and season with salt and pepper.

Cuisson

1. Melt the butter in a large sauté pan over medium heat and add the carrot. Stir to coat and season with salt and pepper. Add the onion and continue to sweat (***suer***) until the onions turn translucent, before adding the garlic. Cook for 2 minutes, stirring well. Mix in the chopped parsley and blanched lardons then place the ***bouquet garni*** in the center. Place a wire rack over them and arrange the stuffed lettuce on top. Cover with a piece of barding fat and, if available, the bacon rind. Cover with the reserved outer leaves. Pour in the hot stock. Cover and transfer the pan to the oven to braise gently for 45 minutes.

Finishing

1. Butter a medium saucepan and prepare a paper cartouche. Remove the pan from the oven. Remove the lettuce leaves, bacon rind, and barding fat and discard. Lift out the wire rack with the braised lettuce, set aside, and keep warm. Remove the ***bouquet garni*** and discard. Strain the garniture through a fine mesh sieve (***chinois***), pressing lightly but being careful not to crush it. Transfer the garniture to the buttered pan. Butter the cartouche and place over the garniture. Keep warm. Reduce the strained cooking liquid (***jus de braisage***) in a medium saucepan over medium heat to a syrupy consistency. Verify and adjust the seasoning, then mount with cold butter.
2. Using a pastry brush, coat the braised lettuce with the reduction until they are evenly covered in a thick glaze.

To Serve

1. Arrange the reserved garniture in the center of a hot serving dish or plate and arrange the braised lettuce on top. Pour some reduction into the dish or plate.
2. *Using caul fat (optional):* Wrapping the lettuce in caul fat ensures that the leaves remain moist and basted during the cooking process, but also maintains their shape. If using, lay out the caul fat and place a finished lettuce in the middle. Tightly wrap in just enough caul fat to cover it. Trim off any excess and tuck in the ends. Proceed as usual for the cooking (***cuisson***).

PETITS FARCIS NIÇOIS

Learning Outcomes

Farce simple
Tronçons
Évider
Blanchir
Sauté
Hacher

Equipment

Knives:
Vegetable peeler (*économe*), paring knife (*office*), slicing knife (*éminceur*)

Tools:
Bowls, wooden spatula, plastic spatula, cutting board, melon baller

Pans:
Sautoir, medium sauce pan (*russe*), baking pan

Serving

6 persons

FYI...

Not only does this recipe celebrate ingredients grown in the countryside around the city of Nice (Niçoise), but it also favors the use of immature produce—when certain vegetables or roots (in this recipe, turnips) are harvested prior to reaching full maturity, they will be sweeter and possess a more delicate flavor.

PETITS FARCIS NIÇOIS

Veal and Pork Stuffed Vegetables

Method

1. Preheat the oven to 400°F (205°C).

Base

1. Wash the vegetables and peel the onions. Peel and trim the turnips so they are evenly shaped.
2. Cut the zucchini and eggplants into 1 ½ in. (4 cm) thick slices (***tronçons***). Empty out the center (***vider***) of the zucchini, eggplants, turnips, tomatoes, and onions using a melon baller (***cuillère parisienne***).
3. Blanch the zucchini, eggplants, and turnips separately in boiling salted water for 1 to 2 minutes each. Refresh immediately in an ice water bath.
4. Place the vegetables upside down to drain on a rack or a baking sheet lined with paper towels. Reserve in the refrigerator until needed.

Farce

1. Soak the bread in the milk. Heat the olive oil in a large sauté pan over low heat and sweat (***suer***) the chopped onion until it begins to soften. Add the ground meats and turn the heat up to medium high. Sauté until the meats begin to color, then add the chopped garlic and continue cooking, stirring from time to time. Season with salt and white pepper to taste. Spread out the mixture on a tray and leave to cool for 5 minutes before transferring to a clean bowl.
2. Squeeze the bread (to extract the milk), then chop finely (***hacher***) and mix it into the meat.
3. Lightly beat the egg and mix it into the farce with the chopped parsley.

Montage

1. Season the inside of the vegetables with salt and pepper and fill the cavities with the farce.

Tip: *To increase efficiency, use a piping bag fitted with a large plain tip.*

1. Brush the stuffed vegetables with olive oil, arrange them on an oiled baking sheet, and bake them in the oven until the vegetables are cooked and beginning to color (*approximately 20 to 25 minutes*).

Optional: *Garlic cloves in their skins, thyme, rosemary, and herbes de provence can also be added to the pan for roasting to add flavor to the dish (**garniture aromatique**).*

Quantity		*Ingredient*
U.S.	*Metric*	*Base*
1 pc	1 pc	Medium zucchini, cut into 6 ***tronçons***
6 pcs	6 pcs	Eggplant, cut into 6 ***tronçons***
6 pcs	6 pcs	Small turnips
6 pcs	6 pcs	Small tomatoes
6 pcs	6 pcs	Medium red onions (*or white*)
		Salt
		Farce
8 oz	250 g	Lean ground veal
8 oz	250 g	Lean ground pork
4 pcs	4 pcs	Slices firm white bread, crusts removed
12 fl oz	360 mL	Milk
½ fl oz	15 mL	Olive oil
½ pc	½ pc	Onion, finely chopped (***hacher***)
6 pcs	6 pcs	Garlic cloves, finely chopped (***hacher***)
1 pc	1 pc	Egg
1 Tbsp	10 g	Parsley, coarsely chopped (***ciseler***)
		Salt and white ground pepper
		To Serve
		Olive oil, for roasting

PETITS GÂTEAUX DE FOIES BLONDS

Learning Outcomes

Mousse de foie de volaille
Réduction
Confire
Concassée de tomates

Equipment

Knives:
Boning knife (***désosseur***), vegetable peeler (***économe***), paring knife (***office***), slicing knife (***éminceur***)

Tools:
Bowls, wooden spatula, plastic spatula, fine chinois, cutting board, skimmer (***écumoire***), ladle (***louche***), cheese grater, food processor, dariole molds

Pans:
Sautoir, small sauce pan (***russe***), bain marie

Serving

8 persons

This recipe is an example of how pastry techniques can cross over into savory preparations. Similar to the execution of a crème brûlée or other set custards, this savory gâteau takes the same amount of care to achieve the desired firmness.

PETITS GÂTEAUX DE FOIES BLONDS

Chicken Liver Creams

Method

1. Preheat the oven to 325°F (150°C).

Base

1. After degorging, pat the livers dry, denerve them, slice them on the bias (***escaloper***), and season them with salt and white pepper.

Note: *When cleaning the livers, make sure to remove and discard any green bile spots (**fiel**). They will give your final dish a bitter flavor.*

2. Place the chicken livers, garlic, and beef marrow into the bowl of a food processor and process until smooth. With the processor running, slowly add (*in the following order*) the flour, whole eggs, egg yolks, cream, and milk. Strain the mixture through a fine mesh sieve (***chinois***) and season with salt, pepper, and nutmeg.
3. Brush 8 ***dariole*** molds with softened butter (***beurrer***), and refrigerate for 5 minutes (*to set the butter*). Once the butter is set, place a round of buttered parchment paper on the bottom of each mold. Line the bottom of a sauté pan with a round of parchment paper. Fill the darioles three-quarters full with the liver mixture and place in the sauté pan. Fill the pan with hot water two-thirds of the way up the sides of the ***darioles***. Bake for about 20 minutes, or until the tip of a knife inserted comes out clean. Remove the molds from the bain marie and leave to cool to room temperature.

Garlic Confit

1. Place garlic cloves (*in their skins*) in olive oil in a small saucepan. Add the bay leaf, thyme sprigs, and peppercorns and cook over very low heat until a knife can be easily inserted into the cloves (*40 minutes to 1 hour*).

Tomate Concassée

1. Heat the olive oil in a sauté pan over medium heat and sweat the diced onion until translucent. Add the garlic and cook for 1 minute, then add the tomatoes and stir well. Season to taste. Mix in the thyme, set aside and keep warm.

Port Reduction

1. In a small pan over high heat, combine the shallots, the port wine, and the demi-glace and reduce until syrupy. Season to taste. Strain through a fine mesh sieve (***chinois)*** and reserve in a bain marie.

Crème d'Ail

1. *Blanch the garlic:* Place the peeled garlic cloves in a small saucepan of cold water and bring to a boil. After 5 minutes of boiling, refresh the garlic under cold running water and drain. Transfer the garlic to a small saucepan and add the cream. Bring the cream to a gentle simmer over low heat and cook the garlic until it is soft and the cream has reduced.
2. Stir in the parmesan and strain through a fine mesh sieve (***chinois***), pressing well (***fouler***).
3. If the cream is too thick, add some fresh cream until the desired consistency is reached. Keep the crème d'ail warm in a bain marie.

To Serve

1. Unmold a gâteau de foie onto a hot plate. Make ***quenelles*** of tomato ***concassée*** and place next to the gâteau. Spoon crème d'ail onto the plate and decorate by drizzling the port reduction around. Finish with a clove of garlic confit (still in its skin) and a sprig of thyme.

Note: *This recipe cannot be unmolded hot. It needs to set in order to come out of the mold cleanly.*

Quantity		Ingredient
U.S.	*Metric*	*Base*
12 oz	340 g	Chicken livers, degorged in milk (***dégorger***)
½ pc	½ pc	Garlic clove, finely chopped (***hacher***)
3 oz	90 g	Beef marrow, degorged in milk (***dégorger***)
1 ¾ oz	50 g	Flour
4 pcs	4 pcs	Whole eggs
4 pcs	4 pcs	Egg yolks
2 fl oz	60 mL	Cream
2 fl oz	60 mL	Milk
Pinch	Pinch	Nutmeg, grated
		Salt and white pepper
		Confit Garlic
2 pcs	2 pcs	Garlic heads, separated, in their skins (***en chemise***)
5 fl oz	150 mL	Olive oil
1 pc	1 pc	Bay leaf
2 br	2 br	Thyme
10 pcs	10 pcs	Black peppercorns
		Tomate Concassée
1 fl oz	30 mL	Olive oil
1 pc	1 pc	Onion, finely diced (***ciseler***)
2 pcs	2 pcs	Garlic
3 pcs	3 pcs	Tomatoes, peeled (***émonder***), seeded (***épépiner***), and roughly diced (***concasser***)
1 br	1 br	Thyme, rubbed (***émietter***)
		Salt and pepper
		Port Reduction
1 ¾ oz	50 g	Shallots, finely diced (***ciseler)***
8 fl oz	250 mL	Port wine
5 fl oz	150 mL	Demi-glace
		Salt and pepper
		Crème d'Ail
1 pc	1 pc	Garlic head
10 fl oz	300 mL	Cream (to reduce to 250 mL (8 oz)
5 oz	150 g	Parmesan cheese, grated
		Additional cream as needed

PISSALADIÈRE

Learning Outcomes

Make a pâte levée
Émincer
Suer
Compoter
Étuver
Lustrer

Equipment

Knives:
Paring knife (*office*), slicing knife (*éminceur*)

Tools:
Bowls, wooden spatula, plastic spatula, rolling pin, cutting board, skimmer (*écumoire*), flan ring or heavy round cake mold

Pans:
Large sauté pan, baking tray

Serving

8 persons

FYI...

The pissalidière is the Provençal version of pizza and is thought to have originated in Nice. It takes its name from the traditional condiment known as pissalat, which is a paste made from salted anchovies. *Pissalat* comes from the local dialect for salted fish.

PISSALADIÈRE
Provençal Onion Pie

Method

1. Preheat the oven to 400°F (205°C).

Pâte Levée

1. Mix the yeast and sugar with the warm water [*around 86°F (30°C)*] and let it sit at room temperature until a light foam appears on the surface. Sift the flour onto the work surface and make a well in the center. Into the well, pour the beaten egg, the salt, and the yeast mixture, then work the ingredients until they combine into a smooth dough. To develop elasticity, knead the dough for 15 to 20 minutes or until it creates a thin diaphragm when stretched between the hands.
2. Pour the oil onto the dough and work it in until it is completely absorbed.
3. Place the dough in a lightly oiled bowl in a warm area and cover with plastic wrap or a humid cloth. Let the dough proof until it has doubled in size.

Garniture

1. Meanwhile, heat the olive oil in a large sauté pan over low heat and sweat (***suer***) the onions until translucent. Add the thyme and rosemary and season lightly. Cover the pan and let the onions cook in their own liquid (***étuver***) until soft and slightly colored (***compoter***). Sprinkle the flour over the onions and cook for 2 to 3 minutes (***singer***), stirring well. Remove the pan from the heat and let the onions cool. Taste and adjust the seasoning.

Montage

1. Once the dough has doubled in volume, punch it down, then transfer to a lightly floured work surface. Roll the dough out (***abaisser***), and place in a 10 in. (26 cm) flan ring or round cake mold. Starting in the center, press the dough outward to create a hollow.

To Finish

1. Fill the hollow with the onions, pressing well. Arrange the anchovies in a lattice pattern. Slice the olives on both sides of the pit and arrange the slices between each lattice. Bake in the oven until the dough is cooked and golden.
2. Brush the pissaladière with olive oil to add shine (***lustrer***). Remove from the mold.
3. Can be served hot or cold.

Quantity		*Ingredient*
U.S.	*Metric*	***Pâte Levée***
½ oz	10 g	Yeast, fresh
¼ oz	5 g	Sugar, granulated
1 oz	30 g	Water
8 oz	250 g	Flour
3 pcs	3 pcs	Eggs, beaten
¾ fl oz	20 mL	Olive oil
¼ oz	5 g	Salt
		Garniture
3 ½ fl oz	100 mL	Olive oil
2 lb	1 kg	Onions (***émincer***)
		Thyme
		Rosemary
1 oz	30 g	Flour
1 ¾ oz	50 g	Anchovies
15 pcs	15 pcs	Black olives
		Salt and pepper
		To Serve
		Olive oil

QUICHE LORRAINE

Learning Outcomes

Pâte brisée
Foncer
Cuisson à blanc
Cooking of a set cream
Tamiser
Sabler
Fontaine
Fleurer
Abaisser
Piquer
Chiqueter

Equipment

Knives:
Knife (*éminceur*)

Tools:
Cheese grater, whisk, rolling pin, baker's brush, corne, pique pâte, baking beads/beans

Pans:
Tart pan with a removable bottom, small saucepan, small sauté pan

Serving

4 persons

FYI...

Lorraine is a province in France bordering on Belgium, Germany, and Switzerland. As pork is a central ingredient in Lorraine cuisine, the lardons are one clue to this preparation's heritage. Even the concept of quiche (an open tart filled with an egg mixture) may be a Lorraine invention, though the word *quiche* comes from the German *küchen,* which means cake.

QUICHE LORRAINE

Savory Custard and Bacon Tart

Method

1. Preheat the oven to 370°F (190°C).

Pâte Brisée

1. Sift (***tamiser***) the flour onto a clean work surface and make a large well (***fontaine***) in the center using a plastic scraper (***corne***).
2. Place the cold, diced butter in the center of the well. Work the butter into the flour using your fingertips while simultaneously cutting through the mixture with the corne. Rub the mixture between your fingers until it resembles fine sand (***sabler***).
3. Gather the flour-butter mixture into a neat pile and make a well in the center using the ***corne***. Dissolve the salt in the cold water and add with the egg yolks to the center of the well. Stir these ingredients together using your fingertips while simultaneously using the ***corne*** to gradually incorporate the dry ingredients from the sides. Continue this process until the mixture in the center of the well resembles a paste.
4. Add the remaining dry ingredients and incorporate by cutting in using the ***corne***. Repeat this process until the mixture forms a homogeneous dough. Using the heel of your palm, firmly smear the dough away from yourself to ensure that no lumps are left (***fraiser***).
5. Shape the dough into a ball, wrap it in plastic, and flatten it into a thick disc. Let the dough rest in the refrigerator for at least 30 minutes (preferably overnight).

Foncage

1. Lightly dust the marble with flour (***fleurer***) and place the dough in the center. Roll out the pastry (***abaisser***), giving it quarter turns as you roll. Continue rolling and turning until the dough is 1.8 in. (3 mm) thick and 3 fingers wider than the tart mold. Prick the dough (***piquer***) with the docker (***pique pâte***) or a fork. Roll the dough onto the rolling pin, and gently lay it on the tart mold. Lift the edges and press the dough into the corners of the mold (***foncer***). To form an even border that can extend above the borders of the mold, you need to leave some extra dough before trimming. Lift the edge of the dough and make a small fold inside the rim. Repeat this process all the way around the mold, then pass the rolling pin over the top of the tart to trim the excess dough. Gently pinch all around the top edge of the tart to create a decorative border (***chiqueter***).
2. Place the lined mold in the refrigerator to rest for at least 20 minutes.

Appareil à Quiche

1. Place the lardons in a pan of cold water, bring the water to a boil, and blanch them for 1 minute. Drain and refresh the lardons under cold running water. Set aside.
2. Heat the oil in a small sauté pan over high heat and sauté the lardons until they are a light golden color. Reserve them on a paper towel until needed.
3. Place the eggs in a mixing bowl, whisk them well, and season with salt, pepper, and nutmeg. Mix in the crème fraîche and set aside until needed.

Cuisson à Blanc

1. Cut a circle of parchment paper that is a few inches larger than the tart shell. Press it into the cavity and fill it with baking beads. Place the tart on a baking sheet in the oven. Once the oven door is closed, reduce the temperature to 350°F (175°C). Cook the tart until the crust is a light golden color. After 15 minutes, remove the shell from the oven, remove the baking beads, and replace the shell in the oven to continue cooking for another 5 minutes.

To Finish

1. Once the shell is cooked, take it out of the oven and sprinkle the bottom with grated cheese and lardons. Ladle half of the custard mixture into the shell and carefully place it in the oven. Allow to cook for 5 minutes, then add the remaining custard mixture.
2. Cook the quiche for 20 to 25 minutes, or until a knife inserted into the center comes out clean. Remove from the oven and allow to rest for 5 minutes.
3. *To remove the quiche from the mold:* Place the mold on top of a small overturned saucepan. Gently push down the outer part of the mold. Slide the quiche off the metal disc onto a serving dish.

Quantity		*Ingredient*
U.S.	*Metric*	*Pâte Brisée*
8 oz	250 g	Flour
4 oz	125 g	Butter, cold, diced
1 tsp	5 g	Salt
2 pcs	2 pcs	Egg yolks
2 tsp	10 mL	Water
		Appareil à Quiche
5 oz	150 g	Smoked pork belly, rind removed, cut into lardons, blanched
2 tsp	10 mL	Vegetable oil
3 pcs	3 pcs	Eggs
8 fl oz	250 mL	Crème fraîche or heavy cream
Pinch	Pinch	Nutmeg
3 ½ oz	100 g	Gruyère cheese, grated
		Salt and pepper

RATATOUILLE NIÇOISE

Learning Outcomes

Émonder
Épépiner
Concassée
Bâtonnet
Saisir
Suer
Mijoter
Ciseler
Chiffonade
Émincer

Equipment

Knives:
Paring knife (*office*), slicing knife (*éminceur*)

Tools:
Bowls, cutting board, wooden spatula

Pans:
Sauteuse, gratin dish

Serving

4–6 persons

FYI...

Ratatouille Niçoise is a dish that is representative of Provençal cuisine. It uses vegetables that grow in profusion in southeastern France, including eggplants, zucchinis, tomatoes, garlic, and green peppers. It was not until the early 20th century that this particular version of ratatouille was formalized as a recipe; before that time, similar preparations existed under the ragoût category. The word *ratatouille* is a cross between *ratouiller* and *tatouiller,* both of which have the suffix *-touiller*, which means "to stir up."

The origin of the word ratatouille comes from the French and Occitan (Langue d'oc) languages.

RATATOUILLE NIÇOISE

Provençal Stewed Vegetables

Method

1. Preheat the oven to 350°F (175°C).

Preparation

1. Cut the eggplant in half lengthwise and score the flesh in a criss-cross pattern. Sprinkle the flesh generously with coarse salt and lay the halves flesh side down in a dish. Leave to degorge for 30 minutes, rinsing well after the process is complete. Cut the bell peppers in half and remove the stem, seeds, and spongy ribs.
2. Cut the eggplant, zucchini, and bell peppers into ***bâtonnets***. Chop the onions and garlic (***hacher***), then cut the tomatoes into a rough dice (***concassée***). Rub the leaves off the thyme (***émietter***) and shred (***chiffonade***) the basil leaves.

Cuisson

1. Heat some olive oil in a large, shallow pan over high heat and sear (***saisir***) the zucchini until lightly colored. Add one-third of the thyme, cayenne, and garlic. Reduce the heat to medium and cook for 1 to 2 minutes. Season to taste, then transfer to a bowl and set aside.
2. Heat some olive oil in the same pan over high heat and sear (***saisir***) the eggplant until lightly colored. Add another one-third of the thyme, cayenne, and garlic. Reduce the heat to medium and cook for 1 to 2 minutes. Season to taste, transfer to a bowl, and set aside.
3. In the same pan, heat some olive oil over low heat and sweat (***suer***) the chopped onions until soft. Add the remaining garlic, cayenne, and thyme. Season to taste. Add the tomato paste and cook 1 to 2 minutes (***pincer la tomate***). Add the red and green peppers, the tomatoes, and the ***bouquet garni***. Season to taste and simmer (***mijoter***) for 25 to 30 minutes. Remove the ***bouquet garni*** and toss in the shredded (***chiffonade***) basil.

To Finish

1. In a deep-sided, ovenproof dish, layer the cooked vegetables in the following order: one-third of the tomato mixture, all of the eggplant, one-third of the tomato mixture, all of the zucchini, then the final one-third of the tomato mixture. Spread a layer of breadcrumbs over the top of the ratatouille and bake in the oven for 45 minutes.

Quantity		*Ingredient*
U.S.	*Metric*	*Base*
1 pc	1 pc	Eggplant, medium
		Coarse salt
5 pcs	5 pcs	Zucchini, medium, cut into ***bâtonnets***
2 pcs	2 pcs	Onion
1 head	1 head	Garlic cloves, chopped
1 pc	1 pc	Red pepper
1 pc	1 pc	Green pepper
8 pcs	8 pcs	Tomatoes, medium
3 brs	3 brs	Thyme
1 bq	1 bq	Basil
4 fl oz	125 mL	Olive oil
1 pc	1 pc	***Bouquet garni***
1 ½ oz	45 g	Tomato paste
		Cayenne pepper
		Salt and pepper
		Finishing
7 oz	200 g	Breadcrumbs

RISSOLES DE VEAU, SAUCE RAVIGOTE

Learning Outcomes

Farce simple
Emulsion instable
Frire

Equipment

Knives:
Paring knife (*office*), slicing knife (*éminceur*)

Tools:
Bowls, cutting board, plastic spatula, rolling pin, araignée, whisk, pastry brush, round pastry cutter

Pans:
Small sauté pan, deep fryer

Serving

8 persons

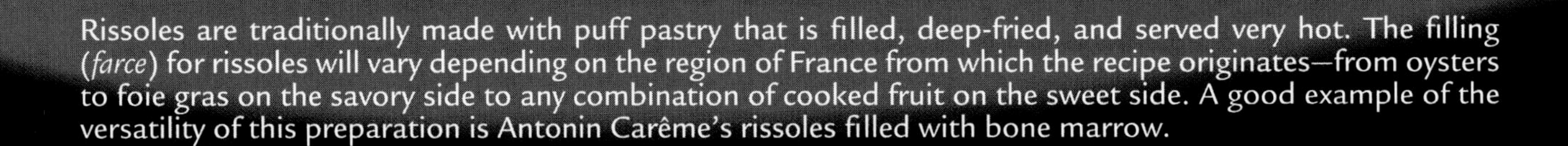

Rissoles are traditionally made with puff pastry that is filled, deep-fried, and served very hot. The filling (*farce*) for rissoles will vary depending on the region of France from which the recipe originates—from oysters to foie gras on the savory side to any combination of cooked fruit on the sweet side. A good example of the versatility of this preparation is Antonin Carême's rissoles filled with bone marrow.

RISSOLES DE VEAU, SAUCE RAVIGOTE

Veal Filled Fried Pastries

Method

1. Heat the deep fryer to 320°F (160°C).

Farce

1. Heat the oil in a small sauté pan over low heat and sweat (***suer***) the shallots and garlic until soft. Set aside to cool.
2. Place the ground veal and fat in a large bowl and mix in the cooled shallots and garlic. Chop the herbs and mix them into the meat. Mix in the egg, season with salt and pepper, and set aside.

Montage des Risoles

1. Roll out the puff pastry to a thickness of ⅛ in. (3 mm). ***Tip:*** *If the dough becomes too soft, place it on a chilled baking sheet that has been kept in the freezer for this purpose.*
2. Using a pastry cutter, cut 3 in. (8 cm) rounds of puff pastry and transfer them to a baking sheet lined with parchment paper. Place the rounds in the refrigerator to rest for 10 minutes.

Assembly

1. Brush the edge of each round with egg wash, place approximately 1 oz (30 g) of the filling in the center of each round, and fold in half. Press them closed, being careful to expel any trapped air. Brush the tops with egg wash and leave the rissoles to rest in the refrigerator on the parchment-lined tray.

Sauce Ravigote

1. Whisk the salt into the sherry vinegar, add the white pepper, and whisk in the oil in a steady stream. Add the capers, shallots, onions, and chopped herbs.

To Finish

1. Cook the rissoles in the deep fryer until golden brown. Drain well on a paper towel and season immediately.

To Serve

1. Serve directly from the deep fryer onto a hot plate with the ***sauce ravigote*** on the side.

Quantity		*Ingredient*
U.S.	*Metric*	
1 lb	500 g	Puff pastry trimmings (***parures de feuilletage***)
1 pc	1 pc	Egg for egg wash (***dorure***)
		Veal Farce
¾ fl oz	20 mL	Vegetable oil
1 oz	30 g	Shallots, finely diced (***ciseler***)
1 pc	1 pc	Garlic clove, finely chopped (***hacher***)
14 oz	400 g	Veal, ground
7 oz	200 g	Pork back fat (***lard***), ground
½ oz	12 g	Parsley
¼ oz	10 g	Sage
1 pc	1 pc	Egg
		Salt and pepper
		Sauce Ravigote
1 fl oz	30 mL	Sherry vinegar
3 ½ fl oz	100 mL	Vegetable oil
1 oz	30 g	Capers, finely chopped (***hacher***)
1 oz	30 g	Shallots, finely diced (***ciseler***)
1 oz	30 g	Onions, finely diced (***ciseler***)
¾ oz	20 g	Tarragon, finely chopped (***hacher***)
¾ oz	20 g	Chives, finely chopped (***hacher***)
¾ oz	20 g	Parsley, finely chopped (***hacher***)
		Salt and white pepper

SOUFFLÉ AU FROMAGE

Learning Outcomes

Sauce Béchamel
Sauce Mornay
Blancs en neige
Appareil
Bain marie
Chemiser

Equipment

Knives:
Knife (***émincer***)

Tools:
Wooden spoon, whisk, rubber spatula, mixing bowls, pastry brush, ramekins

Pans:
Small sauce pan (***russe***), heavy baking sheet, soufflé mold [diameter: 8 in. (18 to 20 cm)]

Serving

6 persons

FYI...

Developed in France during the latter half of the 1700s, the soufflé made its first significant restaurant appearance at *La Grande Taverne de Londres* in Paris. The preparation was executed by Chef Antoine Beauvilliers, who later published a recipe for soufflé in *L'Art du Cuisinier* in 1813. Refining Beauvilliers's recipe, Chef Carême, in the *Pâtissier Royal Parisien,* addressed some problems chefs were having with the tendency for soufflés to fall. During the late 1800s, more diverse soufflé recipes were introduced; recipes started calling for different cheeses, meats, and vegetables. In terms of methodology, the soufflé is closely related to the meringue in that stiffly whisked egg whites are used to create volume.

SOUFFLÉ AU FROMAGE
Cheese Soufflé

Method

1. Preheat the oven to 400°F (205°C).

Chemisage

1. *Prepare the ramekins or molds:* Brush the interiors with softened butter (***pommade***), then mix the flour and breadcrumbs together and coat the inside of the molds with this mixture. Refrigerate until needed.

Sauce Mornay

1. *To make a Mornay sauce, start by making a béchamel sauce:* Melt the butter in a small pan over medium heat. When the butter begins to foam, add the flour and stir well until the mixture begins to bubble. Cook for 1 to 2 minutes to obtain a white roux (*be careful not to let it color*). Transfer to a small plate and set aside to cool.
2. Place the milk, studded onion, and ***bouquet garni*** in a medium saucepan over medium heat and season with salt, white pepper, and nutmeg. When the milk comes to a boil, whisk the cold roux, piece by piece, into the hot milk until completely incorporated and it begins to thicken. Keep the béchamel at a boil for 1 minute stirring constantly; then take it off the heat, remove the ***bouquet garni*** and the studded onion, and strain it through a fine mesh sieve (***chinois***) into a large bowl.
3. To finish the Mornay, gradually stir the grated cheese into the hot béchamel and allow it to cool until warm to the touch. Beat the egg yolks in a small bowl and temper them by mixing in some warm sauce. Pour this mixture back into the Mornay and stir until completely incorporated. Season the sauce with a pinch of cayenne. Keep warm.

To Finish

1. Beat the egg whites to medium peaks and fold one-third of them into the Mornay. This will lighten the mixture and make it easier to incorporate the remaining egg whites. Fold the remaining egg whites into the Mornay until just combined, then fill the molds with the soufflé mixture. Smooth the top with a spatula and run your thumb just inside the rim of the molds to create a groove around the circumference of the soufflé. This will help it rise evenly. Place a lozenge of Gruyère cheese on top if desired. Transfer the molds to a heavy baking sheet and place in the preheated oven. After 2 minutes, lower the temperature to 350°F (185°C) and continue cooking until the soufflé has doubled in height and is golden in color. Remove from the oven and serve immediately.

Tip: *To avoid the soufflé collapsing, the oven door must stay closed until the end of cooking.*

Quantity		*Ingredient*
U.S.	*Metric*	*Chemisage*
1 oz	30 g	Butter, softened (***pommade***)
1 oz	30 g	Flour
1 oz	30 g	Breadcrumbs
		Mornay Sauce
1 ½ oz	45 g	Butter
1 ½ oz	45 g	Flour
12 ½ oz	375 g	Milk
1 pc	1 pc	Onion, studded (***clouté***)
1 pc	1 pc	***Bouquet garni***
Pinch	Pinch	Nutmeg, grated or ground
4 oz	120 g	Gruyère cheese, grated
5 pcs	5 pcs	Egg yolks
Pinch	Pinch	Cayenne pepper
		Salt and white pepper
		Finishing
5 pcs	5 pcs	Egg whites
2 oz	60 g	Gruyère cheese lozenges (*optional*)

Les Oeufs

(Eggs)

Recipes

Les Oeufs (Eggs)

Oeufs brouillés aux herbes
—Scrambled eggs with herbs

Oeufs Boitelle
—Molded eggs with mushrooms

Oeufs à la coque
—Soft cooked eggs

Oeufs à la mexicaine
—Deep fried eggs with rice and tomato sauce

Oeufs cocotte à la crème
—Coddled eggs with cream

Oeufs mollets à la florentine
—Medium cooked eggs with spinach and Mornay sauce

Oeufs en gelée
—Poached eggs in aspic

Oeufs en meurette
– Poached eggs in red wine sauce

Oeufs à la bretonne
—Hard cooked eggs, leek and mushroom casserole

Oeufs chasseur
—Baked eggs with madeira sauce and chicken livers

Omelette fermière farcie aux champignons
—Mushroom stuffed omelet

Omelette aux fines herbes
—Herb omelet

Omelette paysanne
—Rustic potato and bacon omelet

Omelette grandval
— Omelet and tomato sauce

Omelette soufflée
— Sweet omelet with strawberries

OEUFS BROUILLÉS AUX HERBES

Learning Outcomes

Oeufs brouillés
Cooking in a bain marie

Equipment

Tools:
Bowls, whisk, wooden spatula, ramekin

Pans:
Sauce pan (*russe*), bain marie

Serving

4 persons

FYI...

Oeufs brouillés gives another meaning to scrambled eggs. Gentle cooking over the bain marie prevents the eggs from cooking too quickly and becoming rubbery.

Why the ramekin? Breaking eggs individually first into a ramekin allows you to ensure that the egg isn't spoiled (unfortunately, the only true way to find out) and makes it easier as well to remove the stray piece of egg shell.

OEUFS BROUILLÉS AUX HERBES

Scrambled Eggs with Herbs

Quantity		Ingredient
U.S.	*Metric*	
¼ bunch	¼ bunch	Parsley, finely chopped (***hacher***)
¼ bunch	¼ bunch	Chervil, finely chopped (***hacher***)
¼ bunch	¼ bunch	Chives, finely minced (***ciseler***)
6 pcs	6 pcs	Eggs
3 oz	90 g	Butter
4 fl oz	125 mL	Cream
		Salt and pepper

Method

1. Finely chop (***hacher***) the parsley and chervil. Finely mince (***ciseler***) the chives. Set aside.
2. Break the eggs one at a time into a ramekin and transfer to a bowl. Whisk the eggs until combined.
3. Place a saucepan in a gently simmering bain marie.
4. Melt half the butter in the pan, add the eggs, and stir gently with a wooden spoon until just coagulated.
5. Remove the pan from the bain marie and mix in the remaining butter to stop the cooking process.
6. Slowly pour in the cream while gently stirring the eggs. Season to taste. Stir in the chopped herbs.

To Serve

1. Place the oeufs brouillés in a warm soup bowl and sprinkle with fresh chopped herbs or transfer the eggs to a clean pan, pat (***tamponner***) the surface with butter, and reserve them covered in a gentle bain marie.

OEUFS BOITELLE

Learning Outcomes

Oeufs moulés
Sauter à cru
Croûtons
Beurre clarifié
Essence
Bain marie
Napper
Monter au beurre
Chemiser

Equipment

Knives:
Paring knife (*office*), slicing knife (*éminceur*)

Tools:
Twelve ½-cup oval ramekins or custard cups to be used as molds, bowls, cutting board, wooden spatula, sieve, ladle (*louche*), whisk, fine chinois

Pans:
Sauteuse, sautoir, small russe, bain marie, plaque à rôtir

Serving

4 persons

FYI...

Oeufs boitelle is named for *Antoine Boitelle*, a controversial character in a short story written by author Guy de Maupassant. The story was simply titled *Boitelle*.

This is one of the classic preparations for eggs. Oeufs moulés, literally meaning molded eggs, describes eggs that are cooked in a mold, and then unmolded before serving.

OEUFS BOITELLE

Molded Eggs with Mushrooms

Method

1. Preheat the oven to 300°F (150°C).

Montage

1. Brush the interior of the molds with butter, line the bottom with a piece of parchment paper, and place in the refrigerator to set. Clean, peel, and thinly slice (***émincer***) the mushrooms. Reserve the trimmings.
2. Melt some butter in a medium pan over low heat and sweat (***suer***) the mushrooms until soft. Season the mushrooms and set aside to cool.
3. When the mushrooms are cooled, line (***chemiser***) both the bottom and sides of the molds with overlapping slices. Reserve in the refrigerator.

Sauce

1. *Make an essence from the mushroom trimmings:* In a small pan, cover the trimmings with cold water, season lightly, and bring to a gentle simmer over low heat. Cover and cook for 15 to 20 minutes, then strain through a fine mesh sieve (***chinois***). Mount with cold butter, finish with lemon juice, and season to taste. Set aside and keep warm.

Cuisson

1. Cover the bottom of a roasting pan with a sheet of parchment paper. Arrange the mushroom-lined molds in the roasting pan. Break the eggs one at a time into a small ramekin or bowl. Slip an egg into each mushroom-lined mold, being careful not to break the yolk. Fill the pan with boiling water two-thirds of the way up the sides of the molds, then place in the oven to cook until the yolks are nearly set (*10 to 15 minutes*).

Garniture

1. Melt the butter in a medium sauté pan over low heat and sweat (***suer***) the shallots and garlic until soft. Add the wild mushrooms and tomatoes. Increase the heat to medium high and cook until dry, stirring often. Mix in the chopped parsley, season to taste and set aside.
2. Cut the bread into ovals the same size as the opening of the molds.
3. Heat the clarified butter in a large sauté pan and toast the bread slices in the hot butter until golden on both sides. Transfer the finished croûtons to a plate lined with paper towels and keep warm.

To Serve

1. Mix the chopped parsley into the sauce.
2. Run a knife around the edge of the ramekins and unmold the cooked eggs onto the croûtons. Arrange the eggs on a serving platter. Coat (***napper***) each egg with a spoonful of sauce. Spoon the wild mushroom garnish onto the platter.
3. Serve the remaining sauce on the side.

Quantity		*Ingredient*
U.S.	*Metric*	***Base***
4 pcs	4 pcs	Eggs
		Montage
2 oz	60 g	Unsalted butter, softened (***en pommade***)
8 oz	250 g	Button mushrooms, sliced (***émincer***)
		Sauce
7 oz	200 g	Mushroom trimmings
3 ½ oz	100 g	Butter
1 pc	1 pc	Lemon, juice of
		Salt and pepper
		Garniture
2 oz	60 g	Butter
1 ¾ oz	50 g	Shallots, finely chopped (***hacher***)
2 pcs	2 pcs	Garlic cloves, finely chopped (***hacher***)
7 oz	200 g	Wild mushrooms such as chanterelles
2 pcs	2 pcs	Tomatoes, peeled, seeded, and diced (***concasser***)
5 brs	5 brs	Parsley, finely chopped (***hacher***)
4 pcs	4 pcs	White bread, sliced
3 oz	100 g	Butter, clarified
		Salt and pepper
		To Serve
4 brs	4 brs	Parsley, finely chopped (***hacher***)

OEUFS À LA COQUE

Learning Outcomes

Cuisson des oeufs pochés
en coquille
Mouillette

Equipment

Tools:
Araignée coquetier,
egg cutter

Pans:
Sauce pan (*russe*)

Serving

4 persons

OEUFS À LA COQUE
Soft Cooked Egg

Method

1. Remove the eggs from the refrigerator 10 to 15 minutes before cooking.
2. Bring water to a boil in a deep saucepan with some vinegar (*approximately 15% of the volume of the water*).
Add the eggs and bring back to a simmer.
Maintain a steady simmer for 3 minutes for large eggs [*2 oz (55 grams)*].
3. Remove the eggs from the water, dry, and place the eggs in egg cups with the small ends pointed upward. Just before serving, slice off the top with a pair of egg cutters.

To Serve

1. Toast the bread and cut into strips (***mouillettes***) thin enough to dip into the eggs. Serve on the side, wrapped in a warm napkin and decorated with fresh parsley.

Some recipes refer not only to the ingredients used, but to the utensils that are used to consume them. Traditionally, the egg is served in a ***coquetier*** that is made of porcelain, wood, or earthenware. The purpose of the ***coquetier*** is simply to enable eating the contents of the egg without it falling over.

Quantity		*Ingredient*
U.S.	*Metric*	*Base*
4 pcs	4 pcs	Eggs
		White vinegar
		To Serve
7 oz	200 g	Bread
1 br	1 br	Parsley

FYI...

For all egg-based preparations, it is recommended to use only the freshest of eggs.

Oeufs à la coque are soft boiled eggs served in the shell. *À la coque* doesn't refer to the shell as most people assume. The term dates back to the days of sea travel where the mark of a good chef was his ability to cook an egg without the aid of a clock.

OEUFS À LA MEXICAINE

Learning Outcomes

Oeufs frits
Vider
Tomato sauce
Créole rice
Émonder
Dégorger

Equipment

Knives:
Chef knife (*couteau chef*), paring knife (*office*)

Tools:
Bowl, colander, ovenproof dish, deep fryer, 2 wooden spatulas, food mill, fine chinois

Pans:
Sauce pan (*russe*), frying pan

Serving

6 persons

FYI...

Oeufs à la mexicaine is an example of a French dish with an international influence. Recipes that indicate that something should be made à la *mexicaine* are guaranteed to include tomatoes and/or peppers.

OEUFS À LA MEXICAINE

Deep Fried Eggs with Rice and Tomato Sauce

Method

1. Preheat the oven to 375°F (190°C).
2. Prepare the tomatoes for the garnish: Choose 6 evenly shaped tomatoes. Cut the tops off the tomatoes and set aside. Empty (***vider***) the insides of the tomatoes and reserve the pulp for the sauce. Sprinkle the interiors with salt, then arrange the tomatoes upside down on a wire rack (or tray) that is lined with paper towels to absorb excess water (***dégorger***).

Tomato Sauce

1. Roughly chop the peeled and seeded tomatoes (***concasser***).
2. Heat the oil in a small sauté pan and sweat (***suer***) the onions over low heat until soft. Add the crushed garlic, then the tomato paste, and cook for 1 to 2 minutes (***pincer la tomate***). Add the chopped tomatoes, the reserved tomato pulp, the fresh coriander, a ***bouquet garni***, and salt and pepper. Cover and cook over low heat for 25 to 30 minutes. Purée the sauce through a food mill, then strain it through a fine mesh sieve (***chinois***). The consistency should be thick enough to coat the back of a spoon (***à la nappe***).

Créole Rice

1. *Cook the Créole rice:* Place the rice in a large pan and add the water. Place the pan over medium-high heat and once the rice comes to a boil, cook the rice for 14 minutes. Remove from the heat, strain the rice, and refresh with cold water. Drain well, then transfer the rice to a colander lined with a cloth. Allow to drain completely.
2. Mix one-fourth of the tomato sauce with the cooked rice. Season to taste, then fill the tomatoes with a portion of the rice mixture. Grease an ovenproof dish. Make a bed with the remaining rice in the ovenproof dish. Cover the stuffed tomatoes with their tops (which will keep the rice from drying out) and place in the preheated oven for 25 minutes or until thoroughly heated.

Preparing the Fried Eggs (oeufs frits)

1. While the tomatoes are cooking, prepare the hot oil for frying the eggs. Preheat the oil to 180°C (355°F). If not using a deep fryer, fill a large saucepan with oil at least 3 in. (about 7 cm) deep. Have two wooden spatulas ready. Break the eggs one at a time into a small ramekin or bowl. Slip the egg into the hot oil and fry, turning it with the wooden spatulas. Fry until golden while ensuring the yolk remains runny. Drain on paper towels and season immediately.

Tip: *If in doubt about the cooking time, cook a trial egg. Cooking times will vary depending on the size of the eggs.*

To Serve

1. Once the tomatoes are cooked, remove the tops and discard. Trim the fried egg of any crisp bits of egg white (***ébarber***) and place an egg on top of each tomato. Brush with melted butter to add shine (***lustrer***) and decorate with fresh parsley and coriander.
2. Serve the sauce on the side.

Note: *Deep frying is a dangerous cooking technique to apply to eggs. If not handled properly, the eggs can splash, causing serious burns. Be cautious when using this technique.*

Quantity		Ingredient
U.S.	*Metric*	
6 pcs	6 pcs	Eggs, at room temperature
1 qt	1 L	Oil for deep-frying
6 pcs	6 pcs	Tomatoes
		Salt and pepper
		Tomato Sauce
6 pcs	6 pcs	Tomatoes, peeled (***émonder***) and seeded (***épépiner***)
1 fl oz	30 mL	Oil
3 ½ oz	100 g	Onion
2 pcs	2 pcs	Garlic cloves, crushed
1 oz	30 g	Tomato paste
½ bq	½ bq	Fresh coriander, finely chopped (***hacher***)
1 pc	1 pc	***Bouquet garni***
		Salt and white pepper
		Créole Rice
8 oz	250 g	Long grain rice
1 ½ qt	1 ½ L	Water

OEUFS COCOTTE À LA CRÈME

Learning Outcomes

Cuisson des oeufs façon cocotte
Cuisson au bain marie au four

Equipment

Tools:
Twelve ½-cup ramekins or custard cups

Pans:
Small sauce pan (*russe*), roasting pan

Serving

12 persons

OEUFS COCOTTE À LA CRÈME

Coddled Eggs with Cream

Quantity		*Ingredient*
U.S.	*Metric*	
12 pcs	12 pcs	Eggs
12 ½ fl oz	380 mL	Crème fraîche or heavy cream
2 oz	60 g	Unsalted butter, softened, for ramekins
5 brs	5 brs	Parsley, chopped
		Salt and freshly ground pepper

Method

1. Preheat the oven to 320°F (160°C).
2. Heat the cream in a small saucepan over medium heat and allow it to reduce until it thickly coats the back of a spoon (***à la nappe***). Season to taste with salt and pepper and keep warm.
3. Line a baking dish with parchment paper, then butter the ramekins and season with salt and pepper. Divide the reduced cream among the prepared ramekins. Break the eggs one at a time into a small bowl and slip one into each ramekin. Arrange the filled ramekins in the baking pan, making sure to leave space in between them.
4. Fill the tray with boiling water halfway up the height of the ramekins. Carefully place the pan in the hot oven. Bake until the white is just set, about 3 minutes.

To Serve

1. Remove the ramekin from the baking pan, wipe with a kitchen towel, and place on a plate. Sprinkle with chopped parsley.
2. Serve immediately.

This recipe is named for the heatproof dish that is used to both cook and then act as a serving container for the egg.

FYI...

OEUFS MOLLETS FLORENTINE

Learning Outcomes

Cuisson oeuf mollet
(medium cooked egg)
en coquille
Gratiner
Sauce Mornay

Equipment

Tools:
Bowl, spatula, whisk, fine chinois, cheese grater

Pans:
Sauce pan (*russe*), gratin dish, bain marie

Serving

4 persons

FYI...

À la Florentine is a French term that means Italian style, or in the style of Florence. This implies that the dish contains spinach. Florentine dishes are usually made with fish or eggs. As a finishing touch, Florentine dishes is sometimes lightly topped with grated cheese and baked in the oven until golden.

OEUFS MOLLETS FLORENTINE

Medium Cooked Eggs with Spinach and Mornay Sauce

Method

1. Preheat the oven to 450°F (230°C).

Prepare the Eggs

1. Fill a large saucepan with cold water and generously season with coarse salt and vinegar. Add the whole eggs so they are in a single layer and are covered by at least 1 in. of water. Bring the water to a boil, reduce the heat to a low boil, and cook for exactly 5 minutes. Remove from the heat and transfer the eggs to cold water. Leave them in the cold water until completely cooled, changing the water if necessary. Peel (***écaler***) and rinse the eggs. Set aside.

Sauce Mornay

1. *To make a **sauce Mornay**, start by making a **sauce béchamel**:* Melt the butter in a medium pan over medium heat. When it begins to foam, add the flour and stir well until it begins to bubble. Cook for 1 to 2 minutes to obtain a white roux (*being careful not to let it color*). Transfer to a small dish and set aside to cool. Place the milk in a medium saucepan over medium heat and season with salt, pepper, and nutmeg. Once it is just coming to a boil, whisk the roux, piece by piece, into the hot milk until completely incorporated and beginning to thicken. Bring back to a boil over medium heat, and allow to boil for 1 minute, whisking constantly. Remove from the heat.
2. *To make the **sauce Mornay***: Beat the egg yolk in a small bowl. Stir some of the finished ***sauce béchamel*** into the egg yolk, mixing well to temper it. Stir the warmed egg yolk into the ***sauce béchamel*** until completely incorporated.
3. Strain the sauce through a fine mesh sieve (***chinois***) into a clean pan and mix in the grated Gruyère cheese. Finish by mixing in the butter (***monter au beurre***). Pat (***tamponner***) the surface with cold butter on the end of a fork and reserve in a bain marie.

Garniture Florentine

1. Rinse the spinach in several changes of cold water, then dry and remove the stems. In a large saucepan, bring salted water to a boil. Immerse the spinach in the boiling water and cook until wilted, about 1 minute. Transfer immediately to ice water. Once cooled, drain and press well to extract as much water as possible. Coarsely chop the cooked spinach. Melt some butter in a small pan over medium heat. Cook the butter until it begins to brown. Add the spinach and stir until it is warmed through and coated in brown butter. Stir in one-quarter of the ***sauce Mornay*** and adjust the seasoning if needed.

To Serve

1. Butter a gratin dish. Cover the bottom of the dish with half of the spinach mixture. Nestle the cooked eggs on top and cover with the remaining spinach. Evenly cover with the remaining ***sauce Mornay***. Sprinkle the top with the remaining Gruyère cheese and dot with butter.
2. Brown (***gratiner***) in the oven until golden.

Quantity		*Ingredient*
U.S.	*Metric*	
8 pcs	8 pcs	Eggs
		Coarse salt
		White vinegar
		Mornay Sauce
1 oz	30 g	Butter
1 oz	30 g	Flour
1 pt	500 mL	Milk
		Nutmeg
1 pc	1 pc	Egg yolk
3 ½ oz	100 g	Gruyère cheese, grated
1 oz	30 g	Butter
		Salt and pepper
		Garniture Florentine
1 ½ lb	750 g	Spinach
2 oz	60 g	Butter
		Salt and pepper
		Finishing
2 oz	60 g	Gruyère cheese, grated
1 oz	30 g	Butter

Learning Outcomes

Oeuf poché hors coquille
Aspic
Working with gelatin
Tempérer
Émonder
Évider

Equipment

Knives:
Paring knife (*office*), slicing knife (*éminceur*)

Tools:
6 martini glasses, bowls, cutting board, wooden spatula, plastic spatula, fine chinois, skimmer (*écumoire*), ladle (*louche*), coffee filter

Pans:
Sauce pan (*russe*)

Serving

10 persons

FYI...

A traditional cold preparation, this version is a modern update to an old classic. The use of aspic in the 18th century was popularized by Antonin Carême because it lent itself to his extravagant presentations. The original process of fabricating gelatin was by boiling down cartilage and bones. Although this provided a superior texture to the gelatin, it also gave it a distinct flavor. Aspic can also be made using agar, which provides a harder set and a different texture as well.

OEUFS EN GELÉE
Poached Eggs in Aspic

Method

1. Place the martini glasses on a tray in the refrigerator to chill.

Prepare the Poached Eggs

1. Fill a large shallow pan (***russe***) three-fourths full of water and generously season with white vinegar and salt. Bring to a boil over high heat, and then reduce to a simmer. Break the eggs one at a time into a ramekin or small bowl. Slip each egg into the simmering water, poaching 2 to 3 at a time.

Note: *The rising bubbles will prevent the eggs from sticking to the bottom of the pot and will also help the whites envelop the yolks.*

2. When the whites begin to coagulate, use a skimmer to help them wrap firmly around the yolks. Poach for about 3 minutes or until the whites are set but the yolk is still very soft. Gently remove them from the hot liquid with a slotted spoon and transfer them to an ice bath to cool (see page 252).
3. Once cooled, trim off any loose whites (***ébarber***) and pat the eggs dry.

Garniture

1. *Peel (**émonder**) the tomatoes:* Bring a large pan of water to a boil, dip each tomato in the boiling water, and transfer them to an ice bath. The skin will loosen and be easily peeled. Cut the tomatoes into quarters and remove the pulp and seeds with a small knife to obtain petal shapes. Pat dry and set aside on a tray.
2. Lay out a single layer of the smoked salmon slices.

Gelée

1. Make a little bundle of cheesecloth or gauze containing peppercorns and tie closed with string.
2. Place the chicken consommé in a medium pan, add the bundle of peppercorns, and bring to a simmer over medium heat. After 5 to 10 minutes, remove the peppercorns. If the liquid is fatty or cloudy, blot the surface with a paper towel or re-filter through a fine mesh sieve (***chinois***) lined with a damp towel.
3. Ladle 3 fl oz (100 mL) of the hot consommé into a small bowl and cool until just warm to the touch.
4. Stir in the gelatin powder until completely dissolved. Pour this back into the hot consommé. Pour in the port wine, if using, and stir well. Remove from the heat and reserve at room temperature.

Note: *If using gelatin leaves, first soften in cold water. Squeeze out excess water and add to the warm consommé. Stir until melted. Then proceed as above.*

Montage de l'Aspic

1. Place 14 fl oz (400 mL) of the gelée into a bowl over an ice bath. Using a metal spoon, gently stir the consommé (***vanner***). Continue until the liquid becomes like oil in resistance and movement.

   Tip: *Be careful not to stir too vigorously in order to avoid incorporating any air bubbles.*

2. Remove the martini glasses from the refrigerator. Pour a small amount of the gelée into a glass and swirl to form a thin, even layer at the bottom of the glass. Repeat with the remaining glasses. This must be done quickly as the gelée will set once it comes into contact with the chilled glass.
3. Once the gelée is set, divide the tomatoes, salad, herbs, poached eggs, and smoked salmon between the glasses, arranging them so that each ingredient can be seen through the gelée. Pour another layer of gelée on top.
4. When all the gelée is used, temper another 14 fl oz (400 mL) of gelée and pour into the glasses until the ingredients are just covered. Lightly tap the glasses as they are being filled to remove any air bubbles. Place in the refrigerator to set for at least an hour if serving in the glass. If you are serving them unmolded, it is recommended to refrigerate them for at least 3 or 4 hours. To un-mold, warm the exterior of the glass; the aspic will release from the sides.

Note: *The term aspic is used only in reference to a final preparation using gelée in a mold.*

Quantity		*Ingredient*
U.S.	*Metric*	*Base*
10 pcs	10 pcs	Eggs
		White vinegar
		Salt
		Gelée
1 qt	1 L	Chicken consommé, clarified
		Black peppercorns, crushed (***mignonette***)
¾ oz	22 g	Gelatin powder, or leaves
5 fl oz	150 mL	White port wine (*optional*)
		Salt
		Montage
10 pcs	10 pcs	Smoked salmon, sliced [*¾ oz (25 g) each*]
3 pcs	3 pcs	Medium tomatoes, peeled (***émonder***), quartered, and seeded (***épépiner***)
		Truffle, sliced (***émincer***) (*optional*)
		Caviar (*optional*)
		To Serve
1 ¾ oz	50 g	Micro greens, mixed
		Chervil or parsley leaves

To Serve

1. The ***Oeuf en gelée*** can either be unmolded from the glass and served on a chilled plate, or served in the glass.
2. Present with a touch of caviar, truffle, or a sprig of fresh parsley.

Note: *In this recipe, soft poached eggs are used. The recipe can be adapted to hard poached eggs which would allow the aspic to be kept longer (3 days).*

OEUFS EN MEURETTE

Learning Outcomes

Oeuf poché hors coquille
Blanchir
Lardons

Equipment

Knives:
Chef knife (*couteau chef*), paring knife (*office*)

Tools:
Skimmer (*écumoire*), bowl, fine chinois

Pans:
Sauce pan (*russe*), pan, bain marie

Serving

4 to 8 persons

FYI...

The word *meurette* is thought to have first been used in the Franche-comté region of France in the 15th century. It is derived from the French word *muire*, meaning naturally salted water. *Muire* originally comes from the Latin *muria*, meaning salted water, or brine.

A meurette sauce is one that is made with red wine (ideally Bourgogne). Oeufs en meurette are traditionally accompanied by pieces of bread that have been lightly grilled, rubbed with garlic, and buttered.

OEUFS EN MEURETTE

Poached Eggs in Red Wine Sauce

Method

Sauce

1. *Blanch the smoked lardons:* Place in a small pan with cold water. Bring to a boil (***blanchir***), then strain and refresh. Set aside to drain. Heat the oil in a small frying pan over medium heat and sauté the lardons until lightly colored. Drain and set aside.
2. Heat the clarified butter in a small frying pan over medium-high heat. Add the mushrooms and sauté them until lightly colored. Set aside.
3. Melt the butter in a large pan over low heat and sweat (***suer***) the shallots and garlic until soft. Add the red wine and ***bouquet garni***. Allow to reduce by half.
4. Pour the veal stock into a small pan and place over medium heat. Reduce until thick, then add to the wine reduction and bring to a simmer. Add the beurre manié to the sauce in little pieces, stirring constantly to incorporate it. Simmer the sauce gently until it coats the back of a spoon (***à la nappe***). Simmer for 5 minutes, remove from the heat, and stir in the chocolate until melted. Mount the sauce with cold butter, then strain through a fine mesh sieve (***chinois***) into a clean pan. Season to taste. Add the button mushrooms and lardons. Keep warm in a bain marie.

Garniture

1. Cut the baguette into thin slices and lightly rub them with garlic. Heat some clarified butter in a large sauté pan and toast the bread slices in the hot butter until golden on both sides. Set aside and keep warm.

Poach the Eggs

1. Bring the red wine and red wine vinegar to a simmer in a shallow pan over medium heat. Season with salt and pepper. Break the eggs one at a time into a ramekin or small bowl and gently slip into the poaching liquid. Poach the eggs at a low simmer until the white solidifies but remains soft. Carefully remove the eggs from the poaching liquid with a skimmer. Reserve the eggs on a paper towel-lined plate and keep warm.

Note: *For best results, use the freshest possible eggs. Old eggs have a tendency to separate in the poaching liquid, making it difficult to obtain a nice shape.*

To Serve

1. Serve in a dish large enough for 2 eggs or in smaller dishes for individual eggs. Place 1 croûton per egg in the dish. Trim the eggs of any excess egg white (***ébarber***) and place on top of the croûtons. Coat (***napper***) each egg with sauce, decorate with fresh chervil, and serve.

Quantity		*Ingredient*
U.S.	*Metric*	*Base*
8 pcs	8 pcs	Fresh eggs
24 fl oz	700 mL	Red wine
3 ½ fl oz	100 mL	Red wine vinegar
		Salt and pepper
		Sauce
1 oz	30 g	Smoked pork belly, cut into lardons
2 tsp	10 mL	Oil
1 ¾ oz	50 g	Butter, clarified
5 oz	150 g	Button mushrooms
1 ¾ oz	50 g	Butter
4 pcs	4 pcs	Shallots, finely chopped (***hacher***)
2 pcs	2 pcs	Garlic cloves, finely chopped (***hacher***)
1 pt	500 mL	Red wine (***Bourgogne***)
1 pc	1 pc	***Bouquet garni***
1 pt	500 mL	Brown veal stock
1 oz	30 g	Beurre manié
¾ oz	20 g	Dark chocolate
4 oz	120 g	Butter
		Salt and pepper
		Garniture
8 pcs	8 pcs	French baguette slices
1 pc	1 pc	Garlic clove
		Butter, clarified
		To Serve
4 brs	4 brs	Chervil

OEUFS À LA BRETONNE

Learning Outcomes

Oeuf poché en coquille
Béchamel sauce
Émincer
Étuver
Écaler
Caneler mushrooms
Gratiner

Equipment

Knives:
Paring knife (***office***), slicing knife (***éminceur***), tourner knife

Tools:
Bowls, cutting board, wooden spatula, plastic spatula, fine chinois, salamander, araignée, skimmer (***écumoire***), ladle (***louche***), whisk

Pans:
Sauteuse, sautoir, Sauce pan (***russe***), gratin dish

Serving

4 persons

OEUFS À LA BRETONNE

Hard Cooked Eggs, Leek and Mushroom Casserole

Method

1. Preheat oven to 350°F (185°C).

Base

1. *Prepare hard cooked eggs* (***oeufs durs***): Place the eggs in a saucepan large enough for them to fit in a single layer. Cover with cold water and generously season with white vinegar. Bring the water to a boil over high heat. Reduce the temperature to a low boil and cook for 10 minutes. Remove from the heat, drain, and cover with cold water to cool. Once cooled, peel (***écaler***) and set aside.

Appareil

1. Select four of the nicest mushrooms, remove the stems, channel (***canneler***) the caps (see pages 130–133), and reserve the trimmings.
2. Bring a small saucepan of water to a simmer over low heat. Season and add the vinegar. Add the channeled mushroom caps and poach them for 5 minutes. Reserve for presentation.
3. Remove the stems from the remaining mushrooms. Peel and thinly slice the caps. Reserve the sliced caps and trimmings separately.
4. Heat the white stock (***fonds blanc***) in a medium saucepan over low heat and add the mushroom trimmings. Allow to gently simmer for 10 minutes.
5. Meanwhile, melt the butter in a small sauté pan over medium heat. Add the chopped onions and sliced leeks. Cook until they begin to soften. Add the sliced mushrooms and cook for 5 minutes.
6. Strain the mushroom-infused white stock (***fonds blanc***) into the pan and cook until completely dry, stirring from time to time. Season to taste.

Sauce Béchamel

1. Melt the butter in a medium pan over medium heat. When it begins to foam, add the flour and stir well until it begins to bubble. Cook for 1 or 2 minutes to obtain a white roux (*be careful not to let it color*). Transfer to a small dish and set aside to cool.
2. Place the milk in a medium saucepan over medium heat. Season with salt, white pepper, and nutmeg. Whisk the roux, piece by piece, into the hot milk until completely incorporated and beginning to thicken. Bring to a boil and allow to boil for 1 minute, whisking constantly. Remove from the heat and stir in the leek and mushroom mixture.

To Serve

1. Butter an ovenproof dish. Cut the hard cooked eggs in half.
2. Spread half of the leek and mushroom mixture evenly over the bottom of the dish and arrange the hard boiled eggs on top, cut side down. Cover with the remaining leek and mushroom mixture. Place in the oven to heat for 5 minutes. Arrange the channeled mushrooms on top and place under a salamander or broiler to brown (***gratiner***). Just before serving, sprinkle with chopped chervil or parsley.

Quantity		*Ingredient*
U.S.	*Metric*	*Base*
8 pcs	8 pcs	Eggs, at room temperature
		White vinegar
		Appareil
4 pcs	4 pcs	Mushrooms (***canneler***)
½ fl oz	15 mL	White vinegar
3 ½ fl oz	100 mL	White stock (***fonds blanc***)
3 ½ oz	100 g	Butter
7 oz	200 g	Onions, finely chopped (***hacher***)
7 oz	200 g	Leek whites, sliced (***émincer***)
7 oz	200 g	Mushrooms, sliced (***émincer***)
		Salt and white pepper
		Sauce Béchamel
¾ oz	20 g	Butter
¾ oz	20 g	Flour
8 fl oz	250 mL	Milk
Pinch	Pinch	Nutmeg
		Salt and white pepper
		To Serve
1 oz	30 g	Butter
1 oz	30 g	Chervil or parsley leaves, finely chopped (***hacher***)

OEUFS CHASSEUR

Learning Outcomes

Cuisson des oeufs sur plat (miroir) hors coquille
Reduction
Demi-glace

Equipment

Knives:
Vegetable peeler (*économe*), paring knife (*office*), slicing knife (*éminceur*)

Tools:
Bowls, wooden spatula, plastic spatula, fine chinois, ladle (*louche*)

Pans:
Sautoir, Sauce pan (*russe*), bain marie, egg dishes

Serving

4 persons

FYI...

Eggs cooked in this manner are also known as oeufs miroir (mirror eggs) because when the yolk is touched, it should be just warm and liquid inside, and shiny as a mirror.

OEUFS CHASSEUR

Baked Eggs with Madeira Sauce and Chicken Livers

Method

1. Preheat the oven to 350°F (185°C).

Sauce Madère

1. Place the Madeira and ***bouquet garni*** in a small pan over medium heat and reduce by one-third. Add the demi-glace, bring to a simmer, and season to taste. Off the heat, remove the ***bouquet garni*** and mount the sauce with cold butter. Season to taste and keep warm in a bain marie.

Garniture

1. Clean the livers and slice thickly on the bias (***éscaloper***). Pat dry and set aside. Finely chop (***hacher***) the shallots. Heat the clarified butter in a medium pan over medium heat and add the liver and shallots. Sauté the livers with the shallots until the livers are still pink in the middle (***rosé***). Deglaze with 1 ½ oz (50 mL) of Madeira, allow to reduce until almost dry, then add half the sauce. Season to taste with salt and pepper and set aside.

Croûtons

1. Using a serrated knife, cut the bread slices in half on the diagonal, then remove the crusts and trim one end to form a teardrop shape. Heat some clarified butter in a large sauté pan over medium-high heat and toast the bread slices in the hot butter until golden on both sides. Set aside and keep warm.

Cuisson

1. Heat the egg dishes on the stove top with some clarified butter until hot. Break each egg into a small bowl or ramekin, being careful not to break the yolk. Place 2 eggs into each heated dish and transfer to the oven until the egg white is set (about *5 minutes*).

To Serve

1. Arrange the livers on the eggs and pour some of the sauce around. Dip the tip of the croûtons in the sauce and then coat with chopped parsley. Place on the eggs and serve.

Quantity		Ingredient
U.S.	*Metric*	*Base*
8 pcs	8 pcs	Fresh eggs
1 oz	30 g	Butter, clarified
		Sauce Madère
6 fl oz	175 mL	Madeira
1 pc	1 pc	***Bouquet garni***
7 fl oz	200 mL	Demi-glace
3 ½ oz	100 g	Butter, clarified
		Salt and pepper
		Garniture
7 oz	200 g	Chicken livers
2 pcs	2 pcs	Shallots, finely chopped (hacher)
1 oz	30 g	Butter, clarified
2 fl oz	60 mL	Madeira
		Salt and pepper
		Croûtons
4 pcs	4 pcs	White bread slices
		Butter, clarified
		To Serve
¼ oz	10 g	Parsley, finely chopped (***haché***)

OMELETTE FERMIÈRE FARCIE AUX CHAMPIGNONS

Learning Outcomes

Omelette farcie
Sauce duxelle

Equipment

Knives:
Paring knife (*office*), slicing knife (*éminceur*)

Tools:
Bowls, cutting board, fork, wooden spatula, whisk

Pans:
Sauteuse, sautoir, poêle noir or nonstick omelette pan, bain marie

Serving

3 to 4 persons

FYI...

There are four classic French methods for making an omelette: plate (flat), roulée (rolled), farcie (filled), and fourrée (stuffed).

May the best pan win: traditionally a poêle noir, or black iron pan, is used for making omelettes. Like cast iron, the pan requires seasoning and, over time with proper care, it takes on the properties of a nonstick pan. Nonstick pans have certainly made life easier, but some complain that they limit the amount of coloring you would obtain with an iron pan. Commonplace in some countries, nonstick cookware can be quite expensive than in others.

OMELETTE FERMIÈRE FARCIE AUX CHAMPIGNONS

Mushroom Stuffed Omelet

Method

Sauce Duxelle

1. Melt the butter in a small pan over low heat and sweat (***suer***) the shallot until soft. Add the finely chopped mushrooms and cook until dry. Deglaze with the Madeira and reduce until dry (stirring well). Add the demi-glace, season to taste, and keep warm in a bain marie.

Garniture

1. Clean and slice (***émincer***) the mushrooms.
2. Finely dice (***ciseler***) the shallots and chop (***hacher***) the garlic. Melt the butter in a medium pan over low heat and sweat (***suer***) the shallots until soft. Add the garlic and sliced mushrooms. Season with salt and pepper. Cook until all the water evaporates from the mushrooms, then remove from the heat. Select 12 of the nicest slices of mushroom for decoration and set them aside. Stir the chopped parsley into the remaining mushrooms and season to taste with salt and pepper. Set aside and keep warm.

Cuisson

1. Warm a plate and brush some butter down the middle. Season with salt and pepper and set aside in a warm place. Heat the clarified butter in a non-stick omelette pan over medium heat. Beat the eggs in a large bowl, season with salt and pepper, then pour the beaten eggs into the pan. Stir gently with a fork, lifting the bottom to allow the uncooked eggs to flow underneath. The eggs should not set too quickly or take on too much color.

To Serve

1. Once the eggs are almost completely set—that is, they can no longer be stirred—spoon the ***garniture*** into the middle of the omelette. Give the pan a good shake or tap and lift the pan almost vertically. With the aid of the fork, fold the omelette in half, then slip the omelette onto the plate, letting it fold over again onto itself. Cover with a clean kitchen towel and press well, forming points at both ends. Remove the towel. Brush with clarified butter (***lustrer***) (see page 249).
2. Decorate with mushroom slices, pour the sauce around, and serve immediately.

Quantity		*Ingredient*
U.S.	*Metric*	*Base*
12 pcs	12 pcs	Eggs
4 oz	120 g	Butter, clarified
		Salt and pepper
		Sauce Duxelle
1 ¾ oz	50 g	Butter
1 pc	1 pc	Shallot, finely chopped (***hacher***)
3 ½ oz	100 g	Mushrooms, finely chopped (***hacher***)
1 fl oz	30 mL	Madeira
3 ½ fl oz	100 mL	Demi-glace
		Salt and pepper
		Garniture (farcis)
1 lb	500 g	Mushrooms, finely sliced (***émincer***)
2 pcs	2 pcs	Shallots, finely chopped (***hacher***)
1 pc	1 pc	Garlic clove, finely chopped (***hacher***)
1 ¾ oz	50 g	Butter
¼ bq	¼ bq	Parsley, finely chopped (***hacher***)
		Salt and pepper
		Finishing
1 oz	30 g	Butter, clarified (*for shine*, ***lustrer***)

OMELETTE AUX FINES HERBES

Learning Outcomes

Omelette roulée à la Française
Blanchir

Equipment

Knives:
Chef knife (*couteau chef*), paring knife (*office*)

Tools:
Ramekin, bowl, whisk, fork

Pans:
Poêle noir or nonstick omelette pan, bain marie

Serving

4 to 6 persons

OMELETTE AUX FINES HERBES

Herb Omelet

Method

Garniture

1. Pick the leaves off the chervil, parsley, and tarragon, and trim the bottoms off the chives. Blanch the herbs separately in boiling salted water, then refresh immediately in ice water. Squeeze out the excess water and finely chop. Warm a large plate, and brush the center with melted butter. Lightly season with salt and pepper. Set aside in a warm place.

Cuisson

1. Break the eggs into a large bowl, season, and whisk well. Mix in the chopped herbs.
2. Heat the clarified butter in the omelette pan over medium heat. When the butter is hot, pour the egg mixture into the pan. Stir gently with a fork, lifting the bottom to allow the uncooked eggs to flow underneath. The eggs should not set too quickly or take on too much color.

To Serve

1. Once the eggs are almost completely set—that is, they can no longer be stirred—give the pan a good shake or tap. Lift the pan almost vertically; with the aid of the fork, fold the omelette in half and slip it onto the prepared plate, folding it again onto itself. Cover with a clean kitchen towel, and press along the sides, forming points at each end.
2. Brush the top of the omelette with clarified butter before serving (see page 249).

Note: *Traditionally, omelettes are rolled using a cloth. Not only does it assist in giving the omelette its final torpedo shape, but it also absorbs any excess butter from the cooking.*

Quantity		*Ingredient*
U.S.	*Metric*	*Base*
12 pcs	12 pcs	Eggs
4 oz	120 g	Butter, clarified
		Salt and pepper
		Garniture
5 brs	5 brs	Chervil, finely chopped (***hacher***)
5 brs	5 brs	Parsley, finely chopped (***hacher***)
5 brs	5 brs	Tarragon, finely chopped (***hacher***)
5 brs	5 brs	Chives, finely minced (***ciseler***)
		Salt and pepper
		To Serve
1 oz	30 g	Butter, clarified
		Salt and pepper

FYI...

There are different degrees of cooking for omelettes: ***baveuse*** (slightly runny) or ***bien cuite*** (well done). When ordering an omelette, one can ask for it to be cooked either way.

OMELETTE PAYSANNE

Learning Outcomes

Omelette plate (flat)

Equipment

Knives:
Paring knife (*office*), slicing knife (*éminceur*)

Tools:
Bowls, cutting board, fork, wooden spatula, whisk

Pans:
Poêle noir or nonstick omelette pan, bain marie

Serving

5 to 6 persons

OMELETTE PAYSANNE

Rustic Potato and Bacon Omelet

Quantity		Ingredient
U.S.	*Metric*	*Base*
12 pcs	12 pcs	Eggs
3 oz	90 g	Butter
		Salt and pepper
		Garniture Paysanne
1 lb	500 g	Potatoes
3 oz	90 g	Butter
7 oz	200 g	Slab bacon, cut into lardons
		To Serve
1 br	1 br	Parsley

Method

Garniture Paysanne

1. Peel the potatoes and slice them thinly (***émincer***). Rinse in cold water to prevent them from sticking to the pan.
2. Melt the butter in a nonstick pan over medium-high heat and sauté the sliced potatoes until cooked and light brown in color. Drain and set aside.
3. Cut the smoked slab bacon into lardons, place them in a small pan of cold water, and bring to a boil. Blanch for 1 minute, then drain and refresh under cold running water. Add the lardons to the cooked potatoes. Set aside some lardons and slices of potato for decoration.

Cuisson

1. Heat the butter in a nonstick omelette pan over medium heat. Beat the eggs in a large bowl and season with salt and pepper. Add the potato slices and lardons to the hot butter and stir until heated through. Spread evenly. Pour in the beaten eggs and stir gently with a fork, lifting the bottom to allow uncooked egg to flow underneath. The eggs should not take on too much color. When the eggs are no longer runny, give the pan a good shake to loosen the omelette. Place a plate over the pan and flip the omelette onto the plate. Carefully slide the omelette back into the hot pan to finish cooking on the other side, adding more butter if needed. Shake the pan occasionally to ensure it doesn't stick. When cooked, slide onto a warm plate. Brush with some butter and decorate with the reserved lardons, potato slices, and sprigs (***pluches***) of parsley. Serve immediately.

FYI...

In this recipe, *paysanne* means "of the country" or peasant. Therefore it usually features simple fare such as bacon and potatoes. The omelette plate is similiar to the Spanish tortilla and the Italian frittata.

Learning Outcomes

Omelette fourrée
Concassée
Beurre clarifié
Pocher
Cuisson des oeufs durs

Equipment

Knives:
Paring knife (*office*), slicing knife (*éminceur*)

Tools:
Bowls, cutting board, fork, wooden spatula, whisk

Pans:
Sauteuse, sautoir, poêle noir or nonstick omelette pan (see note under Omelette Fermière Farcie)

Serving

4 persons

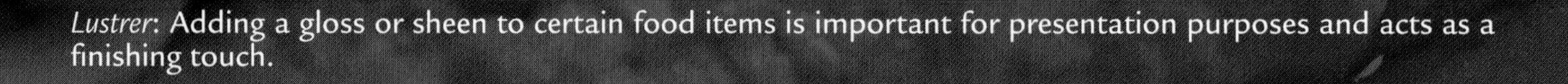

FYI...

Lustrer: Adding a gloss or sheen to certain food items is important for presentation purposes and acts as a finishing touch.

OMELETTE GRANDVAL

Omelet and Tomato Sauce

Method

Sauce Tomate

1. *Skin the tomatoes* (***émonder***): First bring a large pan of water to a boil over a high heat. Remove the stem ends and place 2 to 3 tomatoes at a time in the boiling water. Leave in the boiling water about 30 seconds to 1 minute then immediately transfer to an ice bath. The skin will loosen and be easily peeled.
2. Quarter the tomatoes and cut away the seeds and pulp, reserving for the sauce tomate. Separate the tomatoes for the garnish (***fondue de tomates***) and set aside. Remove and reserve the seeds and the pulp (***épépiner***), then dice the tomatoes (***concasser***). Heat the olive oil in a shallow pan over low heat and sweat (***suer***) the onions or shallots, then add the finely chopped (***hacher***) garlic. Add tomato paste and cook for a few minutes (***pincer la tomate***). Add the diced tomatoes and pulp, the parsley, the ***bouquet garni***, and the salt and pepper. Cook over low heat until all the water has evaporated (*25 to 30 minutes*).
3. Purée the mixture through a food mill and strain it through a fine mesh sieve (***chinois***).

Garniture

1. Heat the olive oil in a medium pan over medium heat and sweat the shallots until soft. Add the tomatoes and the ***bouquet garni*** and cook until thick. Season to taste, then stir in the chopped parsley.
2. Prepare the hard cooked eggs: Place the 2 eggs in a small saucepan and cover with cold water. Generously season the water with white vinegar and salt. Bring the water to a boil, reduce the heat to a low boil, and poach the eggs for 10 minutes. Pour out the hot water and replace with cold water. Peel the eggs (***écaler***) while still hot and reserve.
3. Heat a plate and brush butter down the middle. Sprinkle with salt and pepper and keep warm.

Cuisson

1. Heat the clarified butter in a nonstick omelette pan over medium heat. Beat the eggs in a large bowl and season with salt and pepper. Pour the beaten eggs into the pan and stir gently with a fork, lifting the bottom to allow the uncooked eggs to flow underneath. The eggs should not set too quickly or take on too much color.

To Serve

1. Once the eggs are almost completely set—that is, they can no longer be stirred—give the pan a good shake or tap. Lift the pan almost vertically and, with the aid of the fork, fold the omelette in half, then slip it onto a plate, letting it fold over again onto itself. Cover with a clean kitchen towel and press well, forming points at both ends. Remove the towel and brush the omelette with clarified butter (***lustrer***).
2. Make an incision down the length of the omelette (the incision should be deep enough to reach the center of the omelette), and fill the cavity with the ***concassée*** of tomato. Slice the hard cooked eggs, and lay them on top of the filling. Pour some tomato sauce around the omelette and serve immediately.

Tip: *It is important to use enough clarified butter to ensure that the omelette does not stick. The kitchen towel not only helps give the omelette its final shape, but also blots up any excess butter.*

Quantity		*Ingredient*
U.S.	*Metric*	*Base*
12 pcs	12 pcs	Eggs
4 oz	120 g	Butter, clarified
		Salt and pepper
		Sauce Tomate
4 pcs	4 pcs	Tomatoes, peeled (***émonder***), seeded (***épépiner***), and chopped (***concasser***)
1 fl oz	30 mL	Olive oil
3 oz	100 g	Onions or shallots, finely chopped (***hacher***)
2 pcs	2 pcs	Garlic cloves, finely chopped (***hacher***)
1 oz	30 g	Tomato paste
10 brs	10 brs	Fresh parsley, roughly chopped (***concasser***)
1 pc	1 pc	***Bouquet garni***
		Salt and white pepper
		Garniture
1 ¾ fl oz	50 mL	Olive oil
1 ¾ oz	50 g	Shallots, finely diced (***ciseler***)
6 pcs	6 pcs	Tomatoes, peeled (***émonder***), seeded (***épépiner***), and chopped (***concasser***)
1 pc	1 pc	***Bouquet garni***
5 brs	5 brs	Parsley, finely chopped (***hacher***)
		To Serve
2 pcs	2 pcs	Eggs, hard cooked
8 fl oz	250 mL	Tomato sauce (see above)

OMELETTE SOUFFLÉE

Learning Outcomes

Omelette Soufflée
Separating eggs (clarifier)
Simple meringue
Flambéing
Caramelizing

Equipment

Knives:
Paring knife (*office*)

Tools:
Pastry brush, mixing bowls, whisks, rubber spatula, plastic scraper, metal skewers, metal spatula, kitchen towels

Pans:
Nonstick omelette pan, small pan

Serving

4 to 5 persons

FYI...

In Ancient Rome, eggs and dairy were combined with other ingredients to make dishes that were both sweet and savory. An egg recipe written by *Apicius*, called *"ova sfongia ex lactem,"* included milk, oil, honey, and pepper, and would have been cooked much like a pancake. Omelette souflée, in the long history of cooking with eggs, distinguishes itself as a sweet omelette made light and airy with meringue.

OMELETTE SOUFFLÉE

Sweet Omelet with Strawberries

Quantity		Ingredient
U.S.	*Metric*	*Base*
6 pcs	6 pcs	Egg yolks
4 oz	120 g	Sugar
6 pcs	6 pcs	Egg whites
4 oz	120 g	Sugar
½ fl oz	15 mL	Kirsch (optional)
		Butter
		Sugar
		Filling
5 oz	150 g	Strawberries, quartered
8 oz	250 g	Strawberry jam
		Finishing
1 oz	30 g	Powdered sugar
1 oz	30 g	Kirsch (*optional*)
		To Serve
1 br	1 br	Fresh mint

Method

1. Preheat the oven to 355°F (180°C).
2. Prepare all the elements ahead of time.
3. Separate the eggs.
4. Brush a serving platter and a large nonstick omelette pan with softened butter. Pour granulated sugar onto the platter and into the pan, then shake them to form a thin coating of sugar. Tap off the excess sugar.
5. Place the metal skewers on an open gas flame (set to low) and heat until red hot.
6. Lay a clean kitchen towel out flat on the work surface and dust it generously and evenly with powdered sugar.

Filling

1. Stem the strawberries and cut into quarters.
2. Heat the jam in a small pan over low heat until just liquid. Allow to cool slightly, then add to the strawberries. Stir until they are completely coated. Set aside.

Base

1. *Blanch the sugar and egg yolks:* Beat the sugar 4 oz (120 g) into the egg yolks using a whisk. Whisk vigorously until the yolks have thickened and turned a pale yellow and the wires of the whisk leave trails when lifted. Set aside.
2. Whisk the egg whites in a large mixing bowl until soft peaks form. Gradually incorporate the sugar into the egg whites while beating. Continue beating until they form a stiff and glossy meringue.
3. If using, whisk the kirsch into the blanched egg yolks.
4. Add one-third of the meringue to the egg yolks and stir with a whisk until just incorporated. Gently fold in the remaining meringue with a rubber spatula until the mixture is homogenous. Pour the mixture into the sugar-coated pan and spread it out using the spatula to form an even layer. Work quickly, without overworking the mixture.
5. Immediately place the pan in the oven and bake until the omelette is puffed up and nicely colored *(check for doneness after 15 minutes of cooking)*.
6. To check for doneness, open the oven door and insert a trussing knife into the omelette. If it comes out clean, the omelette is cooked. If not, continue baking for 5 more minutes before checking for doneness again.

Montage

1. Remove the omelette from the oven and immediately turn over onto the prepared kitchen towel. Working quickly, make a small indentation across the middle of the omelette with the back of a spoon, and fill with the strawberries. Lift one side of the towel and fold one side of the omelette onto the strawberries. Gently press to secure. Then lift the other side of the towel to bring the other side of the omelette over the first. Using the towel, roll the omelette onto its seam.
2. Using two long palette knives or narrow metal spatulas, slip one underneath the omelette while securing the side with the other palette knife. Transfer the omelette to the prepared platter. Lay a clean towel over the omelette and press lightly along the sides to give it an even form and to secure the filling.
3. Sprinkle with powdered sugar, then take the red-hot skewers and mark the top in a criss-cross pattern. Heat the remaining kirsch in the saucepan until hot *(optional)*. Pour over the omelette and light to flambé.

To Serve

4. When the flames die down, cut the omelette into slices and place them on hot plates. Decorate with fresh mint.

Les Volailles
(Poultry)

Recipes

Les Volailles (Poultry)

Ballotine de volaille
—Roasted boneless stuffed chicken

Canard à l'orange, pommes gaufrette
—Duck in orange sauce, waffle potatoes

Canard poêlé aux navets
—Pan roasted duck with turnips

Oie confite, haricots blancs
—Confit of goose with white beans

Coq ou chapon au vin
—Rooster or capon stewed in red wine

Coquelet en crapaudine
—Spatchcock game hen

Filet de dinde sauté chasseur
—Sautéed turkey breast in mushroom and tomato sauce

Jambonnette de volaille farcies au bulgar, sauce Riesling
—Deboned chicken legs stuffed with bulgur in a Riesling sauce

Lapin à la graine de moutarde, pommes rissolées
—Rabbit in whole-grain mustard sauce with sautéed potatoes.

Pintade cocotte grand-mère
—Guinea fowl with potatoes, mushrooms, and bacon

Poularde pochée sauce suprême, riz pilaf
—Poached chicken with a white cream sauce and boiled rice

Poulet rôti, pommes mignonette
—Roast chicken with mignonette potatoes

Poulet sauté Boivin
—Sautéed chicken and artichokes

BALLOTINE DE VOLAILLE

Learning Outcomes

Rôtir
Making a ballotine
Boning a large bird
Farce simple
Poêler
Pommes Anna

Equipment

Knives:
Boning knife, paring knife (*office*), chef knife (*couteau chef*)

Tools:
Tray, food processor, plastic scraper, bowls, ice bath, rubber spatula, kitchen twine, wooden spatula, chinois, meat grinder, trussing needle

Pans:
Small saucepan, 8-inch cake pan, medium sauté pan, small sauté pan, roasting pan

Serving

4 to 6 persons

FYI...

Ballotine stems from the Old French *bale,* meaning "a rolled-up bundle."

BALLOTINE DE VOLAILLE

Roasted Boneless Stuffed Chicken

Method

1. Preheat the oven to 350°F (175°C).
2. *Clean the chicken* (***habiller***): See pages 333–336 in *Cuisine Foundations.*
3. *Debone the chicken:* See pages 339–343 in *Cuisine Foundations.*

Tip: *Any discarded chicken bones may be used to add flavor to the veal stock that will be added to the preparation during the "finishing" step.*

Farce Simple

1. Cut the chicken, pork, and veal into a large dice and grind them together in a meat grinder fitted with a fine grinding disk. Mix the eggs into the ground meat one at a time with a rubber spatula. Keeping the bowl over the ice bath, beat in the cream, little by little, then mix in the pistachio, ham and sage. To taste and adjust the seasoning, first cook a spoonful of farce in a small sauté pan over medium heat. It will now be safe to taste. Reserve the seasoned farce in a covered bowl in the refrigerator until ready to use.

Montage

1. Lay the chicken out flat horizontally to form a neat rectangle. Season with salt and pepper. Mound the farce down the middle of the flattened chicken. Fold the bottom flap up, then fold the top over being sure it overlaps. Tuck any loose skin underneath. Thread a trussing needle with kitchen twine. Starting at one end, pierce through the seam and tie a knot. Sew along the long seam to the other end and cut the string and knot. Spread out the caul fat and wrap the ballotine. Trim the ends and tuck underneath. Slip a length of string under the ballotine from end to end. Pull the string up and tie snugly but not too tightly. Trim the string. Tie a string around the middle, snugly but not too tight. In order to maintain an even shape, continue tying at regular intervals, but alternating the left side and the right side. This ensures that the ties are evenly placed and tied.

Pommes Anna

1. Wash and peel the potatoes. Cut them into ***bouchons*** and slice them thinly. Coat the bottom of an 8-inch cake pan with clarified butter over medium-high heat. To gauge the temperature of the butter, place a potato slice in the pan. When it begins to bubble, the pan is ready. Reduce the heat to medium and line the bottom of the pan with a layer of potato slices arranged so that they overlap in concentric circles. Cook the potatoes until they begin to brown. Season them with salt and white pepper, then arrange a second layer of potatoes on top, in the same pattern, overlapping each other in the opposite direction. Pour in enough clarified butter to just cover the potatoes. Continue layering the potatoes in opposing directions, seasoning and adding butter until the cake pan is about ½ in. (1 cm) from being

Quantity		*Ingredient*
U.S.	*Metric*	
1 pc	1 pc	Chicken (*1.6 kg / 2 1.4 lb*)
4 tsp	20 mL	Cognac
1 pc	1 pc	Caul fat (***crépine***), degorged in water and vinegar
5 oz	150 g	Butter, clarified
		Salt and white pepper
		Farce Simple
3 ½ oz	100 g	Chicken
3 ½ oz	100 g	Pork
3 ½ oz	100 g	Veal
2 pcs	2 pcs	Eggs
½ pt	250 mL	Cream
3 ½ oz	100 g	Ham, cooked, cut into small dice
1 ¾ oz	50 g	Pistachio, blanched
10 pcs	10 pcs	Sage leaves, blanched, coarsely chopped (***ciseler***)
		Salt and white pepper
		Pommes Anna
2 lb	1 kg	Potatoes
10 oz	300 g	Butter, clarified
		Salt and white pepper

Quantity		Ingredient
U.S.	*Metric*	*Garniture de Poelage*
0.5 pc	0.5 pc	Onion, medium, cut into ***paysanne***
0.5 pc	0.5 pc	Carrot, medium, cut into ***paysanne***
0.5 pc	0.5 pc	Shallot, cut into ***paysanne***
0.5 pc	0.5 pc	Leek, white, cut into ***paysanne***
0.5 pc	0.5 pc	Celery stalk, cut into ***paysanne***
5 oz	150 g	Lardons
1 pc	1 pc	***Bouquet garni***
		Salt and white pepper
		Finishing
1 pint	500 mL	Veal stock
1 ¾ fl oz	50 mL	Meat glaze (***glace de viande***)

Method

full (*5 to 6 layers*). Place a circle of parchment paper over the potatoes. Transfer the cake pan to the oven to cook until the potatoes are easily pierced with the tip of a knife (*45 minutes*) and the surface has reached a golden color. Remove the pommes Anna from the oven and allow to rest, uncovered, for 30 minutes, then pour out the excess butter and press the potatoes down to compress them.

Tip: *To avoid the surface browning before the center is cooked, cover the cake pan with aluminum foil for the first 20 minutes of cooking.*

Cuisson

1. Pat the ***ballotine*** dry and season it with butter, salt and pepper. Place the ***ballotine*** on a wire rack over a roasting pan, top with some pieces of cold butter and place in the oven to roast for approximately 35 minutes. Once the butter has melted, baste the ***ballotine*** frequently throughout the roasting period. Once roasted, set the ***ballotine*** aside on its cooking rack. Cover with foil and keep warm.
2. Place the roasting pan over medium heat and sauté the lardons until they take on a blond color. Add the ***paysanne*** of vegetables, the bouquet garni and stir to coat. Season. Place the ***ballotine*** and cooking rack on top of the garniture and finish cooking in the oven for 15 minutes. Remove from the oven and set the ***ballotine*** aside to rest for approximately 15 minutes. Cover with foil to keep warm.

Finishing

1. Strain the garniture from the roasting pan into a fine mesh sieve (***chinois***) in order to remove the excess fat (***dégraisser***). Transfer the drained garniture to a clean saucepan, wet with the brown veal stock and allow it to reduce over medium heat for 5 minutes. Check for consistency, adjust the seasoning, and reserve the sauce. Brush the surface of the ***ballotine*** with meat glaze (***glace de viande***) and place in the oven just long enough to create a shiny coating on the surface of the meat (***glacer***). Repeat if needed.
2. *Finishing the pommes Anna:* Check that the potato slices are not stuck to the bottom of the pan by placing it over medium heat and gently shaking it to see if the contents slide around. Drain again to remove any excess fat, then flip the pommes Anna onto a serving plate.

To Serve

1. Pour some sauce and garniture on the bottom of the serving platter.
2. Place the ***ballotine*** on the serving platter with two or three slices cut for presentation. Serve the pommes Anna on the side.

CANARD À L'ORANGE, POMMES GAUFRETTE

Learning Outcomes

Rôtir
Habiller
Brider
Détailler à cuit
Gastrique
Sauce bigarade
Peler à vif and suprêmes of citrus
Glazing a roast bird
Pommes gaufrettes

Equipment

Knives:
Boning knife, paring knife (*office*), chef knife (*couteau chef*), cleaver

Tools:
Trussing needle, pastry brush, soup spoon, wire rack, sieve, skimmer (*écumoire*), chinois, mixing bowls, wooden spatula, citrus juicer, mandolin

Pans:
Small saucepan, roasting pan, medium saucepan, deep fryer, baking sheet

Serving

4 persons

FYI...

This recipe was first published in the *Libro di Cucina* in 1835 by Francesco Zambrini. The recipe was adapted from a 14th-century manuscript that was written by an unknown Tuscan author.

Quantity		Ingredient
U.S.	*Metric*	
1 pc	1 pc	Duck [*5 lb (2.5 kg)*]
2 pcs	2 pcs	Oranges
4 brs	4 brs	Thyme
3 pcs	3 pcs	Bay leaves
5 oz	150 g	Butter
		Salt and pepper
		Gastrique
3 ½ oz	100 g	Sugar, granulated
3 ½ oz	100 g	Red wine vinegar
2 pcs	2 pcs	Orange, juice of
		Sauce Bigarade
0.5 pc	0.5 pc	Carrot, cut into ***mirepoix***
0.5 pc	0.5 pc	Onion, cut into ***mirepoix***
1 pc	1 pc	Celery stalk, cut into ***mirepoix***
1 pc	1 pc	Garlic head, crushed
2 tsp	10 g	Tomato paste
2 pc	2 pc	Tomatoes, quartered
1 pc	1 pc	***Bouquet garni***
1 pt	500 mL	Veal stock or brown duck stock
1 ¾ oz	50 g	Butter
		Salt and pepper

CANARD À L'ORANGE, POMMES GAUFRETTE

Duck in Orange Sauce, Waffle Potatoes

Method

1. Preheat the oven to 400°F (200°C).
2. Preheat the deep fryer to 285°F (140°C).
3. *Clean the duck* (***habiller***): See pages 333–336 in *Cuisine Foundations.* Chop off the tip of the drumsticks using a cleaver. Reserve the trimmings. Pat the duck dry and season the inside and outside with salt and pepper. Cut the oranges in half or quarters and place in the cavity of the duck with the thyme and bay leaves. With a small knife, remove the glands on the top of the tail, then bend it back into the carcass and truss (***brider***) the duck. (See pages 337–339 in *Cuisine Foundations.*)

Cuisson

1. Place the duck, breast side up, on a wire rack over a roasting pan. Cut the butter into small pieces and spread over the top. Place the duck in the oven to roast for 15 minutes; start basting once the fat is melted. Reduce the oven temperatures to (350°F). Roast for 1 hour, basting frequently. When the duck is cooked through, the juices will run clear when the thickest part of the thigh is pierced with a trussing needle. Once cooked, transfer the duck on its wire rack onto a plate. Cover with foil and keep warm. Turn off the oven.
2. Pour off the excess fat and place the pan on the stove top over medium-high heat to brown the cooking juices (***pincer les sucs***). Add the duck trimmings, mirepoix, bouquet garni, and crushed garlic to the hot pan. Stir well until it begins to color. Add the tomato paste and cook 1 to 2 minutes (***pincer la tomate***), then the tomatoes (this is to reinforce the color of the jus). Deglaze with the veal stock, stirring well to dissolve the ***sucs***, and add the bouquet garni. Add the gastrique, reduce the heat and reduce until it nicely coats the back of a spoon. Strain through a fine mesh sieve (***chinois***) into a clean pan. Skim off any fat from the surface (***dégraisser***). Mount with butter and keep warm in a bain-marie.

Sauce Bigarade

1. *Prepare a gastrique:* Melt the sugar with a little water in a medium sauce pan over medium heat. Cook the sugar until it begins to color. Continue cooking to a medium brown, then deglaze with the vinegar and reduce until almost dry. Wet (***mouiller***) with the orange juice. Allow to reduce over medium heat until syrupy in consistence.

Note: *When deglazing with vinegar, be careful not to inhale the resulting steam as it will be quite strong.*

Quantity		*Ingredient*
U.S.	*Metric*	*Garniture*
2 pc	2 pc	Oranges, zest, suprêmes
1 fl oz	30 mL	Cointreau liqueur
1 bq	1 bq	Parsley (***hacher***)
		Pommes Gaufrette
		Potatoes
		Salt

Method

Garniture

1. Using a knife, cut away the peel following the natural curve of the oranges, exposing the flesh of the fruit (***peler à vif***). Holding the peeled orange in one hand over a small bowl, cut between the membranes to free the segments (***suprêmes***). Set aside.
2. Cut the pith off the zest (*discard the pith*) and cut the zest into ***julienne***.
3. *Blanch the **julienne** of zest:* Place the zest in a small saucepan of cold water and bring it to a boil over high heat. Blanch for 2 or 3 minutes, then drain the zest and refresh it under cold running water. Drain and place the zest in fresh water with some Cointreau and a pinch of salt. Cook gently for 10 to 15 minutes. Set aside.

Glaçage

1. Remove the string from the duck. Place the duck on a rack over a baking sheet and evenly coat it with the sauce using a pastry brush. Place the duck in the warm oven for 20 to 30 seconds to dry out the glaze. Repeat 2 to 3 times, or until the duck is coated in an even, shiny glaze. Keep warm.

Pommes Gaufrette

1. Wash and peel the potatoes. Using the fluted blade of a mandolin, slice the potatoes, alternating hand position 90° with every swipe over the blade. **Note**: *The first slice will have one flat side, but succeeding slices should have a waffle pattern on both sides.*
2. Rinse the ***pommes gaufrette*** in a large bowl of cold water and pat them dry. Immerse the chips in the deep fryer and stir gently to ensure even coloring and to prevent sticking. Cook the ***pommes gaufrette*** to an even golden color (***blond***) and drain them on a paper towel. Season and serve the ***pommes gaufrette*** immediately.

To Serve

1. Carve the duck into 8 pieces. Gently reheat the sauce and mount with cold butter. Drain the julienned orange zest and stir into the sauce. Reserve some for the decoration. Pour the sauce into a hot serving platter. Arrange the duck pieces on top and decorate with the ***julienne*** of orange zest, the orange segments (***suprêmes***), and chopped parsley. Serve the ***pommes gaufrette*** on the side.

Note: *Because the duck is glazed, it can be presented whole and carved at the table.*

CANARD POÊLÉ AUX NAVETS

Learning Outcomes

Poêler
Habiller
Brider
Saisir
Glacer des légumes
Tourner
Glacer

Equipment

Knives:
Boning knife, paring knife (*office*), vegetable peeler (*économe*), chef knife (*couteau chef*), turning knife

Tools:
Colander, wooden spatula, kitchen twine, wire rack, chinois, skimmer (*écumoire*), pastry brush, trussing needle

Pans:
Cocotte, medium saucepan, large sauté pan, baking sheet

Serving

4 persons

FYI...

In 1873, nine Peking ducks were exported from China to Long Island, New York. It is believed that all domesticated ducks in the United States are offspring of these first few fowl.

A preparation that is ***poêlé*** necessarily involves the use of the cocotte, a versatile oval cast-iron pot with a lid. Prior to the 19th century, the pot was called a *coquasse* or *cocasse*—which is where we get the term *casserole*. In the following recipe, the cocotte is used on top of the stove for searing the duck as well as in the oven to complete the cooking process. Also used in braising, this pot may even be used to serve a preparation such as a stew.

CANARD POÊLÉ AUX NAVETS

Pan Roasted Duck with Turnips

Quantity		Ingredient
U.S.	Metric	Base
1 pc	1 pc	Duck [*5 lb (2.5 kg)*]
4 oz	120 g	Duck fat
		Salt and pepper
		Garniture de Cuisson
4 oz	120 g	Smoked pork belly, cut into ***lardons,*** blanched
2 oz	60 g	Butter
1 pc	1 pc	Carrot, cut into large ***paysanne***
1 pc	1 pc	Onion, cut into large ***paysanne***
1 pc	1 pc	***Bouquet garni***
		Sauce et Glaçage
8 fl oz	250 mL	White wine
1 pt	500 mL	Veal stock
1 ¾ oz	50 g	Butter
		Salt and pepper
		Navets Glacés à Brun
28 pcs	28 pcs	Turnips, turned, ***cocotte*** size
1 oz	30 g	Sugar, granulated
2 oz	60 g	Butter
		Salt
		To Serve
1 bq	1 bq	Fresh parsley, chopped

Method

1. Preheat the oven to 400°F (200°C).
2. *Clean the duck* (***habiller***): See pages 333–336 in *Cuisine Foundations.* Bend the tail back into the carcass and truss (***brider***) the duck. (See pages 337–339 in *Cuisine Foundations.*)

Cuisson

1. Place the duck in the dutch oven (***cocotte***). Cover with the duck fat and season well with salt and pepper, then transfer to the hot oven. When it begins to color (***saisir***), baste frequently. Once colored, turn the duck onto one side. Baste until colored, then turn the duck onto its other side. Continue turning and basting until the duck is evenly colored on all sides. Reduce the heat to 325°F (160°C) and leave to cook about 25 minutes, basting frequently.

Garniture de Cuisson

1. *Blanch the lardons:* Place the lardons in a medium saucepan of cold water and bring it to a boil over high heat. Refresh the lardons under cold running water, strain them, and pat them dry.
2. Melt some butter in a shallow pan over medium heat and color the blanched lardons. Add the carrot and sweat (***suer***) until it begins to soften, then add the onion and cook until soft. Remove the duck from the cocotte and place on a wire rack. Pour off the excess fat, and add the garniture and the bouquet garni. Place the wire rack with the duck on top, cover the cocotte, and finish cooking for 15 to 20 minutes.

Sauce

1. Remove the cocotte from the oven and check for doneness of the duck (*when done, the juices will run clear from the thickest part of the thigh when pricked with the tip of a knife*). Remove the wire rack with the duck, cover with foil, and keep warm. Place the cocotte over medium high in order to brown any cooking juices (***pincer les sucs***). Strain the vegetables and set aside. Place the cocotte back on the heat and deglaze with white wine. Scrape the bottom to dissolve any of the ***sucs***. Reduce until syrupy, then add the veal stock and allow to reduce until thick enough to coat the back of a spoon (***à la nappe***). Keep warm.

Navets Glacés à Brun

1. Place the turnips in a small pan that is large enough to hold them in a single layer. Pour in enough cold water so that the turnips are two-thirds immersed. Add the sugar and butter and season with salt. Cover with a parchment-paper lid (***cartouche***) and cook the turnips over low heat until all the water has evaporated. Remove the cartouche and roll the turnips in the resulting syrup to glaze and evenly coat. Continue cooking until they take on a deep golden brown color, shaking the pan to color evenly. Set aside.

Tip: *Old or off-season turnips can be bitter. To remove any bitterness, blanch the turnips in salted boiling water for 2 minutes before glazing them. If blanching the turnips, use less water when glazing them.*

Glaçage

1. Remove the kitchen twine from the duck (***débrider***). Place the duck back on the wire rack over a baking sheet. Using a pastry brush, completely coat the top and the sides of the duck in sauce. Place the duck in the oven for 20 to 30 seconds to dry out the glaze. Repeat 4 or 5 times, or until the duck is coated in an even, thick shiny glaze. Turn off the heat of the oven and reserve the duck in the oven with the door ajar. Mount the remaining sauce with butter, then strain into the reserved garniture, and gently reheat if needed. Stir in some chopped parsley.

To Serve

1. Arrange the turnips around the edge of a hot serving dish and sprinkle with chopped parsley if desired. Place the garniture in the middle. Place the duck on top and fill the rear cavity with a bouquet of fresh watercress (optional).

OIE CONFITE, HARICOTS BLANCS

Learning Outcomes

Confire
Cured meat (*salage*)
Détailler à cru
Manchonner
Tremper

Equipment

Knives:
Boning knife, cleaver, paring knife (*office*), chef knife (*couteau chef*), vegetable peeler (*économe*)

Tools:
Bowls, colander, skimmer (*écumoire*), immersion blender, chinois, wooden spatula, whisk, kitchen twine

Pans:
Large pot, baking sheet, small saucepan, large saucepan, medium saucepan, large sauté pan

Serving

8 persons

FYI...

Confit comes from the French verb *confire,* which means to preserve. This stems from the Latin *conficere*, which means to do, to produce, to make, or to prepare. A confit is a food item that can have a long shelf life if it is packaged and stored correctly. Confire can be applied to fruits (in syrups, honey, or alcohol) and meats (salted and covered by fat).

OIE CONFITE, HARICOTS BLANCS
Confit of Goose with White Beans

Quantity		Ingredient
U.S.	*Metric*	
1 pc	1 pc	Goose (*whole, with heart and gizzards*)
4 lb	2 kg	Coarse sea salt
1 br	1 br	Rosemary, roughly chopped (***concasser***)
5 brs	5 brs	Thyme, rubbed (***émietter***)
2 brs	2 brs	Sage, roughly chopped (***concasser***)
2 pcs	2 pcs	Bay leaves
		Peppercorns, crushed (***mignonette***)
		To Confire
4 lb	2 kg	Duck fat
3 pcs	3 pcs	Shallots
5 pcs	5 pcs	Garlic cloves
1 pc	1 pc	Onion, halved
1 pc	1 pc	***Bouquet garni***
5 brs	5 brs	Parsley stems
		Tomates Confites
8 pcs	8 pcs	Tomatoes, peeled (***émonder***), seeded (***épépiner***), and quartered
1 bq	1 bq	Basil, roughly chopped (***ciseler***)
3 ½ oz	100 g	Anchovies, finely chopped (***hacher***)
4 pcs	4 pcs	Garlic cloves, finely chopped (***hacher***)
3 brs	3 brs	Thyme
1 pc	1 pc	Bay leaf
10 pcs	10 pcs	Black olives, pitted, finely chopped (***hacher***)
5 fl oz	150 mL	Olive oil
		Salt and pepper
		Crème d'Ail
1 pc	1 pc	Garlic head, peeled, blanched, refreshed
5 fl oz	150 mL	Cream
		Salt and pepper

Method

1. Preheat oven to 320°F (160°C).
2. Soak the beans in cold water overnight.
3. *Clean the goose* (***habiller***): See pages 333–336 in *Cuisine Foundations.* Make an incision (*right to the bone*) around the end of the drumstick. Scrape the flesh and tendons away from the bone. Chop off the tip of the drumsticks using a cleaver. Push the flesh back to expose the bone (***manchonner***).
4. Cut the bird into 8 pieces: separate the wings, thighs, drumsticks, and breasts. Then cut each into two. Place the goose with its heart and gizzards in a bowl and toss with the herbs and coarse salt. Leave this to marinate for 48 hours in the refrigerator, stirring occasionally.

Oie Confite

1. Remove the meat from the salt, rinse it well under cold running water, drain it, and pat it dry.
2. Place the goose pieces (*including the gizzard and heart*) in a large pot with the shallots, garlic, onion, parsley, ***bouquet garni***, and duck fat. Bring the fat to a gentle simmer over medium heat. Cover the pot and place it in the oven to cook for 2 to 4 hours (***confire***). When the meat is tender enough to break apart under light pressure, remove the pot from the heat and set aside (*still covered*). Reduce the oven to 200°F (95°C).

Tomates Confites

1. Peel (***émonder***), seed (***épépiner***), and quarter the tomatoes. Mix them with the chopped basil, anchovies, garlic, herbs, and olives. Toss the mixture in olive oil. Spread the tomato mixture on a baking sheet and place it in the oven. Leave to cook until soft (***confire***) (*about 1 hour*). Drain the tomatoes and cut them into a small dice (***brunoise***). Season to taste. Set aside in a covered container with a few tablespoons of the cooking oil. Strain the rest of the cooking oil through a fine mesh sieve (***chinois***) and reserve for the vinaigrette. Increase the oven temperature to 400°F (205°C).

Crème d'Ail

1. *Blanch the garlic:* Place the peeled garlic cloves in a small saucepan of cold water and bring to a boil. After 5 minutes of boiling, refresh the garlic under cold running water and drain. Transfer the garlic to a small saucepan and add the cream. Bring the cream to a gentle simmer over low heat and cook the garlic until it is soft. Purée the garlic and cream with an immersion blender, then strain the mixture through a fine mesh sieve (***chinois***). Taste, season, and set aside.

Garniture

1. Drain the white beans. Place them in a large saucepan with the carrot, onion, leek, celery, garlic, and ***bouquet garni***. Add in the water and bring it to a boil over high heat. Reduce the heat to medium and simmer gently until the beans are tender to the bite and their skins are slightly split [*skim* (***écumer***) *the surface of the water regularly*].

Quantity		Ingredient
U.S.	Metric	Haricots
13 ½ oz	400 g	White beans (***haricots blancs***), dry, (*soaked for 24 hours*)
1 pc	1 pc	Carrot
1 pc	1 pc	Onion
1 pc	1 pc	Leek white
½ pc	½ pc	Celery stalk
½ pc	½ pc	Garlic head
1 pc	1 pc	***Bouquet garni***
2 qt	2 L	Water
		Garniture
5 oz	150 g	Smoked pork belly, cut into ***lardons***
3 ½ oz	100 g	Duck fat (*left over from goose* ***confit***)
3 pcs	3 pcs	Shallots, thinly sliced (***émincer***)
4 pcs	4 pcs	Garlic cloves, finely chopped (***hacher***)
3 brs	3 brs	Thyme, rubbed (***émietter***)
3 ½ fl oz	100 mL	***Glace d'agneau***
3 ½ oz	100 g	Butter
1 ½ oz	40 g	Parsley, finely chopped (***hacher***)
		Salt and pepper
		Salade
1 pc	1 pc	Curly endive, washed
1 fl oz	30 mL	Red wine vinegar
1 pc	1 pc	Shallot, finely chopped (***hacher***)
3 pcs	3 pcs	Garlic cloves, finely chopped (***hacher***)
2 ½ fl oz	80 mL	Olive oil, leftover from tomato (***confit***)
		Salt and pepper

Season the beans 10 minutes into cooking. Once cooked, remove from the heat and set aside to cool in its cooking liquid.

2. *Blanch the lardons:* Place the lardons in a medium saucepan of cold water and bring to a boil over high heat. Refresh the lardons under cold running water, then drain.
3. Pour off 3 ½ oz (100 g) of the goose fat. Heat this fat in a large sauté pan over medium-high heat, add the lardons, and sauté them until lightly golden (***blond***). Reduce the heat to low, add the sliced shallots and chopped garlic, and sweat (***suer***) them until soft. Drain the beans and remove the vegetables and the ***bouquet garni***. Add the beans to the pan of lardons and stir them in. Add the thyme and ***glace d'agneau*** and cook gently for 5 minutes, stirring carefully. Mix in the tomato ***brunoise*** (***tomates confits***) and garlic cream and let the mixture simmer gently for 10 to 15 minutes. Mix in the cold butter until it is completely melted (***lier au beurre***) and stir in the chopped parsley. Taste and adjust the seasoning.

Salade

1. Tear the lettuce into bite-sized pieces. Rinse well in cold water and drain thoroughly.
2. *Prepare the vinaigrette:* Season the vinegar with salt and pepper and stir in the chopped shallot and garlic. Pour in the reserved oil from cooking the tomatoes in a thin stream (*whisking continuously*) to create an emulsion. Set aside.

To Serve

1. Remove the meat from the duck fat. Slice the heart and gizzards and set aside. Place the duck meat on a parchment paper-lined baking sheet in the oven until golden and crispy.
2. Place the bean mixture in a hot serving dish, arrange the goose ***confit*** on top, and sprinkle the dish with chopped parsley.
3. Toss the lettuce in the vinaigrette and place it in a cold serving bowl. Top the lettuce with the sliced gizzards and heart.

Learning Outcomes

Mariner
Habiller
Détailler
Liaison au beurre manié
Croûtons
Mirepoix

Equipment

Knives:
Boning knife, chef knife (*couteau chef*), paring knife (*office*), serrated knife, vegetable peeler (*économe*)

Tools:
Mixing bowls, wire rack, wooden spoon, whisk, chinois, butcher's twine, kitchen twine

Pans:
Cocotte, small saucepan, sauté pan, large saucepan

Serving

8 persons

FYI...

The traditional thickening agent for coq au vin is blood. Although protein-enriched sauces are both rich and nutritious, caution should be taken during and after the preparation of a sauce of this variety. Harmful microbes that could cause botulism could present themselves, so it is best to either keep the temperature below 40°F (5°C) or above 140°F (60°C).

Once a rooster goes through the process of castration, it becomes less aggressive and spends less time fighting and defending its territory—it also spends more time eating. Without sex-producing hormones, the male instincts are curtailed, but there are also considerable changes to the meat. Castrated males possess higher dark to light meat ratios, produce a higher percentage of body fat, and tend to have more cooking juices to baste with.

Quantity		Ingredient
U.S.	Metric	
1 pc	1 pc	Rooster or capon [*10 lb (5 kg)*]
3 ½ oz	100 g	Duck fat or oil
1 qt	1 L	Brown chicken stock
		Salt and pepper
		Marinade
3 qt	3 L	Dry red wine (*preferably Bordeaux*)
2 pcs	2 pcs	Carrots, large, cut into ***mirepoix***
2 pcs	2 pcs	Onions, large, cut into ***mirepoix***
2 pcs	2 pcs	Tomatoes, cut into ***mirepoix***
1 pc	1 pc	Garlic head, halved
10 brs	10 brs	Parsley
1 pc	1 pc	Celery stalk, cut into large ***mirepoix***
1 pc	1 pc	***Bouquet garni***
1 pc	1 pc	Leek, cut into large ***mirepoix***
20 pcs	20 pcs	Peppercorns, crushed (***mignonette***)
4 fl oz	125 g	Red wine vinegar
4 fl oz	125 mL	Cognac
4 fl oz	125 mL	Vegetable oil
1 Tbsp	15 g	Coarse salt

COQ OU CHAPON AU VIN

Rooster or Capon Stewed in Red Wine

Method

1. Preheat the oven to 375°F (190°C).
2. *Clean the bird* (***habiller***):See pages 333–336 in *Cuisine Foundations.* Make an incision (*right to the bone*) around the end of the drumstick. Scrape the flesh and tendons away from the bone. Chop off the tip of the drumsticks using a cleaver.
3. Cut the bird (***détailler***) into 16 pieces. Separate the wings, thighs, drumsticks, and breasts, then cut each in two.

Marinade

1. Place the pieces in a large recipient with the carrots, onions, tomatoes, garlic, parsley, celery, leek, ***bouquet garni***, black peppercorns, red wine vinegar, cognac and red wine. Add the oil last, which forms a protective barrier on the surface, then cover the recipient and leave to marinate from 6 to 12 hours in the refrigerator.

Cuisson

1. Remove the meat from the marinade (***décanter***). Pat dry and season with salt and pepper. Strain the marinade through a fine mesh sieve (***chinois***) into a large saucepan. Reserve the vegetables and bring the liquid to a boil over high heat. Skim (***écumer***) the surface and let the liquid reduce by one-third. Strain it through a fine mesh sieve (***chinois***) and reserve.
2. Heat the duck fat or oil in a Dutch oven (***cocotte***) over medium-high heat. Add the fowl pieces (a few at a time) and sear them in the ***cocotte*** until browned (***saisir à brun***). When done, transfer the pieces to a wire rack. Pour the fat out of the ***cocotte*** (***dégraisser***) and place it back over medium-high heat. Add the marinade vegetables to the ***cocotte*** and sauté until colored. Add the seared fowl pieces and stir well. Wet (***mouiller***) with the reduced marinade and the brown chicken stock. Increase the heat to high and bring the liquid to a boil. Skim the surface (***écumer***), cover the ***cocotte***, and braise the fowl pieces in the oven until the meat is easily pierced with a knife (*approximately 2 ½ to 3 ½ hours*).
3. *Prepare a beurre manié:* Combine the softened butter (***beurre pommade***) with the flour and mix until the ingredients form a smooth paste. Set aside.

Garniture

1. ***Glacer à brun*** *the pearl onions:* Place the onions in a sauté pan that is large enough to hold them in a single layer. Pour in enough cold water so that the onions are two-thirds immersed. Add the sugar and butter and season with salt. Cover with a buttered parchment paper lid (***cartouche***). Cook the onions over low heat until all the water has evaporated and the sugar and butter mixture begins to caramelize.

Method

Roll the onions in the resulting syrup to evenly coat. Reserve in the sauté pan at room temperature.

2. *Blanch the lardons:* Place the lardons in a medium saucepan of cold water and bring to a boil over high heat. Refresh the lardons under cold running water, then drain.
3. Heat a tablespoon of oil over medium-high heat and sauté the lardons until they begin to brown (***blond***). Reduce the heat to medium, add the mushrooms, and cook them until soft. Set aside in a clean bowl.

Croûtons

1. Cut the bread slices in half diagonally. Melt the clarified butter in a medium sauté pan over medium heat. Add the bread shapes and cook them until golden on both sides. Drain the croûtons on a plate lined with a paper towel.

Finishing

1. Remove the chicken pieces from the braising liquid (***décanter***) and strain the cooking liquid through a fine mesh sieve (***chinois***). Discard the vegetables and reduce the braising liquid to concentrate the flavors.
2. Measure out 1.5 qt (1.5 L) of the braising liquid (***jus de braisage***) into a large saucepan and bring the liquid to a boil over medium-high heat. Whisk in the ***beurre manié,*** piece by piece, until incorporated. Continue whisking until the mixture is thick enough to coat the back of a spoon (***à la nappe***). Strain the sauce through a fine mesh sieve (***chinois***) into a clean saucepan and mount the sauce with cold butter. Taste the sauce and adjust the seasoning.

To Serve

1. Combine the lardons, the mushrooms, the sauce, and the fowl pieces in the sauté pan with the glazed onions. Reheat all the ingredients over medium heat, then arrange in a hot serving dish. Lightly dip the narrow tip of the croûtons in the sauce, then the chopped parsley and place them around the edge of the dish.

Optional: *Serve with freshly made pasta.*

Tip: *In Europe, this sauce is finished with blood. If choosing to make a liaison with blood, cool 3 ⅓ oz (100 mL) of the sauce to room temperature and mix in 5 oz (150 mL) of blood. Pour the blood mixture into the finished dish and stir gently over low heat to thicken. If using blood to finish the sauce, reduce the amount of* ***beurre manié*** *used in this recipe from 5 oz (150 g) to 4 oz (120 g).*

2. *This recipe is adaptable to different types of fowl, and cooking times vary depending on the animal used.*

Quantity		Ingredient
U.S.	*Metric*	*Beurre Manié*
2 ½ oz	75 g	Butter, unsalted, softened (***pommade***)
2 ½ oz	75g	All-purpose flour
		Garniture
40 pcs	40 pcs	Pearl onions, peeled
1 ¾ oz	50 g	Sugar, granulated
4 oz	120 g	Butter
		Salt
24 pcs	24 pcs	Mushrooms, peeled, halved
10 oz	300 g	Smoked pork belly, cut into ***lardons***, blanched
1 ¾ oz	50 g	Butter
		Croûtons
8 slices	8 slices	Firm white bread, crusts removed
2 fl oz	60 mL	Butter, clarified
1 oz	30 g	Parsley, finely chopped (***hacher***)
		To Serve
2 ½ oz	75 g	Butter
		Salt and pepper

COQUELET EN CRAPAUDINE

Learning Outcomes

Crapaudine
Habiller poultry
Griller
Sauter
Reduction
Sauce américaine

Equipment

Knives:
Cleaver, désosseur, scissors, vegetable peeler (***économe***), office, slicing knife (***émninceur***)

Tools:
Wooden spatula, fine chinois, skewers, bowls, pastry brush, cutting board, wire rack

Pans:
Grill, large sauté pan, medium saucepan, bain marie

Serving

4 persons

FYI...

The term *en crapaudine* comes from *crapaud*, the French word for toad. The end product in this recipe is a small chicken that is partially deboned, flattened, breaded, and grilled—and ends up looking uncannily like a frog. Accentuating this resemblance, the traditional presentation for crapaudine was done with makeshift eyes that were made from cooked egg whites and either caper or truffle pieces. The technique was originally applied to cooking pigeons in the 1740s.

COQUELET EN CRAPAUDINE
Spatchcock Game Hen

Method

1. Preheat the oven to 400°F (205°C).

Crapaudine

1. *Clean the birds* (***habiller***): See pages 333–336 in *Cuisine Foundations.* Make an incision (*right to the bone*) about three-quarters of an inch from the end drumstick. With the back of the knife, scrape the flesh and tendons toward the end of the drumstick. Chop off end using a cleaver. Push the flesh back to expose the bone (***manchonner***). Scrape off any remaining meat.
2. Place the bird on its back on the cutting board. If you are right-handed, hold the shoulders of the bird with your left hand, then place the knife through the opening under the breastbone and cut the sides to just above the wing joints. Grasp the breastbone with your left hand while securing the legs, and pull up and back until you hear the joints snap, and the interior of the bird is exposed. Remove the rib cage and other small bones. Turn the bird over, cover with plastic, and flatten using a meat pounder, or the bottom of a heavy saucepan. In order to keep its flat shape during cooking, secure the wing and leg together with a skewer. Repeat on the other side.

Sauce Américaine

1. Combine the shallots, white wine vinegar, white wine, black peppercorns, and thyme in a medium saucepan over low heat and reduce until almost dry. Add the tomatoes and simmer gently until almost all the liquid has evaporated. Add the veal stock and reduce the sauce until it is thick enough to coat the back of a spoon (***à la nappe***). Taste and season. Strain the sauce through a fine mesh sieve (***chinois***) and reserve in a bain marie.

Cuisson

1. Preheat the grill.
2. Mix the mustard with an equal amount of clarified butter and brush both sides of the bird with the mixture. Season with salt and pepper. Mix the breadcrumbs with a little clarified butter and spread over the skin side of the bird. Once the grill is very hot, lay the prepared birds skin-side down at a 45° angle. Grill until marked by the hot grill, then turn them in the opposite direction to create a criss-cross pattern. Turn over and grill the other side. Transfer to a rack and place in the oven to finish cooking, about 20 minutes. Set aside on a wire rack.

Pommes de Terre à la Lyonnaise

1. Peel and wash the potatoes. Cut the potatoes into 1 cm (½ in.) pieces (***coin de la rue***).
2. Heat 5 oz (150 g) clarified butter in a large sauté pan over high heat and sauté the potatoes until golden (*tossing them every 5 to 10 minutes to cook them evenly*).
3. Meanwhile, peel and slice the onion. Heat 1 ¾ oz (50 g) clarified butter in a medium sauté pan over medium heat and sauté the onions until golden. Add the onions to the potatoes and continue cooking until the potatoes are cooked through and golden. Pour the clarified butter out of the pan, replace it with the cold butter, and toss until melted. Season to taste.

To Serve

1. Remove the skewers from the birds. Spoon some sautéed potatoes into the center of the plate. Rest the ***crapaudine*** against the potatoes and decorate with fresh herbs. Serve the sauce on the side.

Quantity		*Ingredient*
U.S.	*Metric*	
2 pcs	2 pcs	Young roosters or Cornish game hens *[500 g (1 lb each)]*
1 ½ oz	40 g	Dijon mustard
3 ½ fl oz	100 mL	Butter, clarified
1 ½ oz	40 g	Breadcrumbs
		Salt and white pepper
		Sauce Américaine
5 oz	150 g	Shallots, finely chopped (***hacher***)
3 ½ fl oz	100 mL	White wine vinegar
3 ½ fl oz	100 mL	White wine
1 tsp	5 g	Black peppercorns
5 brs	5 brs	Fresh thyme
2 pcs	2 pcs	Tomatoes, peeled (***émonder***), seeded (***épépiner***), and diced (***concasser***) (*see p. 51)*
6 ½ fl oz	200 mL	Brown veal stock
		Salt and white pepper
		Pommes de Terre à la Lyonnaise
4 pcs	4 pcs	Potatoes, ½ in. (1 cm) dice (***coin de la rue***)
5 oz	150 g	Butter, clarified
1 pc	1 pc	Onion, finely sliced (***émincer***)
1 ¾ oz	50 g	Butter, clarified
1 ¾ oz	50 g	Butter, cold
		Salt and white pepper

FILET DE DINDE SAUTÉ CHASSEUR

Learning Outcomes

Pommes soufflés
Habiller a turkey breast
Ficeler
Sauté a turkey breast

Equipment

Knives:
Boning knife, chef knife (*couteau chef*), paring knife (*office*), vegetable peeler (*économe*)

Tools:
Twine, wooden spatula, tongs, skimmer (*écumoire*), chinois, spider, wire rack, trussing needle

Pans:
Large sauté pans, bain marie, large saucepan, deep fryer

Serving

8 persons

FYI...

Duke Philippe de Mornay (November 5, 1549–November 11, 1623) is credited with creating the chasseur sauce.

FILET DE DINDE SAUTÉ CHASSEUR

Sauteed Turkey Breast in Mushroom and Tomato Sauce

Quantity		*Ingredient*
U.S.	*Metric*	
1 pc	1 pc	Turkey breast [*3 lbs (1.5 kg)*]
2 fl oz	60 mL	Vegetable oil
2 fl oz	60 g	Butter, clarified
		Salt and white pepper
		Sauce Chasseur
1 oz	30 g	Tomato paste
¾ oz	20 g	Flour
3 ½ fl oz	100 mL	White wine
8 fl oz	250 mL	Brown veal stock
1 pc	1 pc	***Bouquet garni***
		Mushroom peels (*from garniture*)
		Tomato peels (*from garniture*)
1 pc	1 pc	Garlic clove
2 tsp	10 g	Parsley, finely chopped (***hacher***) reserve some for decor
2 tsp	10 g	Tarragon, finely chopped (***hacher***)
2 tsp	10 g	Chervil, finely chopped (***hacher***)
4 ½ oz	125 g	Butter
		Salt
		Garniture and Finish
1 ¾ oz	50 g	Butter
½ lb	250 g	Mushrooms, peeled, quartered
2 pcs	2 pcs	Shallots, finely chopped (***hacher***)
7 oz	200 g	Tomatoes, peeled (***émonder***), seeded (***épépiner***), and diced (***concasser***) (*see page 51*)
		Pommes Pailles
4 pcs	4 pcs	Potatoes, large
		Salt

Method

1. Preheat the deep fryer to 320°F (160°C).
2. *Clean the turkey breast* (***habiller***): Trim off any fat, silver skins, tendons, and surplus skin. Truss (***ficeler***) the breast to obtain an even cylindrical shape. Set aside.

Garniture

1. Melt the butter in a large sauté pan over medium-high heat and sauté the quartered mushrooms until they are lightly colored (***blond***). Reduce the heat to low, add the chopped shallots, and sweat (***suer***) them until soft (*5 minutes*). Stir in the tomato ***concassée*** and cook gently for 10 minutes. Cover the pan and set aside.

Cuisson

1. Pat the turkey dry and season well. Heat the oil and clarified butter in a large sauté pan over medium-high heat. Add the turkey breast, skin-side down, and sauté it until the skin is golden. Turn the breast and continue until the entire exterior is golden. When the turkey is cooked (*see note*), transfer it to a wire rack to rest, covered in foil.

Note: *The breast is cooked if the juices run clear from the center when it is pricked with a trussing needle.*

Sauce

1. Pour the fat (***dégraisser***) out of the pan used to cook the turkey. Place the pan on medium heat for 1 minute to concentrate (*without burning*) the cooking residues (***pincer les sucs***). Add the tomato paste and cook for 1 to 2 minutes, stirring well (***pincer la tomate***). Add the flour and cook for 2 to 3 minutes, stirring well (***singer***). Deglaze with the white wine and let it reduce until almost dry. Wet the pan with veal stock and add the ***bouquet garni***, mushroom and tomato peels, and the garlic clove. Simmer the mixture over medium heat, skimming the surface (***écumer***) until the sauce is thick enough to coat the back of a spoon (***à la nappe***). Strain the sauce through a fine mesh sieve (***chinois***) and reserve covered in a bain marie.

Pommes Pailles

1. Wash the potatoes and cut them into ***pommes paille*** (***julienne size***).
2. Rinse the ***pommes paille*** in fresh water and pat them dry. Immerse the ***pommes paille*** in the deep fryer and cook them until they are crispy, golden, and cooked through. Strain the potatoes and drain them on a dish lined with a paper towel. Sprinkle the ***pommes paille*** generously with salt and serve immediately.

Note: *Unlike **pommes Pont-Neuf**, which are blanched in the deep fryer and finished later, **pommes paille** are cooked at a lower temperature in one cooking. Due to their thinness, they are cooked by the time they are crisp and golden.*

Persil Frit

1. Wash the reserved parsley and pat it dry. Toss it in flour and pat off any excess. Deep fry the sprigs for a few seconds, being careful not to burn the leaves. Drain the fried parsley on a paper towel and sprinkle with salt.

To Serve

1. Remove the string from the turkey and slice the breast. Mount the sauce with cold butter and mix in the garniture. Spoon the sauce and garnish onto a heated serving platter and arrange the sliced turkey breast on top. Decorate with fresh herbs. Serve the pommes pailles and the fried parsley in a dish that is lined with a folded napkin.

JAMBONNETTES DE VOLAILLE FARCIES AU BULGUR, SAUCE RIESLING

Learning Outcomes

Making jambonnette
Farcis
Brunoisette
Désosser a thigh
Glacer à blanc

Equipment

Knives:
Cleaver, désosseur, vegetable peeler (***économe***), office, slicing knife (***éminceur***), scissors

Tools:
Bowls, cutting board, fork, wooden spatula, fine chinois, colander, skimmer (***écumoire***), ladle (***louche***), whisk, trussing needle, kitchen twine

Pans:
Sauteuse, sautoir, sauce pan (***russe***), bain marie

Serving

4 persons

Jambonnette translates from the French as "little ham" and, in this recipe, refers to the shape of the chicken leg after it has been deboned and stuffed.

JAMBONNETTES DE VOLAILLE FARCIES AU BULGUR, SAUCE RIESLING

Deboned Chicken Legs Stuffed with Bulgur in a Riesling Sauce

Quantity		Ingredient
U.S.	*Metric*	***Base***
4 pcs	4 pcs	Chicken legs [*200 g (7 oz each)*]
1 ½ fl oz	50 mL	Vegetable oil
1 ½ oz	50 g	Butter
		Salt and white ground pepper
		Farcis
3 ½ oz	100 g	Bulgur
3 ½ oz	100 g	Carrots, cut into ***brunoisette***
3 ½ oz	100 g	Celery, cut into ***brunoisette***
1 ¼ oz	40 g	Butter
1 ¼ oz	40 g	Chives, sliced (***émincer***)
13 fl oz	400 mL	White chicken stock
3 ½ fl oz	100 mL	White wine, Riesling
1 pc	1 pc	Egg
		Salt and white ground pepper
		Sauce Riesling
1 ½ oz	40 g	Butter
1 ½ oz	40 g	Flour
1 pt	500 mL	White wine, Riesling
7 fl oz	200 mL	White chicken stock
1 oz	30 g	Chives, sliced (***émincer***)
5 fl oz	150 mL	Cream
2 pcs	2 pcs	Egg yolks
		Salt and white ground pepper
		Garniture
10 oz	300 g	Oyster mushrooms
3 ½ oz	100 g	Butter
2 pcs	2 pcs	Shallots
½ oz	15 g	Parsley
20 pcs	20 pcs	Pearl onions, peeled
¾ oz	25 g	Sugar, granulated
1 ½ oz	40 g	Butter
4 brs	4 brs	Sage, sliced (***émincer***)
4 brs	4 brs	Tarragon, sliced (***émincer***)
		Salt and white ground pepper

Method

1. Preheat the oven to 360°F (180°C).

Farcis

1. Melt the butter in a medium saucepan over low heat and sweat (***suer***) the carrots and celery until soft. Deglaze with white wine and let it reduce until it is almost dry. Wet (***mouiller***) with white chicken stock and mix in the bulgur. Increase the heat to medium and season. Simmer until the bulgur is soft (*25 to 30 minutes*) and drain the excess liquid through a strainer. Mix the chives into the bulgur, then taste and adjust the seasoning. Let the mixture cool to room temperature, mix in the egg and reserve.

Jambonnette

1. Remove all the thigh bones in the legs leaving the drumstick bone below the joint. Reserve the bones and trimmings. Make an incision (*right down to the bone*) around the end of the drumstick. Scrape the flesh and tendons away from the bone. Chop off the tip of the drumstick bone using a cleaver. Push back the meat to expose the bone (***manchonner***). Season the legs on the inside and outside and fill them with ***farcis***. Sew the opening shut using a trussing needle and kitchen twine. Pat the skin dry and season. Sear the legs in oil and butter in a large pan over high heat until colored on all sides. Add the meat trimmings and place the pan in the oven to roast for 40 minutes, basting the legs frequently. The legs are cooked when a trussing needle inserted into the center comes out too hot to touch. Transfer the legs to a wire rack.

Tip: *Some chickens will have thin skins where the skin will rip while being sewed. If the skin is thin and fragile, either use toothpicks to hold the skin closed or wrap the legs tightly in caul fat, which can be removed before serving.*

Sauce Riesling

1. Melt the butter in a medium pan over medium heat. When the butter begins to foam, add the flour and stir well until the mixture begins to bubble. Cook for 5 to 10 minutes over low heat to obtain a roux blond (*be careful not to let it color*). Transfer to a small plate and set aside to cool.
2. Pour the excess fat out of the cooking pan (***dégraisser***).
3. Place the cooking pan used for the jambonnettes on medium heat for 1 to 2 minutes to concentrate (*without burning*) the cooking juices (***pincer les sucs***). Deglaze the pan with the Riesling wine and reduce by half. Wet (***mouiller***) the pan with the chicken stock and bring it to a simmer. Mix in the cold roux, piece by piece, and let the sauce cook for 10 to 15 minutes over low heat. Strain the sauce through a fine mesh sieve ***(chinois)*** into a small saucepan and mix in the herbs. Mix the cream into the egg yolks. Ladle some of the hot sauce into the yolks to temper. Off the heat, incorporate the tempered yolks back into the sauce and mix well. Season to taste. Pat (***tamponner***) the surface of the sauce with cold butter on the end of a fork to prevent a skin from forming. Reserve the sauce in a bain marie.

Garniture

1. Cut the mushrooms into bite-sized pieces.
2. Melt the butter in a large sauté pan over low heat and sweat (***suer***) the shallots until soft. Increase the heat to medium high and add the mushrooms. Sauté the mushrooms until they begin to color, then season with salt and pepper. Set aside.
3. ***Glacer à blanc*** *the pearl onions:* Place the onions in a sauté pan that is large enough to hold them in a single layer. Pour in enough cold water so that the onions are two-thirds immersed. Add the sugar and butter and season with salt. Cover with a parchment paper lid (***cartouche***) and cook the onions over low heat until all the water has evaporated. Roll the onions in the resulting syrup to glaze them. Mix the onions with the oyster mushrooms.

To Serve

1. Remove the kitchen twine from the legs (***déficeler***). Trim the base of the legs to help them stand upright.
2. Arrange the legs on a hot serving dish with the garnish in the center. Coat with the sauce and decorate with fresh herbs.

LAPIN À LA GRAINE DE MOUTARDE, POMMES RISSOLÉES

Learning Outcomes

Braiser
Détailler
Mariner
Saisir
Pincer les sucs
Écumer
Réduire
Sauter à cru
Monter au beurre

Equipment

Knives:
Cleaver, désosseur, vegetable peeler (***économe***), office, slicing knife (***éminceur***), scissors

Tools:
Bowls, cutting board, fork, wooden spatula, fine chinois, colander, skimmer (***écumoire***), ladle (***louche***), whisk, wire rack

Pans:
Sauteuse, sautoir, sauce pan (***russe***), bain marie

Serving

4 persons

FYI...

The French call whole grain mustard, *moutarde à l'ancienne.* The recipe for whole grain mustard predates the 18th century and remains popular even today. The texture is mealy and the flavor is similar to that of Dijon, only it is less intense. Prepared mustards are made up of vinegar, spices, and a variety of mustard seeds ranging from ground to semi ground.

LAPIN À LA GRAINE DE MOUTARDE, POMMES RISSOLÉES

Rabbit in Whole-Grain Mustard Sauce with Sauteed Potatoes

Method

1. Preheat the oven to 350°F (180°C).
2. Cut the rabbit into 8 pieces. Mix the Whole-grain and the Dijon mustards together and rub the mixture onto the rabbit pieces to coat them. Leave to marinate up to 12 hours in a covered recipient in the refrigerator.

Cuisson

1. Season the rabbit with salt and pepper. Heat the oil butter in a large sauté pan over high heat and sear the rabbit until it is a golden brown. Transfer the pieces to a bowl and reserve. Pour off the excess fat (***dégraisser***) and place the pan over medium heat for 1 to 2 minutes to concentrate (*but not burn*) the cooking residues (***pincer les sucs***). Add the shallots, leek, and celery and sweat (***suer***) them until they begin to soften. Deglaze the pan with white wine, scraping the pan well to dissolve the cooking juices (***sucs***) and reduce it by half. Wet (***mouiller***) the pan with the chicken stock and add the ***bouquet garni***. Return the rabbit to the pan and roll it in the sauce until coated. Increase the heat to medium high, bring the liquid to a simmer, season lightly and skim (***écumer***) the surface. Cover the pan and place it in the oven to braise until the rabbit is cooked through (*45 minutes*).

Sauce Moutarde

1. Transfer the rabbit to a wire rack and cover it in foil to keep it warm.
2. Strain the braising liquid (***jus de braisage***) through a fine mesh sieve (***chinois***). Add the cream and reduce the sauce over medium heat until it is thick enough to coat the back of a spoon (***à la nappe***). Taste the sauce and adjust the seasoning. Reserve in a bain marie.

Pommes Rissolées

1. Melt the clarified butter in a large pan over medium-high heat and sauté the diced potatoes until they are golden on all sides (*toss them every 5 to 10 minutes to ensure even coloring*). Drain the grease out of the pan and toss the potatoes in fresh butter and parsley.

To Serve

1. Stir the whole-grain mustard into the sauce and mount it with cold butter. Adjust seasoning as needed. Warm the rabbit by placing it in the sauce over low heat for 5 minutes. Transfer the rabbit to a hot serving dish and coat it in the sauce (***napper***). Sprinkle the dish with chopped parsley and serve the ***pommes rissolées*** on the side in a hot serving dish.

Quantity		*Ingredient*
U.S.	*Metric*	
4 lb	2 kg	Whole rabbit
3 ½ oz	100 g	Whole-grain mustard
3 ½ oz	100 g	Dijon mustard
		Salt and white pepper
		Garniture de Cuisson
3 ½ fl oz	100 mL	Oil
4 pcs	4 pcs	Large shallots, finely chopped (***hacher***)
1 pc	1 pc	Leek, finely chopped (***hacher***)
1 pc	1 pc	Celery stalk, cut into ***mirepoix***
8 fl oz	250 mL	Dry white wine
1 pc	1 pc	***Bouquet garni***
1 pt	500 mL	White stock
5 oz	150 g	Butter
		Salt and white pepper
		Sauce
8 fl oz	250 mL	Cream
2 ½ oz	75 g	Butter
		Whole-grain mustard to taste
		Dijon mustard (*optional*)
		Salt and white pepper
		Pommes Rissolées
5 fl oz	150 g	Butter, clarified
4 pcs	4 pcs	Potatoes, cut into dice
2 ¾ oz	80 g	Fresh butter
2 ¾ oz	80 g	Parsley (***hacher***)
		Salt and white pepper

PINTADE COCOTTE GRAND-MÈRE

Learning Outcomes

Habiller
Brider
Pommes cocotte
Garniture grand-mère
Braising
Déglacer
Poêler une volaille entière

Equipment

Knives:
Vegetable peeler (***économe***), office, slicing knife (***éminceur***), cleaver, turning knife, boning knife

Tools:
Wooden spoon, trussing needle, wire rack, mixing bowls, chinois, kitchen twine

Pans:
Cocotte

Serving

4 persons

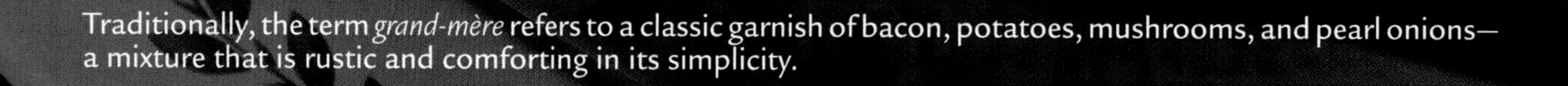

TIP...

Traditionally, the term *grand-mère* refers to a classic garnish of bacon, potatoes, mushrooms, and pearl onions—a mixture that is rustic and comforting in its simplicity.

PINTADE COCOTTE GRAND-MÈRE

Guinea Fowl with Potatoes, Mushrooms, and Bacon

Quantity		Ingredient
U.S.	*Metric*	
1 pc	1 pc	Guinea fowl [*3 lbs (1.5 kg)*]
3 ½ fl oz	100 mL	Butter, clarified
2 pcs	2 pcs	Apples
5 brs	5 brs	Thyme
1 pc	1 pc	Bay leaf
		Salt and pepper
		Garniture Grand-Mère
4 oz	125 g	Smoked pork belly, cut into lardons, blanched
5 oz	150 g	Mushrooms
4 oz	125 g	Pearl onions
2 lb	1 kg	Potatoes, turned to ***cocotte***, blanched
1 ¾ oz	50 g	Butter
1 ¾ fl oz	50 mL	Oil
		Salt
		Fond de Braisage
3 ½ fl oz	100 mL	White wine
7 fl oz	250 mL	Brown veal stock
3 ½ fl oz	100 mL	Thickened brown veal stock
1 ¾ oz	50 g	Butter
		To Serve
3 brs	3 brs	Parsley (***hacher***)

Method

1. Preheat oven to 400°F (205°C).
2. *Clean the guinea fowl* (***habiller***): See pages 333–336 in *Cuisine Foundations.* Make an incision (*right to the bone*) around the end of the drumstick. Scrape the flesh and tendons away from the bone. Chop the tip off the drumsticks using a cleaver. Push the flesh back to expose the bone (***manchonner***).
3. Pat dry and season the bird inside and out. Place the apples, thyme, and bay leaf inside the carcass.
4. Truss the bird (***brider***) using a trussing needle and kitchen twine. See pages 337–339 in *Cuisine Foundations.*
5. Set aside.

Garniture Grand-Mère

1. *Blanch the lardons:* Place the lardons in a medium saucepan of cold water and bring it to a boil over high heat. Refresh the lardons under cold running water, then drain.
2. Peel the mushrooms and cut off the stems, then cut into quarters.
3. Place the cocotte potatoes in cold salted water and bring to a boil. Blanch 1 to 2 minutes, then drain. Set aside.

Cuisson

1. Heat the oil in a Dutch oven (***cocotte***) over high heat. Place the guinea fowl on its side and sear (***saisir***) until golden. Turn the guinea fowl onto its other side and sear until golden. Place the guinea fowl on its breasts and sear until golden, basting all the while. Turn the guinea fowl on its back, cover the cocotte, and place it in the oven. Leave to cook for 10 to 15 minutes, basting often. Add the potatoes, blanched bacon, quartered mushrooms, and pearl onions and stir to coat. Cover and place back in the oven to finish cooking, about 15 to 20 minutes or until the juices run clear when the thickest part of the thigh is pierced with a trussing needle. Transfer the pintade to a wire track, cover with foil, and keep warm. Decant the garniture grand-mère, cover, and keep warm.
2. Pour off the fat and place the cocotte over medium-high heat to brown the cooking juices (***pincer les sucs***). Deglaze with white wine and scrape well to dissolve the sucs. Reduce by half, then add the veal stock and thickened veal stock. Leave to gently simmer until thick enough to coat the back of a spoon. Season to taste, then strain through a fine mesh sieve (***chinois***) into a clean pan. Mount with butter and set aside.

To Serve

1. Add the garniture grand-mère back into the cocotte and mix in the chopped parsley. Pour the sauce over and stir well. Remove the kitchen twine from the pintade and place on top.

POULARDE POCHÉE SAUCE SUPRÊME, RIZ PILAF

Learning Outcomes

Pocher
Habiller and brider poultry
Riz pilaf
Sauce suprême
Égrener
Nacrer

Equipment

Knives:
Boning knife, chef knife (*couteau chef*), paring knife (*office*), scissors

Tools:
Kitchen twine, bowls, wire rack, skimmer (*écumoire*), chinois, wooden spatula, whisk, carving fork, ladle (*louche*), trussing needle

Pans:
Medium stock pot, large saucepan, small saucepan, medium saucepan, bain marie

Serving

4 persons

FYI...

When cream is added to a velouté, it is transformed into something extraordinary. However, adding cream will thin out your *sauce suprême*—this is why the cream is reduced beforehand.

POULARDE POCHÉE SAUCE SUPRÊME, RIZ PILAF

Poached Chicken with a White Cream Sauce and Boiled Rice

Method

1. Preheat the oven to 350°F (180°C).

Base

1. *Clean the chicken* (***habiller***): See pages 333–336 in the *Cuisine Foundations.*
2. Truss the chicken (***brider***) using a needle and kitchen twine. See pages 337–339 in *Cuisine Foundations.*
3. Set aside.

Cuisson

1. Place the chicken in a medium stock pot with the carrots, celery, studded onion, leek, garlic, and the ***bouquet garni***. Cover with cold water and bring to a gentle simmer over medium heat. Lightly season the liquid, cover the stock pot, and poach (***pocher***) the chicken for 45 minutes to 1 hour. Once cooked, decant the chicken from the pot, place it on a wire rack, and cover it with foil to keep it warm. Skim (***écumer***) the poaching liquid and strain it through a fine mesh sieve (***chinois***) into a clean saucepan. Reduce the poaching liquid down to 1 quart (1 liter) over medium heat.

Sauce Suprême

1. Reduce the cream by half in a small saucepan over low heat.
2. Melt the butter in a medium pan over medium heat. When the butter begins to foam, add the flour and stir well until the mixture begins to bubble. Cook for 1 or 2 minutes to obtain a white roux (*be careful not to let it color*). Transfer to a dish and set aside to cool.
3. Whisk the cold roux, piece by piece, into the reduced poaching liquid until smooth. Turn the heat up to medium high and bring the liquid to a boil, stirring constantly. Reduce the heat to low and simmer the ***sauce suprême*** gently for 20 minutes. Strain through a fine mesh sieve (***chinois***) into a clean saucepan. Taste and adjust the seasoning. Add the reduced cream to the sauce and simmer gently until it is thick enough to coat the back of a spoon (***à la nappe***). Adjust the seasoning and finish the sauce with lemon juice to taste. Pat (***tamponner***) the surface with cold butter (*on the end of a fork*) to prevent a skin from forming and reserve the sauce in a bain marie.

Riz Pilaf

1. Melt the butter in a medium saucepan over low heat and sweat (***suer***) the onion until transparent. Add the rice and cook it (*stirring gently*) until it turns opaque and shiny (***nacrer***). Add the cold water and the ***bouquet garni***. Increase the heat to medium high and, just before it begins to boil, remove the pan from the heat. Cover the pan with a parchment paper lid (***cartouche***) and place in the oven. Cook the rice in the oven for 17 minutes, then remove the pan, lift the cartouche, and dot the rice with butter. Replace the cartouche until the butter has melted. Gently separate the grains with the prongs of a carving fork.

To Serve

1. Remove the string from the chicken, then carefully remove the skin. Fill the neck cavity with rice pilaf.
2. Make a bed of rice on the bottom of a hot serving platter. Place the chicken on top (*breast side up*) and coat the bird with sauce. Fill the rear cavity with a bouquet of fresh parsley. Serve the remainder of the rice on the side.

Quantity		*Ingredient*
U.S.	*Metric*	***Base***
1 pc	1 pc	Chicken [*1.5 kg (3 lb), giblets removed*]
		Salt and white ground pepper
		Garniture Aromatique
2 pcs	2 pcs	Carrots, unpeeled, whole
1 pc	1 pc	Celery stalk, whole
1 pc	1 pc	Leek
1 pc	1 pc	Onion, studded with one clove
1 pc	1 pc	Garlic head, halved
1 pc	1 pc	***Bouquet garni***
		Sauce Suprême
1 ¾ oz	50 g	Flour
1 ¾ oz	50 g	Butter
1 qt	1 L	Cooking liquid
12 fl oz	350 g	Cream
1 pc	1 pc	Lemon, juice of
		Salt and pepper
		Riz Pilaf
1 ¼ oz	40 g	Butter
3 ½ oz	100 g	Onion, finely chopped (***hacher***)
½ lb	250 g	Long grain rice
12 ½ fl oz	375 mL	Water
2 ½ oz	80 g	Butter
		Bouquet garni
		Salt and white ground pepper
		To Serve
1 bunch	1 bunch	Fresh parsley

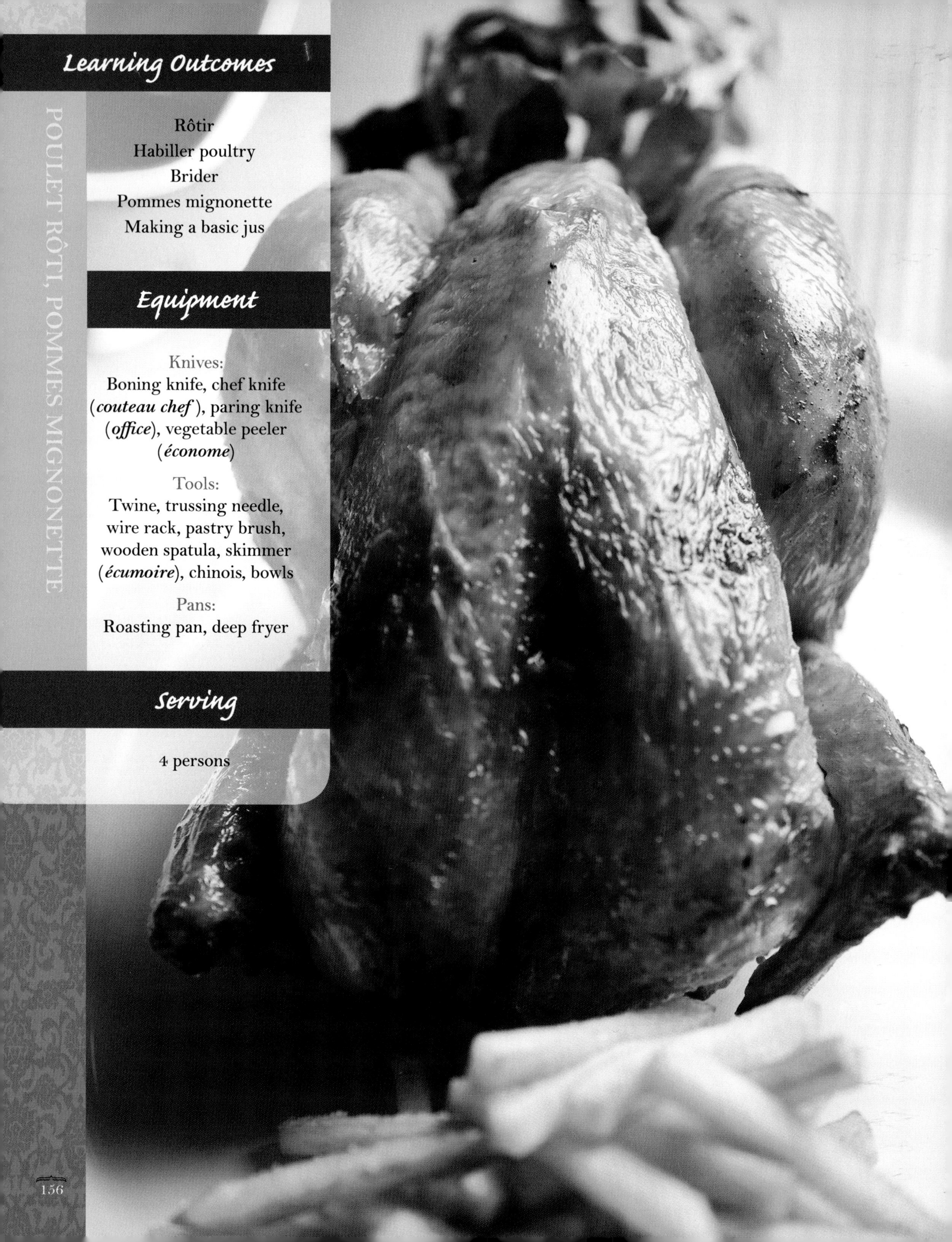

Learning Outcomes

Rôtir
Habiller poultry
Brider
Pommes mignonette
Making a basic jus

Equipment

Knives:
Boning knife, chef knife (*couteau chef*), paring knife (*office*), vegetable peeler (*économe*)

Tools:
Twine, trussing needle, wire rack, pastry brush, wooden spatula, skimmer (*écumoire*), chinois, bowls

Pans:
Roasting pan, deep fryer

Serving

4 persons

POULET RÔTI, POMMES MIGNONETTE

Roast Chicken with Mignonette Potatoes

Quantity		*Ingredient*
U.S.	*Metric*	*Base*
1 pc	1 pc	Chicken [*3 lb (1.5 kg)*]
3 ½ fl oz	100 g	Unsalted butter, clarified
		Salt and white pepper
		Jus
0.5 pc	0.5 pc	Garlic head
1 pc	1 pc	***Bouquet garni***
0.5 pc	0.5 pc	Carrot, cut into ***mirepoix***
0.25 pc	0.25 pc	Onion, cut into ***mirepoix***
As needed	As needed	Water
		Salt
		Pepper
		Salt and pepper
		Pommes Mignonette
4 pcs	4 pcs	Potatoes, large
		Salt
		To Serve
1 bq	1 bq	Watercress

Method

1. Preheat the oven to 450°F (230°C).
2. Preheat the deep fryer to 320°F (160°C).

Base

1. *Clean the chicken* (***habiller***): See pages 333–336 in *Cuisine Foundations.*
2. Truss the chicken (***brider***) using a needle and kitchen twine. See pages 337–339 in *Cuisine Foundations.*

Cuisson

1. Place the chicken on its back on a wire rack set in a roasting pan. Cut the butter into pieces and arrange on top of the chicken and season with salt and pepper. Place in the hot oven, and reduce the temperature to 375°F (190°C). Roast until golden brown, basting frequently. The chicken is cooked when the juices run clear when pierced at the thigh joint or when lifted from the pan. Once cooked, transfer the chicken on its wire rack onto a plate, cover with foil, and reserve in a warm place.

Jus

1. Place the roasting pan on the stovetop over medium heat to brown the cooking juices (***pincer les sucs***). Add the vegetables and trimmings and stir well until nicely colored. Carefully pour off any excess oil (***dégraisser***) and add enough water and the *bouquet garni* to just cover the bottom of the pan. Scrape well to dissolve the juices and cook until almost dry. Repeat with more water 4 or 5 times. The final time, add water to three-quarters the height of the solids and allow to reduce by 25%. Strain into a clean pan, then mount with browned butter (***beurre noisette***). Season to taste with salt and pepper and keep warm.

Pommes Mignonette

1. Heat the deep fryer to 320°F (160°C). Peel and trim the potatoes and cut into sticks 8 cm (3 ¼ inches) long and ¼" square. Rinse in cold water to remove excess starch and drain well. Deep fry until cooked but not colored (***blanchir***). Drain and set aside on a paper towel. When ready to serve the pommes mignonette, increase the heat of the deep fryer to 360°F (182°C). Plunge the blanched potatoes into the hot oil and cook until golden. Drain and transfer to paper towel and season with salt. Serve immediately.

To Serve

1. Remove the kitchen twine from the chicken and return the chicken to its cooking rack (*over a pan to catch the drippings*). Brush it with its cooking fat and return it to the oven for 5 minutes (*to crisp up the skin and reheat the bird*).
2. Place the roast chicken on a hot platter and brush it with clarified butter (***lustrer***). Fill the rear cavity with a bouquet of watercress. Serve the ***pommes mignonette*** on the side and the ***jus*** in a gravy boat.

Note: *Traditionally, the maître d'hôtel would cut and serve the chicken to the guests to ensure that each plate contains a piece of white and dark meat, with the bones.*

POULET SAUTÉ BOIVIN

Learning Outcomes

Habiller poultry
Détailler poultry à cru
Making a white stock
Pommes cocotte
Étuver

Equipment

Knives:
Cleaver, boning knife, chef knife (*couteau chef*), turning knife, paring knife (*office*), vegetable peeler (*économe*)

Tools:
Bowls, wooden spatula, skimmer (*écumoire*), chinois, wire rack

Pans:
Medium stock pot, large sauté pans, large saucepan

Serving

8 persons

FYI...

Classically, the backbone and ribs left over from cutting up the chicken are named le bateau (the boat). Poulet boivin is classically served reconstructed around the roast bateau.

There is no equivalent to the term *étuver*. *Étuver* is usually applied to vegetables that are cooked in some fat with perhaps the addition of a little liquid, covered over low heat until tender.

POULET SAUTÉ BOIVIN

Sautéed Chicken and Artichokes

Method

Fond de Volaille

1. Degorge the veal knuckles for 24 hours in several changes of cold water (*keep refrigerated*).
2. *Blanch the veal knuckles:* Place them in a large pot of cold water and bring to a boil over high heat. Boil the veal for 10 minutes, skimming the surface of impurities and fat (***écumer, dégraisser***). Drain the veal knuckles and refresh them under cold running water.
3. Place the chicken and veal knuckles in a medium stock pot. Add the carrots, leeks, celery, peppercorns, ***bouquet garni***, and the studded onion. Cover the ingredients in cold water and bring to a simmer over high heat. Reduce the heat to medium and let the stock cook for 2 to 4 hours, regularly skimming (***écumer***) the surface of impurities and fat (***dégraisser***). Strain the stock through a fine mesh sieve (***chinois***) and reserve the meat and fowl, for another use.

Optional: *For extra flavor and texture, the stock can be reduced to the amount needed for the recipe.*

Cuisson

1. *Clean the chicken* (***habiller***): See pages 333–336 in *Cuisine Foundations.* Make an incision (*right to the bone*) around the end of the drumstick. Scrape the flesh and tendons away from the bone. Chop off the tip of the drumsticks using a cleaver. Push the flesh back to expose the bone (***manchonner***). Cut each chicken into 8 pieces. Separate the wings, thighs, drumsticks and breasts, then cut each in two. Pat the chicken pieces dry and season them with salt and pepper. Heat the clarified butter in 2 large sauté pans over medium-high heat. In one pan, sauté the breasts and wings; in the other, the thighs and drumsticks. Sauté the chicken pieces until the skin is golden and the juices run clear from the thickest parts. Transfer to a wire rack to rest.

Sauce

1. Pour the excess fat (***dégraisser***) out of the pans. Place the pans on medium heat for 1 to 2 minutes to concentrate (*without burning*) the cooking residues (***pincer les sucs***). Deglaze each pan with a liter (a quart) of the stock. Scrape well to dissolve the browned cooking juices (***sucs***), then pour the stock from one pan into the other. Reduce the combined liquid to 1 quart (1 liter) over medium heat, skimming (***écumer***) the surface regularly. Add the cream and simmer gently until the sauce is thick enough to coat the back of a spoon (***à la nappe***). Taste and adjust the seasoning. Finish the sauce with lemon juice to taste.

Garniture

1. ***Glacer à brun*** *the pearl onions:* Place the onions in a sauté pan large enough to hold them in a single layer. Pour in enough cold water so that the onions are two-thirds immersed. Add the sugar and butter and season with salt. Cover with a parchment paper lid (***cartouche***). Cook the onions over low heat until all the water has evaporated and the sugar and butter mixture has begun to caramelize. Roll the onions in the resulting syrup to glaze and evenly coat. Reserve the onions in a clean bowl at room temperature.
2. *Cook the* ***pommes cocotte***: Turn the peeled potatoes to ***cocotte*** size (*5 cm long, about 20 g*). Place the potatoes in a large saucepan of cold, salted water and bring to a boil over high heat. Blanch the potatoes for 5 minutes and drain. Heat the clarified butter in a large sauté pan over medium-high heat. Sauté the potatoes, tossing them every few minutes, until they are golden on all sides and soft in the center. Season them to taste and set aside.
3. *Cook the artichoke bottoms* ***à l'étuvée***: Turn the artichokes and rub them with lemon to prevent discoloring (see pages 160–164 in *Cuisine Foundations*). Using a melon baller or spoon, scrape out the choke and discard (***évider à cru***). Reserve the artichoke bottoms in a mixture of lemon juice and water.
4. Drain the artichoke bottoms and cut into wedges. Heat the clarified butter in a large sauté pan over low heat. Add the lemon juice, the artichokes, and season lightly. Stir well and cover the pan and cook the artichokes (***étuver***) over low heat. Cook the artichokes until they are soft enough to be easily pierced with the tip of a knife (*15 to 20 minutes*). Adjust the seasoning, cover the pan, and set aside.

To Serve

1. Reheat the chicken with the onions, potatoes, and artichokes. When hot, cover with the sauce. Transfer to a heated serving platter and sprinkle with chopped parsley. Mount the sauce with cold butter.

Quantity		*Ingredient*
U.S.	*Metric*	*Fond de Volaille*
1 pc	1 pc	Chicken [*3 lb (1.5 kg)*]
7 pcs	7 pcs	Veal knuckles [*½ lb (250 g each)*]
1 pc	1 pc	Onion (*clove studded*)
3 pcs	3 pcs	Cloves
2 pcs	2 pcs	Carrots, large
1.5 pcs	1.5 pcs	Leek
2 pcs	2 pcs	Celery stalks
20 pcs	20 pcs	Black peppercorns
1 pc	1 pc	***Bouquet garni***
5 qt	5 L	Water
		Cuisson
2 pcs	2 pcs	Chickens [*3 lb (1.5 kg each)*]
5 fl oz	150 g	Butter, clarified
		Salt and white pepper
		Sauce
1 qt	1 L	Fond de volaille
8 ½ fl oz	250 mL	Cream
1 pc	1 pc	Lemon, juice of
		Salt and white pepper
		Garniture
40 pcs	40 pcs	Pearl onions
1 ¾ oz	50 g	Butter
3 tsp	15 g	Sugar, granulated
40 pcs	40 pcs	Potatoes, turned to ***cocotte*** size
5 oz	150 g	Butter, clarified
8 pcs	8 pcs	Artichokes, medium
2 ½ oz	75 g	Butter, clarified
1 pc	1 pc	Lemon, juice of
		Salt and white pepper
		To Serve
½ lb	250 g	Butter
5 brs	5 brs	Parsley

Les Poissons et Coquillages

(Fish and Shellfish)

Recipes

(Fish and Shellfish)

Beignets de gambas
—Shrimp fritters

Crabe farci à la thermidor
—Stuffed crab thermidor

Cuisses des grenouilles, sauce poulette
—Frog legs with white sauce

Darne de saumon au beurre-blanc
—Salmon steak with white butter sauce

Escabèche de maquereaux
—Marinated mackerel

Escalope de saumon en papillotte
—Salmon escalope baked in a paper parcel

Escargots bourguignons
—Snails with garlic and parsley butter

Filets de sole bonne-femme
—Sole fillet in velouté sauce

Homard à l'armoricaine
—Lobster in a white wine and tomato sauce with tarragon

Lotte rôtie, fenouil braisé
—Roasted monkfish with braised fennel

Matelote de doré au vin rouge
—Yellow pike stewed in red wine

Merlan frit Colbert
—Breaded and fried whiting

Moules marinières and Derivatives
—Steamed mussels

Mouclade
—Steamed mussels with curry

Moules a la crème
—Steamed mussels with cream

Petit coulibiac de saumon frais et fumé
—Fresh and smoked salmon baked in a leavened dough

Quenelles de brochet, sauce cardinal
—Pike dumplings with a lobster sauce

Saint-Jacques à la provençales
—Scallops with tomatoes

Saumon cru à l'aneth
—Raw marinated salmon with dill

Sole et coquillages en nage
—Sole and shellfish in their cooking broth

Sole meunière
—Sauteed sole with lemon and parsley butter

Terrine chaude de poisson, sauce Américaine
—Hot fish mousse with shellfish and tomato sauce

Truite en Bellevue
—Cold poached trout in aspic

Tronçons de turbotin à la Dugléré
—Turbot steaks with a tomato and white wine sauce

Vol-au-vent marinière
—Seafood with white sauce in a puff pastry case

Learning Outcomes

Pâte à frire
Marinating seafood
Hacher
Deveining prawns

Equipment

Knives:
Paring knife (*office*), slicing knife (*éminceur*)

Tools:
Mixing bowls, whisk, tamis, araignée, deep fryer

Serving

2 persons

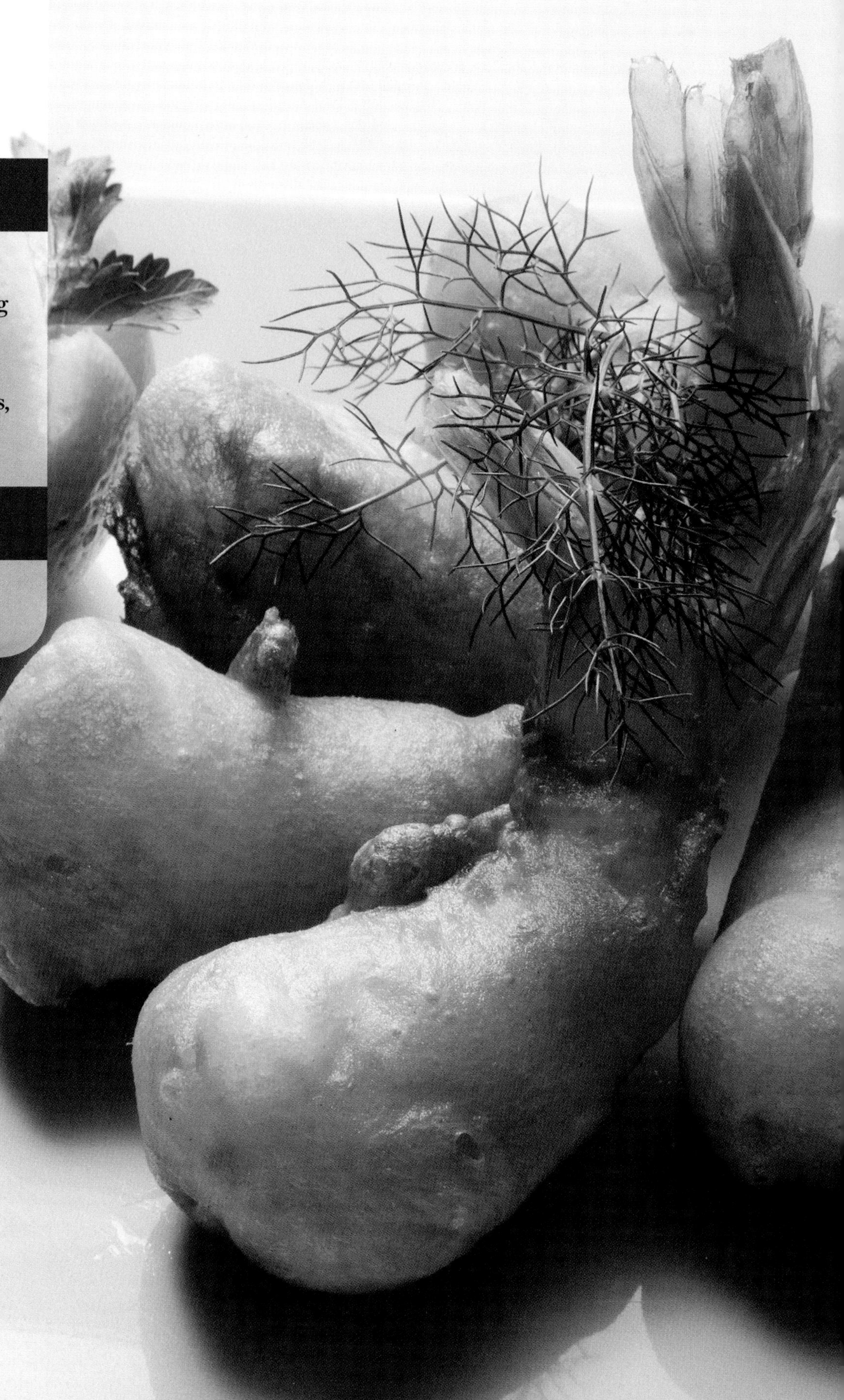

BEIGNETS DE GAMBAS

Shrimp Fritters

Method

1. Preheat the deep fryer to 370°F (190°C).
2. Remove the prawn heads and peel and devein the tails. Reserve the heads and shells for another use.

Marinade

1. Mix the garlic, basil, and olive oil together. Season the prawns and place them in a recipient with a cover, pour the marinade over the prawns, cover, and place the container in the refrigerator for a minimum of 2 hours (preferably overnight).

Pâte à Frire

1. Combine and sift the flour, cornstarch, and baking powder through a sieve (***tamis***). Mix in the salt and pepper, then slowly pour in the water and mix with a whisk until just incorporated. Let the batter rest for 30 minutes to a few hours in the refrigerator.

Sauce

1. Mix the parsley, basil, dill, and garlic into the mayonnaise. Add curry powder and cayenne pepper to taste, then reserve the sauce covered in the refrigerator.

Cuisson

1. Decant the prawns from the marinade and pat them dry. One by one, hold the prawns by the tail and dip them into the ***pâte à frire***. Dip the battered part in the deep fryer and drag the prawn back and forth in the hot oil before letting go.
2. When the tails float to the top and have a nice golden color, transfer them to a paper towel to drain and salt them immediately.

To Serve

1. Arrange the prawns on a heated plate or platter on a white napkin or paper. Decorate with a carved lemon and fresh parsley. Serve the sauce on the side.

Quantity		Ingredient
U.S.	*Metric*	*Marinade*
20 pcs	20 pcs	Prawns
2 pcs	2 pcs	Garlic cloves, finely chopped (***hacher***)
1 ¾ fl oz	50 mL	Olive oil
10 pcs	10 pcs	Basil leaves, finely chopped (***hacher***)
		Salt, pepper, cayenne pepper
		Pâte à Frire
3 ¼ oz	100 g	Flour
1 ¾ oz	50 g	Cornstarch
2 tsp	8 g	Baking powder
		Salt and pepper
10 ¼ fl oz	300 mL	Water
		Sauce
9 oz	250 g	Mayonnaise (*see page 236 in Cuisine Foundations*)
½ oz	15 g	Parsley, finely chopped (***hacher***)
½ oz	15 g	Basil, finely chopped (***hacher***)
¼ oz	10 g	Dill, finely chopped (***hacher***)
2 pcs	2 pcs	Garlic cloves, finely chopped (***hacher***)
Pinch	Pinch	Curry powder
Pinch	Pinch	Cayenne pepper
		To Serve
2 pcs	2 pcs	Lemons
5 brs	5 brs	Parsley or dill

FYI...

Do you follow proper shrimp etiquette? Fried shrimp should be eaten using a fork and knife. If served on a skewer or toothpick, either of these two items should be used as your utensil. Shrimp that are served with an accompaniment of cocktail sauce should be dipped with a fork (no double dipping, though). However, if shrimp are served with their tails intact, it is perfectly acceptable to hold the tail and eat using your fingers. "Pinch the tail, suck the head," as is said in New Orleans.

CRABE FARCI À LA THERMIDOR

Learning Outcomes

Preparing a crab
Preparation "à la thermidor"
Gratiner

Equipment

Knives:
Chef knife (*couteau chef*), scissors

Tools:
Araignée, wooden spoon, whisk, cheese grater, cutting board, tweezers

Pans:
Large stock pot, medium saucepan, bain marie, baking sheet, large sauté pan

Serving

2 persons

FYI...

From the Greek *thermos* meaning "heat" and *doron* meaning "gift," the word *thermidor* was adopted into the French revolutionary (Republican) calendar and referred to the period between July 19 and August 17. Later the term was used as the title of Victorian Sardou's scandalous 1891 play *Thermidor*. The preparation known as thermidor was subsequently named at the height of this scandal. Generally speaking, to prepare something "thermidor" the shell of the crustacean (crab or lobster) must be large enough to act as a serving container for the prepared meat and sauce.

CRABE FARCI À LA THERMIDOR
Stuffed Crab Thermidor

Quantity		Ingredient
U.S.	*Metric*	*Base*
2 pcs	2 pcs	Crabs [*1 lb 10 oz (750 g) each*]
		Salt
		Sauce
2 oz	60 g	Butter
2 pcs	2 pcs	Shallots, finely chopped (***hacher***)
1 pt	500 mL	White wine
1 qt	1 L	Cream
3 ½ fl oz	100 mL	Béchamel sauce
5 brs	5 brs	Chervil, finely chopped (***hacher***)
Pinch	Pinch	Cayenne pepper
Pinch	Pinch	Mustard powder
1 ¾ oz	50 g	Gruyère cheese, grated
		Salt and white pepper
		To Serve
1 ¾ oz	50 g	Gruyère cheese, grated
		Coarse salt

Method

Base

1. Bring a large pot of salted water to a boil over high heat. Add the live crabs, and poach for 15 minutes, then refresh under cold running water. Twist off the legs, then extract and reserve the meat. Twist off the claws and reserve them intact for the final presentation.
2. Separate the top shell from the abdomen and reserve both parts separately. Blanch the top shell in boiling water for 10 minutes (to clean and prepare it for presentation), then scrape off the hairs around the edge of the shell. Reserve the shell for the ***montage***.
3. *Extract the meat from the abdomen:* Cut along the line that runs down the middle of the abdomen (*this will produce two halves*). To gain access to the flesh, lay the two halves cut side down on the cutting board and cut in half again. With the aide of tweezers, pull the flesh out of the cartilage cells and set aside. Discard the carcass.

Note: *Be careful not to use any of the gills. They resemble long, gray fingers and are inedible.*

Sauce

1. Melt the butter in a large sauté pan over medium heat and sweat (***suer***) the shallots until they begin to color. Deglaze the pan with white wine and reduce by half. Add the cream and reduce by half. Add the béchamel sauce and cook for 10 minutes until creamy. Mix in the crab meat, chervil, and cayenne pepper. Season to taste and let the sauce cook gently for 5 minutes to thicken it slightly and to develop its flavor. Remove the pan from the heat and mix in the mustard powder and the grated cheese.

Finishing

1. Bring a small pot of salted water to a boil over high heat (***chauffante***).
2. Make a small mound of wet, coarse salt on a baking sheet and set the crab shell, cavity side up, on the salt (the salt provides a base to keep the shell from spilling). Fill the shell with the crab mixture, sprinkle the surface with grated cheese, and place it in the oven for 10 to 15 minutes or until hot. Finish under the salamander to brown the cheese (***gratiner***). Meanwhile, reheat the crab claws in the ***chauffante*** for 3 to 4 minutes.

To Serve

1. Make a small mound of wet, coarse salt on a hot plate. Transfer the crab to the salt mound (to keep it from rocking) and arrange the claws under the crab to make it appear whole.

CUISSES DES GRENOUILLES SAUCE POULETTE

Learning Outcomes

Réduire
Pocher
Lier
Monter au beurre
Pluches

Equipment

Knives:
Paring knife (*office*), slicing knife (*éminceur*)

Tools:
Bowls, fine chinois, colander, skimmer (*écumoire*), ladle, whisk

Pans
Sautoirs, Sauce pan (*russe*), bain marie

Serving

4 persons

FYI...

Frog legs have been part of the human diet since the Neolithic era. They were made popular in France in the 10th century when monks, who were forced to observe up to 200 days of fasting a year, got frogs classified as fish in order to vary their diet. Classifying frog legs as fish applies in the culinary world to this day. In the 19th century, when Escoffier was head chef at the Savoy in London, he decided to trick British guests at a dinner held for the Prince of Wales. He euphemistically named frogs legs, *cuisses de nymphes à l'aurore* (nymph thighs aurora) and served them to 600 unsuspecting diners who devoured them greedily—much to the amusement of the Prince, who, being partial to frog legs, was aware of the hoax.

CUISSES DES GRENOUILLES SAUCE POULETTE

Frog Legs with White Sauce

Method

Preparing the Frog Legs

1. Cut down the middle to separate the legs.
2. *Degorge the legs:* Cover the frog legs with cold milk in a deep container with a cover and place in the refrigerator to degorge for up to 12 hours.

Cuisson

1. Butter a shallow dish. In a medium saucepan over medium heat, bring the fish ***fumet*** and mushroom ***fumet*** to just under a simmer (*the liquid should be as hot as it can be without bubbling*). Add the ***bouquet garni,*** season to taste, and poach the frog legs for 10 minutes. When cooked through, transfer the frog legs to the butter dish. Keep the frog legs covered and warm.

Sauce Poulette

1. Remove the poaching liquid from the heat, pour in 6 ¾ fl oz (200 mL) of cream, and return to medium heat. Reduce the liquid by one-half.
2. Lightly beat the egg yolks with the remaining 4 tsp (20 mL) of cream. Add 1 or 2 ladles of hot liquid into the yolk mixture and stir well to temper. Off the heat, pour the tempered egg yolks into the hot liquid, stirring continuously with a whisk. Turn the heat down to low and stir the sauce in a figure 8 motion using a wooden spoon. Cook the sauce until it is thick enough to coat the back of the spoon (***à la nappe***), then strain it through a fine mesh sieve (***chinois***).
3. Mount the sauce with cold butter and mix in lemon juice and chopped parsley. Taste and adjust the seasoning.

To Serve

1. Transfer the frog legs to the sauce to reheat them and decant them into a warm serving dish.
2. Coat them (***napper***) with the ***sauce poulette*** and decorate the dish with a sprig of fresh parsley.

Quantity		*Ingredient*
U.S.	*Metric*	*Preparation*
24 pcs	24 pcs	Frog legs
1 pt	500 mL	Milk
		Cuisson
¾ oz	20 g	Butter
8 fl oz	250 mL	Fish ***fumet***
4 fl oz	125 mL	Mushroom ***fumet***
1 pc	1 pc	***Bouquet garni***
		Salt and pepper
		Sauce Poulette
8 fl oz	250 mL	Cream
2 pcs	2 pcs	Egg yolks
¾ oz	25 g	Butter
1 pc	1 pc	Lemon juice
¼ oz	10 g	Parsley, finely chopped (***hacher***)
		To Serve
1 br	1 br	Parsley

DARNE DE SAUMON AU BEURRE-BLANC

Learning Outcomes

Griller
Détailler une darne
Émulsion
Quadriller
Beurre monté

Equipment

Knives:
Vegetable peeler (*économe*), paring knife (*office*), slicing knife (*éminceur*), serrated knife

Tools:
Cutting board, ladle (*louche*), whisk, wire rack, metal spatula

Pans:
Sauce pan (*russe*), bain marie, grill

Serving

4 persons

FYI...

Thicker than a *tronçon*, a *darne* is a thick cut that includes the central bone and is suited to larger round fish such as salmon and tuna. The word *darne* is an evolution of the Breton word "*dam*" which simply means "piece."

DARNE DE SAUMON AU BEURRE-BLANC

Salmon Steak with White Butter Sauce

Method

1. Prior to preheating, clean and grease the grill.
2. Cut the salmon into ***darnes***.

Marinade Sèche

1. Mix all the marinade ingredients together. Place the salmon in a shallow dish and pour the marinade over the top. Cover the dish in plastic wrap and leave to marinate in the refrigerator for a minimum of 2 hours.

Mousseline de Fenouil

1. *Prepare a* ***blanc de cuisson:*** Mix the flour and lemon juice into enough cold water to cover the fennel bulb. Season with a large pinch of salt.
2. Bring the liquid to a boil in a small saucepan over medium heat and cook the fennel until it is easily pierced by a knife (10 to 15 minutes). Refresh the fennel under cold running water.
3. Peel off the outer layer of the fennel, cut the bulb in half, and remove and discard the root. Finely slice the fennel (***émincer***).
4. Melt the butter in a medium sauté pan over low heat and sweat (***suer***) the fennel until it begins to soften. Season lightly with salt and pepper, cover the pan, and cook the fennel until soft (***étuver***). Wet (***mouiller***) the pan with the fish ***fumet*** and turn the heat up to medium. Cook covered until the fennel absorbs all the liquid, then pour in the cream. Once the cream comes to a boil, press (***fouler***) the contents of the pan through a fine mesh sieve (***chinois***) to obtain a fine purée / mousseline. Stir in the chopped dill and parsley. Reserve the mousseline in a bain marie.

Cuisson

1. Pat the salmon steaks dry and sprinkle them generously with salt and pepper. Place the presentation side of the fish diagonally on a hot grill and cook until the grill leaves dark lines on the flesh. Turn the ***darnes*** 90° and cook again until the grill leaves dark lines on the flesh (***quadriller***). Flip the ***darnes*** and repeat the same operation. Remove the fish from the grill and place it on a wire rack by the grill to rest for 5 minutes.

Note: *Keep the fish warm.*

Beurre-blanc

1. Season the lemon juice and reduce it by half in a medium pan over medium heat. Increase the heat to high and whisk in the cold butter, piece by piece, until melted and incorporated.

To Serve

1. Using a carving fork, pierce the central bone of the ***darnes***, and pull up to remove. Place a ***darne*** on a hot plate and quickly shape the mousseline into quenelles (*using two equally sized spoons*) and arrange on the side of the plate. Spoon some ***beurre-blanc*** onto the plate, brush the fish with melted butter to add shine (***lustrer***), and decorate the plate with a sprig of parsley. If desired serve with fresh lemon.

Quantity		*Ingredient*
U.S.	*Metric*	***Base***
4 pcs	4 pcs	Salmon, cut into ***darnes*** [*5 oz (150 g) each*]
1 fl oz	30 mL	Oil
		Marinade Sèche
3 ½ fl oz	100 mL	Vegetable oil
3 brs	3 brs	Thyme, rubbed (***émietté***)
1 pc	1 pc	Bay leaf
½ pc	½ pc	Lemon, juice of
		Salt and pepper
		Mousseline de Fenouil
1 pc	1 pc	Fennel bulb [*12 oz (350 g)*]
¼ oz	10 g	Flour
2 tsp	10 mL	Lemon, juice of
1 ¾ oz	50 g	Butter
4 fl oz	125 mL	Fish ***fumet***
8 fl oz	250 mL	Cream, reduced by half
5 brs	5 brs	Dill (***hacher***)
2 brs	2 brs	Parsley (***hacher***)
		Salt and pepper
		Beurre-blanc
2 pcs	2 pcs	Lemon, juice of
7 oz	200 g	Butter
		Salt and pepper
		To Serve
2 brs	2 brs	Parsley or dill
1 pc	1 pc	Lemon (*optional*)

ESCABÈCHE DE MAQUEREAUX

Learning Outcomes

Marinade cuite
Habiller and filleting a round fish

Equipment

Knives:
Paring knife (*office*), filet de sole, slicing knife (*éminceur*), scissors

Tools:
Wooden spoon, mixing bowls, zester

Pans:
Medium pan, hotel pan

Serving

4 persons

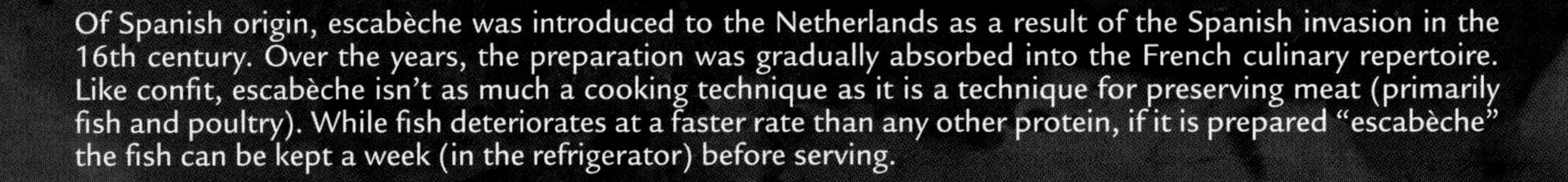

FYI...

Of Spanish origin, escabèche was introduced to the Netherlands as a result of the Spanish invasion in the 16th century. Over the years, the preparation was gradually absorbed into the French culinary repertoire. Like confit, escabèche isn't as much a cooking technique as it is a technique for preserving meat (primarily fish and poultry). While fish deteriorates at a faster rate than any other protein, if it is prepared "escabèche" the fish can be kept a week (in the refrigerator) before serving.

ESCABÈCHE DE MAQUEREAUX

Marinated Mackerel

Method

1. ***Habiller*** *the mackerel:* Using a pair of scissors, cut all the fins and trim the tail fin.
2. Fillet the fish using a filleting knife (discard the carcasses).

Marinade Cuite

1. Heat the olive oil in a medium pan over low heat and sweat (***suer***) the leeks and shallots until soft. Season lightly. Add the celery and carrot and sweat until soft. Add the garlic, rosemary, thyme, bay leaf, chili peppers, and lemon zest and sweat for 10 minutes. Wet (***mouiller***) the pan with the white wine and vinegar, season to taste, and simmer the marinade for 5 more minutes. Set aside, covered to keep it hot.

Cuisson

1. Pat the fillets dry and season with salt and pepper.
2. Heat the olive oil in a large sauté pan over high heat. Add the fillets, skin side down, and sear them just enough to color the skin but not enough to cook the flesh. Transfer the fillets to a deep dish and cover them with lemon slices. Being careful not to dislodge the lemon slices, gently pour the hot marinade into the dish, covering the fish completely. Allow to cool to room temperature, then cover and refrigerate for at least 12 hours before serving.
3. This dish can be stored for 1 week in the refrigerator.

To Serve

1. Serve cold.
2. Remove the fish from the marinade and cut into individual portions. Arrange the fish on a plate, skin side down. Cover the fish with fresh lemon slices and spoon the marinade on top. Decorate with fresh herbs.

Quantity		*Ingredient*
U.S.	*Metric*	*Cuisson*
2 pcs	2 pcs	Fresh mackerel [*8 oz (250 g) each*]
2 Tbsp	30 mL	Olive oil
1 pc	1 pc	Lemon, thinly sliced (***rondelles***)
		Salt and pepper
		Marinade Cuite
8 fl oz	250 mL	Olive oil
1 pc	1 pc	Leek whites, cut into ***brunoise***
2 pcs	2 pcs	Shallots, finely chopped (***hacher***)
½ pc	½ pc	Celery stalk, cut into ***brunoise***
1 pc	1 pc	Carrot, cut into ***brunoise***
2 pcs	2 pcs	Garlic cloves, finely chopped (***hacher***)
1 br	1 br	Rosemary
2 brs	2 brs	Thyme
1 pc	1 pc	Bay leaf
2 pcs	2 pcs	Small chili peppers, cut into ***brunoise***
2 pcs	2 pcs	Lemon zest, grated
8 fl oz	250 mL	Dry white wine
4 fl oz	125 mL	White wine vinegar
		Salt and pepper
		Finishing
1 pc	1 pc	Lemons, channeled (***canneler***), sliced (***rondelles***)
		Fresh herbs

ESCALOPE DE SAUMON EN PAPILLOTE

Learning Outcomes

Cuisson en papillote
Tomato concassée
Marinade sèche
Tapenade

Equipment

Knives:
Paring knife (***office***), slicing knife (***éminceur***), filet de sole

Tools:
Zester, ice bath, mixing bowl, mortar and pestle, pastry brush

Pans:
Large russe, large pan, small russe, medium russe, baking sheet

Serving

4 persons

FYI...

Escalope is a regional word from northeastern France that refers to either a shell from a nut or a snail. As a culinary term (meaning a slice cut on the bias), *escalope* dates back to the 17th century and referred strictly to a cut of veal. Today, an escalope can be cut from chicken, pork, turkey, veal, or fish.

The first *papillote* was made to honor a Brazilian balloonist at a banquet. The food was carefully wrapped in a parchment envelope and put in the oven to bake. Steam trapped inside the parchment caused the envelope to inflate like a balloon. The purpose of a *papillote* is to gently (not to mention healthily) steam the food inside the parchment. However, there is an entertainment factor in this preparation as well; when guests "pop" the parchment, the envelope deflates like a balloon, releasing the vaporous aroma of the food inside.

ESCALOPE DE SAUMON EN PAPILLOTE

Salmon Escalope Baked in a Paper Parcel

Method

1. Preheat the oven to 400°F (205°C).

Portugaise

2. Heat the olive oil in a medium sauté pan over low heat and sweat (***suer***) the chopped shallots and onion until translucent. Season lightly, add the chopped garlic, and sweat (***suer***) until dry. Add the tomato ***concassée*** and sweat (***suer***) until dry. Turn off the heat, add the basil, and adjust the seasoning. Set aside.

Marinade Sèche

1. Combine the garlic, the peppercorns, the blanched lemon zest, and stir in the lemon juice. Pour in the olive oil in a thin stream, whisking continuously to emulsify.
2. To infuse the flavors, heat the marinade in a small saucepan over medium heat until just hot to the touch [*104°F (40°C)*]. Taste and adjust the seasoning, then remove the marinade from the heat and let it cool to room temperature. In an appropriate-sized recipient, arrange the salmon in a single layer. Cover the fish in the marinade and leave it to marinate in the refrigerator for at least 1 hour.

Tapenade

1. Crush the olives, the garlic, the anchovy fillet, and the basil leaves with a mortar and pestle. Once these ingredients have formed a dark paste, mix in the olive oil. Taste, season, and set aside.

Sauce

1. Heat the olive oil in a medium pan over low heat and sweat (***suer***) the garlic until it is soft. Add the tomato pulp and peels and gently cook the mixture for 5 minutes. Wet (***mouiller***) with the fish ***fumet*** and simmer to reduce the liquid by half. Strain the sauce through a fine mesh sieve (***chinois***) and mount it with the ***tapenade***. Taste and adjust the seasoning and set aside.

Montage

1. Lay out a large piece of parchment paper or foil on a tray and spoon 2 ½ oz (70 g) of the *Portugaise* in the middle. Place a piece of marinated salmon on top of the *Portugaise* and place two branches of thyme on top of the salmon. Drizzle with a spoonful of marinade and season with salt and pepper.
2. *To seal the papillote using parchment paper:* With a pastry brush, wet the edges with egg white and seal, keeping the edges even. Fold the edges over about (¾ in.). Brush again with egg white and fold again. Make sure that the folds remain secure. If using foil, even the edges and fold 2 or 3 times.

Cuisson

1. Transfer the ***papillote*** to the hot oven to bake for 10 to 15 minutes. The ***papillote*** should puff up and be served straight out of the oven on a hot plate with the tapenade sauce on the side. The diner or server opens the ***papillote*** at the table.

Quantity		*Ingredient*
U.S.	*Metric*	
4 pcs	4 pcs	Salmon, cut into [5 ¼ oz (150 g)]
		Salt and pepper
		Portugaise
4 fl oz	120 mL	Olive oil
3 pcs	3 pcs	Shallots, finely chopped (***hacher***)
½ pc	½ pc	Onion, finely chopped (***hacher***)
2 pcs	2 pcs	Garlic cloves, finely chopped (***hacher***)
8 pcs	8 pcs	Tomatoes, peeled (***émonder***), seeded (***épépiner***), and diced (***concasser***)
1 bq	1 bq	Basil, shredded (***chiffonade***)
		Salt and pepper
		Marinade Sèche
3 pcs	3 pcs	Garlic cloves, finely chopped (***hacher***)
5 pcs	5 pcs	Peppercorns, crushed (***mignonette***)
1 pc	1 pc	Lemon zest, blanched, cut into ***brunoise***
1 pc	1 pc	Lemon, juice of
5 fl oz	150 mL	Olive oil
		Salt and pepper
		Tapenade
5 oz	150 g	Black olives (***Kalamata***)
2 pcs	2 pcs	Garlic cloves
1 pc	1 pc	Anchovy fillet
5 brs	5 brs	Basil
2 ½ oz	75 g	Olive oil
		Salt and pepper
		Sauce
4 tsp	20 mL	Olive oil
1 pc	1 pc	Garlic clove, finely chopped (***hacher***)
		Tomato peels, left over from ***concasser***
		Tomato pulp, left over from ***concasser***
1 ½ pt	750 mL	Fish ***fumet*** [reduced to 12 ½ fl oz (375 mL)]
		Salt and pepper
		Montage
2 pcs	2 pcs	Egg whites
8 brs	8 brs	Thyme

ESCARGOTS BOURGUIGNONS

Learning Outcomes

Beurre Bourguignon

Equipment

Knives:
Paring knife (*office*), slicing knife (*éminceur*), small offset spatula

Tools:
Pastry bag, small pastry tip

Pans:
Escargotière, frying pan

Serving

4 persons

FYI...

Evidence of grilled snails has been found in archeological digs that are dated as far back as 3000 BC. The first written evidence of snails as a food source dates back to ancient Rome where they were reared in special gardens called *cochleariae*. In these gardens, they were fattened on milk and grain (sometimes even wine) until they were too big to fit into their shells. The Romans were so fond of snails that they brought back exotic varieties from the four corners of their empire. Under Roman reign, the Gauls of ancient France also took a liking to these mollusks and ate them as a sweetened dessert. Over the centuries, snails have gone in and out of fashion. Their status as a gastronomic classic was sealed when Antonin Carême cooked them for the Czar of Russia in 1814. He prepared them in the manner he had learned in the Bourgogne region, with hot butter, parsley, and garlic. Ever since then, "*escargots bourguignons*" has been emblematic of French cuisine throughout the world.

ESCARGOTS BOURGUIGNONS

Snails with Garlic and Parsley Butter

Method

1. Preheat the oven to 400°F (205°C).

Beurre Bourguignon

1. Mix all the ingredients into the softened butter (***beurre pommade***) and transfer the mixture to a pastry bag fitted with a small plain tip.

Escargots

1. Melt the butter in a large sauté pan over medium heat and sweat (***suer***) the garlic and shallots until they are soft. Increase the heat to high and sauté the escargots for 2 to 3 minutes. ***Flambé*** the pan with cognac (be careful not to overcook the escargots; otherwise, they will turn rubbery). Set aside and cool to room temperature.
2. Pipe a portion of butter the size of a large pea into each shell. Fit an escargot into the buttered shell and cover with more butter, smooth the opening of the shell with a small offset spatula. Repeat for the remaining escargots. Cover all the shell openings with breadcrumbs and place them in an ***escargotière*** (*a special snail dish*).

To Serve

1. Place the escargots in the oven for 7 to 8 minutes.
2. Serve immediately.

Quantity		Ingredient
U.S.	*Metric*	*Beurre Bourguignon*
8 oz	250 g	Butter, softened (***beurre pommade***)
2 pcs	2 pcs	Shallots, finely diced (***ciseler***)
2 pcs	2 pcs	Garlic cloves (***puréed***)
1 ¾ oz	50 g	Parsley, finely chopped (***hacher***)
2 oz	60 g	Breadcrumbs, dried
½ tsp	3 g	Salt
Pinch	Pinch	Pepper
		Escargots
25 pcs	25 pcs	Snails (***escargots***), with shells
1 ¾	50 g	Butter
1 pc	1 pc	Garlic clove, finely chopped (***hacher***)
2 pcs	2 pcs	Shallots, finely diced (***ciseler***)
1 fl oz	30 mL	Cognac, to ***flambé***

FILETS DE SOLE BONNE-FEMME

Learning Outcomes

Habiller a sole
Lever les filets on a flat fish
Making a velouté
Making a liaison with egg yolks
Glacer

Equipment

Knives:
Filleting knife, paring knife (*office*), slicing knife (*éminceur*)

Tools:
Whisk, sieve, chinois, mixing bowls,

Pans:
Ovenproof dish, medium pan, large pan, roasting pan

Serving

4 persons

FYI...

Bonne femme is French for "good wife." It indicates to the reader that the dish should have a simple presentation—as if it were served at home. Traditionally, this dish is served in the same vessel that it was cooked in.

Method

1. Preheat the oven to 350°F (180°C).
2. Skin the fish and remove the fillets using a filleting knife. See pages 282–285 in *Cuisine Foundations.* Place the fillets in milk to degorge. Reserve.

Potatoes

1. Place the washed, unpeeled potatoes in a large pot of cold salted water over high heat. Bring to a boil and cook until the tip of a knife can be easily inserted. Drain and keep hot in a covered container.

Cuisson

1. Reduce the fish ***fumet*** to 10 fl oz (300 mL) in a small pan over medium heat. Reduce the white wine to 300 mL (10 fl oz). Combine the reduced liquids and set aside.
2. Generously butter a roasting pan. Sprinkle the bottom with finely chopped shallots. Sprinkle with thinly sliced mushrooms, then parsley. Fold the fillets in three and arrange on top of the ***aromates***. Heat the ***fumet*** and white wine reduction and, once hot, pour over the sole. Cover with a buttered piece of parchment paper and place in the oven. Cook the fish about 10 minutes or until they turn a solid white.
3. Remove the roasting pan from the oven. Decant the sole fillets and reserve on a rack (*covered in foil to keep warm*). Gently strain the cooking liquid through a fine mesh sieve (***chinois***). Reserve the cooking liquid and vegetables separately.

Velouté

1. Stir the ***beurre manié***, piece by piece, into the hot cooking liquid until combined. Bring the velouté to a boil over medium heat, stirring constantly until thickened. Reduce the heat to low and simmer the velouté gently for 10 minutes, then remove it from the heat and stir it back and forth until it stops steaming (***vanner***).
2. In a small bowl, temper the egg yolk by mixing in a little velouté sauce. Off the stove and whisking continuously, pour the tempered yolk into the velouté. Stir the sauce over low heat until it is thick enough to coat the back of the spoon (***à la nappe***). Strain the velouté sauce through a fine mesh sieve (***chinois***), mount with cold butter, and mix in the chopped parsley.

To Serve

1. Peel and slice the cooked potatoes.
2. Butter a clean gratin dish and cover the bottom with the reserved garniture in an even layer. Arrange the cooked sole fillets on top and spoon the ***velouté*** over the top to completely coat the contents of the dish (***napper***). Line the edge of the dish with the potato slices. Place the gratin dish under the salamander until brown (***glacer***).
3. Serve immediately.

Quantity		*Ingredient*
U.S.	*Metric*	
12 pcs	12 pcs	Dover sole fillets
		Velouté
1 pt	500 mL	Fish ***fumet***
1 pt	500 mL	White wine
1 ¾ oz	50 g	***Beurre manié*** (*see page 225 in Cuisine Foundations*)
		Cuisson
1 ¾ oz	50 g	Butter
1 ¾ oz	50 g	Shallots, finely chopped (***hacher***)
8 oz	250 g	Mushrooms, peeled, sliced (***émincer***)
1 oz	30 g	Parsley, finely chopped (***hacher***)
		Garniture
5 oz	150 g	Small potatoes, with their skins
		Coarse salt
		Salt and white pepper
		To Serve
3 pcs	3 pcs	Egg yolks
3 ½ oz	100 g	Butter

HOMARD À L'ARMORICAINE

Learning Outcomes

Détailler a lobster
Saisir
Cardinaliser
Mirepoix
Suer
Pincer
Déglacer
Mouiller
Mijoter
Coral and tomalley
Lier with a beurre manié

Equipment

Knives:
Cleaver, vegetable peeler (*économe*), paring knife (*office*), slicing knife (*éminceur*), scissors

Tools:
Bowls, fork, wooden spatula, plastic spatula, regular chinois, fine chinois, colander, cutting board, skimmer (*écumoire*), ladle, rolling pin

Pans:
Sauce pan (*russe*), bain marie, rondeau, large sauté pan

Serving

4 persons

FYI...

By some accounts, both the name and nature of this 19th-century preparation were the invention of necessity: One night around closing time in the Paris restaurant called Noël Peters, a party came in and demanded to be fed. There was little time and little left in the kitchen other than a few lobsters and some standard kitchen ingredients. Pierre Fraise, chef and owner of the restaurant, invented the recipe on the spot. This much of the story most food historians can agree upon; when the subject of the name of the recipe comes up, nationalistic sparks begin to fly. The term *armoricaine* refers to Armoric, which is the ancient Celtic name for Britanny—where Chef Fraise's lobsters may indeed have been caught. Today the dish is still prepared to order, and a well-prepared homard l'amoricaine should take a good kitchen 20–25 minutes to prepare once the order is placed.

HOMARD À L'ARMORICAINE

Lobster in a White Wine and Tomato Sauce with Tarragon

Quantity		*Ingredient*
U.S.	*Metric*	*Chemisage*
4 pcs	4 pcs	Live lobsters [*1 lb 2 oz (500 g) each*]
4 fl oz	120 mL	Olive oil
Pinch	Pinch	Cayenne pepper
		Salt and pepper
		Sauce Armoricaine
2 ¾ oz	80 g	Tomato paste
1 pc	1 pc	Onion, cut into ***mirepoix***
3 pcs	3 pcs	Shallots, cut into ***mirepoix***
1 pc	1 pc	Carrot, cut into ***mirepoix***
4 pcs	4 pcs	Tomatoes, peeled (***émonder***), seeded (***épépiner***), and diced (***concasser***)
1 pc	1 pc	***Bouquet garni***
6 pcs	6 pcs	Garlic cloves
10 pcs	10 pcs	Stems of fresh tarragon, finely chopped (***hacher***)
10 pcs	10 pcs	Stems of fresh chervil, finely chopped (***hacher***)
10 pcs	10 pcs	Stems of fresh parsley, finely chopped (***hacher***)
3 ½ fl oz	100 mL	Cognac
1 ½ pt	750 mL	Dry white wine, reduced to 8 fl oz (250 mL)
1 qt	1 L	Fish ***fumet***
		Tomalley and coral from the lobsters
		Cream or crème fraîche
1 ¾ oz	50 g	Butter
Pinch	Pinch	Cayenne pepper
8 oz	250 g	Salt and ground white pepper
4 oz	120 g	*Beurre Manié*
2 oz	60 g	Butter, softened (***beurre pommade***)
2 oz	60 g	Flour
		To Finish
¼ oz	10 g	Fresh tarragon, blanched, finely chopped (***hacher***)
¼ oz	10 g	Fresh parsley, finely chopped (***hacher***)
¼ oz	10 g	Fresh thyme, finely chopped (***hacher***)

Method

1. Using a pair of scissors, remove the eyes, antennae, and little hairs from the lobsters. Break down (***détailler***) the live lobsters by separating the tail, claws, legs, and knuckles from the head. Cut the head in half, remove the tomalley, and, if the lobster is a female, remove the coral. Refrigerate the tomalley and coral separately.
2. Cut the tail in four pieces.
3. Lightly crack the claws by tapping them firmly with the back of a heavy knife.

Cuisson and Sauce Armoricaine

1. Heat some olive oil in a large sauté pan or a Dutch oven (***cocotte***) over medium-high heat. Add the lobster and sear until it turns red (***cardinaliser***). Add the tomato paste and cook 1 to 2 minutes (***pincer la tomate***), add the mirepoix, herbs, tomato concassée, bouquet garni, and garlic. Stir well. Flambé with the cognac, then deglaze with the white wine. Reduce the wine by half, then add enough fish fumet to come up three-fourths the height of the solids. Cook 5 to 10 minutes. Decant the lobster into a heated serving bowl or copper pan. Remove the lobster heads and set aside. Reduce the liquid by one-third. Mix the tomalley and coral (*optional*) with the beurre manie. Mix into the cooking liquid and cook 5 minutes. Add the cream and reduce until thick enough to coat the back of a spoon. Mount the sauce with butter, then season to taste with salt and pepper and a little cayenne pepper. Strain the sauce through a fine mesh sieve (***chinois***) over the lobster. Add the chopped blanched tarragon and chopped parsley. If desired, finish with a dash of cognac before serving.

To Serve

1. Clean the reserved lobster heads inside and out under running water; round off the edges of the heads with scissors. Clean and oil the head shells.
2. Decorate with the cleaned head shells and fresh herbs (*optional*).

LOTTE RÔTIE ET FENOUIL BRAISÉ

Learning Outcomes

Rôtir
Habiller
Piquer
Blanchir
Blanc de cuisson
Réduire
Monter
Mirepoix
Braiser
Glacer
Dépouiller

Equipment

Knives:
Vegetable peeler (*économe*), paring knife (*office*), slicing knife (*éminceur*), scissors, filet de sole

Tools:
Bowls, cutting board, small larding needle, wooden spatula, plastic spatula, fine chinois, skimmer (*écumoire*), ladle (*louche*), whisk, mortar and pestle, wire rack, tamis

Pans:
Sautoir, small russe, bain marie, roasting pan, medium sauce pan, shallow pan

Serving

4 persons

FYI...

Lotte is French for "monkfish." Monkfish are known by many different names in English: frogfish, sea devil, anglerfish, and as "the poor man's lobster." Lotte has a large and meaty tail that possesses a sweet lobster-like flavor—at a fraction of the cost. It is almost always sold with the head removed, understandably as it has a wide mouth filled with long, sharp teeth.

LOTTE RÔTIE ET FENOUIL BRAISÉ

Roasted Monkfish with Braised Fennel

Method

1. Preheat the oven to 400°F (205°C).
2. Degorge the anchovy fillets in cold milk (***dégorger***) that covers the meat.
3. Skin the monkfish and remove the venous membrane (***dépouiller***). Leave the flesh on the spine. Using a larding needle, stud (***piquer***) the tail with the anchovy fillets, the ***julienne*** of truffle, and garlic.
4. *Blanch the fennel:* Prepare a ***blanc de cuisson***: Mix the flour, lemon juice, and salt into a medium saucepan full of water. Bring the liquid to a boil over high heat and cook the fennel until it is just soft enough to be pierced with the tip of a knife. (The fiber in the fennel should still make a crunching sound as the knife goes in.) Refresh the fennel under cold water, then drain and reserve in a covered container.

Aioli

1. Cook the potato (*skin on*) in a large pan of salted water over high heat until the point of a knife can easily be inserted into the center. Refresh under cold water and peel.
2. Press the potato flesh through a drum sieve (***tamis)*** using a plastic scraper (***corne***). Pound the garlic until smooth using a mortar and pestle. Transfer the paste to a large bowl and, using a whisk, mix it with the potato purée, then beat in the egg yolk. Season the mixture. Slowly pour in olive oil in a thin stream, mixing continuously with a whisk to form an emulsion. Mix in the herbs and transfer the aioli to a covered container. Reserve in the refrigerator.

Cuisson

1. Pat the monkfish dry, sprinkle it with salt and white pepper, and place it in a sauté pan. Top it with cubes of cold butter and transfer it to the hot oven to roast for 10 to 15 minutes. Once the butter has melted, baste and turn the monkfish frequently to color evenly. When cooked, transfer it to a wire rack. Cover with a piece of foil and keep warm.
2. The monkfish is roasted on the bone for two reasons: (1) roasting with bone adds flavor and (2) it allows the fillets to keep their shape for a nicer presentation.

Fenouil Braisé

1. Place the ***mirepoix*** in the pan used to roast the monkfish. Stir until coated, then roast the ***mirepoix*** in the hot oven, stirring periodically until soft. Remove the pan from the oven, pour out the fat, and add the bay leaves. Wet (***mouiller***) with the ***glace de poisson*** and season lightly. Add the fennel bulbs and bring the liquid to a boil over high heat. Cover the pan and place it in the hot oven to braise for 25 minutes (*until the fennel bulbs are easily pierced with the tip of a knife*). When soft, remove the fennel from the braising liquid. Strain the liquid through a fine mesh sieve (***chinois***) and turn off the oven.
2. Reduce the braising juice by half in a small pan over medium heat. Place the fennel on a wire rack in a shallow pan, brush the bulbs with the reduction, and place in the warm oven for 30 seconds. Repeat until covered in a thick shiny glaze (***glaçage***).

Sauce

1. Heat the ***glace de poisson*** in a small saucepan over medium heat. Temper the ***aioli*** by pouring in a little hot ***glace de poisson*** while whisking. Remove the ***glace de poisson*** from the heat and pour in the tempered ***aioli,*** stirring constantly. Return the pot to low heat and stir until the sauce is thick enough to coat the back of a spoon (***à la nappe***).

To Serve

1. Cut the bulbs into 2 or 4 pieces, depending on their size.
2. Remove the filets and place on a heated platter and coat with the sauce. Place the fennel on the plate and decorate with 1 bouquet parsley.

Quantity		*Ingredient*
U.S.	*Metric*	
1 pc	1 pc	Monkfish tail [*5 ½ lb (2.5 kg)*]
		Salt and pepper
		Pour Piquer
8 pcs	8 pcs	Anchovy fillets
3 ½ fl oz	100 mL	Milk
1 pc	1 pc	Truffle, cut into ***julienne***
8 pcs	8 pcs	Garlic cloves, cut into ***julienne***
		Aioli
1 pc	1 pc	Potato
¾ oz	20 g	Garlic, finely chopped (***hacher***)
1 pc	1 pc	Egg yolk
7 fl oz	200 mL	Olive oil
1 ¼ oz	35 g	Basil, finely minced (***ciseler***)
1 ¼ oz	35 g	Parsley, finely chopped (***hacher***)
1 ¼ oz	35 g	Tarragon, finely chopped (***hacher***)
1 ¼ oz	35 g	Thyme, finely chopped (***hacher***)
		Salt and pepper
		Cuisson
5 ¼ oz	150 g	Butter
		Salt and white pepper
		Fenouil Braisé
4 pcs	4 pcs	Small fennel bulbs
		Or
2 pcs	2 pcs	Large fennel bulbs
2 pcs	2 pcs	Fennel trimmings (***parures***), roughly chopped
1–2 oz	30–60 g	Flour
½ pc	½ pc	Lemon
1 pc	1 pc	Carrot, cut into ***mirepoix***
1 pc	1 pc	Onion, cut into ***mirepoix***
½ pc	½ pc	Celery stalk, cut into ***mirepoix***
3 pcs	3 pcs	Shallots, cut into ***mirepoix***
		Bay leaves
8 fl oz	250 mL	Fish ***fumet,*** reduced to a glaze (***glace de poisson***)
		Salt and pepper
		Sauce
1 ¾ oz	50 g	Aioli
		Fish ***fumet,*** reduced to a glaze (***glace de poisson***)

MATELOTE DE DORÉ AU VIN ROUGE

Learning Outcomes

Habiller a fish
Marinade crue
Preparation of lardons
Glacer à brun
Croûtons
Lier with a beurre manié

Equipment

Knives:
Filet de sole, vegetable peeler (*économe*), paring knife (*office*), slicing knife (*éminceur*), serrated knife

Tools:
Mixing bowls, wooden spoons, sieve, chinois, whisk

Pans:
Large shallow pan, medium shallow pan, large deep pan, large pot, large sauce pan, large sauté pan

Serving

4 persons

FYI...

This dish was originally called *plat de matelots*, or *sailor's dish*. Contrary to its name, this recipe is typically made with freshwater fish such as walleye, but can just as easily be made with sautéed veal, or brains.

MATELOTE DE DORÉ AU VIN ROUGE

Yellow Pike Stewed in Red Wine

Method

Marinade

1. ***Habiller** the fish:* Remove the scales (***écailler***) and cut off the fins.
2. Cut the fish into thick slices (***tronçons***), so that the backbone is visible. Each slice should be ¾ in. (1.5 cm) thick.
3. Place the fish in a large recipient and cover it with the marinade ingredients. Cover the container and let the fish marinate in the refrigerator for 24 hours.

Garniture

1. Cut the smoked pork belly into lardons, place them in a pan of cold water, and bring to a boil over high heat. Blanch the lardons for 1 minute, then refresh them under cold running water and drain.
2. Melt 3 ½ oz (100 g) of butter in a large sauté pan over medium heat and sauté the lardons until they are golden on all sides. Transfer the lardons to a large buttered pan, cover them with a circle of buttered parchment paper (***cartouche***), and reserve at room temperature.
3. Melt another 1 ¾ oz (50 g) of butter in the same pan over medium heat and sauté the mushrooms until lightly browned. Reserve with the lardons.
4. ***Glacer à brun** the pearl onions:* Place the onions in a sauté pan that is large enough to hold them in a single layer. Pour in enough cold water so that the onions are two-thirds immersed. Add the sugar and butter and season with salt. Cover with a buttered parchment paper lid (*with a hole in the center*). Cook the onions over low heat until all the water has evaporated and the sugar and butter mixture begins to carmelize. Roll the onions in the resulting syrup to glaze them. Reserve in the saute pan at room temperature.

Croûtons

1. Cut the bread slices into heart shapes. Toast them in clarified butter in a large shallow pan over high heat, until nicely colored on both sides. Drain the croûtons on a paper towel and reserve them uncovered at room temperature.

Cuisson

1. Decant the fish from the marinade and pat the flesh dry. Strain the marinade through a fine mesh sieve (***chinois***) into a large saucepan. Bring this liquid to a boil over medium heat and skim off any impurities (***écumer***). Reduce the liquid by one-third and add the fish ***fumet***. Taste and reduce if necessary. Sauté the marinade vegetables in oil in a large saucepan over high heat until lightly colored. Deglaze with the marinade liquid, add the mushroom trimmings, and bring to a boil. Lower the heat to a gentle simmer and season to taste. Add the fish pieces to the liquid, cover the pot, and reduce the heat to low. Cook the fish gently for 15 to 20 minutes, depending on the size of the ***tronçons***.
2. When the fish is cooked, decant the ***tronçons*** from the liquid.
3. Strain the liquid through a fine mesh sieve into a clean saucepan and discard the vegetables.

Finishing

1. Bring the liquid to a boil over medium heat. Add the ***beurre manié***, piece by piece, whisking constantly until incorporated. Let the sauce boil for 1 minute while stirring, then reduce the heat to low and simmer until it is thick enough to coat the back of a spoon (***à la nappe***). Mount the sauce with butter and mix in the chopped parsley. Taste and adjust the seasoning and return the fish ***tronçons*** to the sauce.

To Serve

1. Transfer the fish, garniture, and sauce to a heated serving dish. Dip the tips of the croûtons in the sauce, then in chopped parsley, and arrange them around the edge of the dish. Decorate with a sprig of fresh parsley.

Quantity		*Ingredients*
U.S.	***Metric***	
1 pc	1 pc	Walleye [*2 lb 10 oz (1.2 kg)*]
1 qt	1 L	Fish ***fumet*** (*poaching element*)
		Marinade
1 ½ qt	1.5 L	Red wine
2 ¾ oz	80 g	Carrots, cut into ***mirepoix***
2 ¾ oz	80 g	Onions, cut into ***mirepoix***
1 ¼ oz	40 g	Celery, cut into ***mirepoix***
1 ¼ oz	40 g	Leeks, cut into ***mirepoix***
10 pcs	10 pcs	Black peppercorns
½ pc	½ pc	Garlic head, halved
10 brs	10 brs	Parsley stems
¼ oz	10 g	Coarse sea salt
½ oz	10 mL	Red wine vinegar (*optional*)
½ oz	10 mL	Peanut oil
¾ oz	20 mL	Cognac
1 pc	1 pc	***Bouquet garni***
		Rind from smoked pork belly
		Garniture
7 oz	200 g	Smoked pork belly, cut into lardons
3 ½ oz	100 g	Butter
10 oz	300 g	Mushrooms, peeled, quartered (*reserve the trimmings*)
2 oz	60 g	Butter
24 pcs	24 pcs	Pearl onions, peeled
1 oz	30 g	Sugar, granulated
		Salt, pinch of
2 oz	60 g	Butter
8 slices	8 slices	White bread, sliced, cut into hearts
3 ½ fl oz	100 g	Butter, clarified
		Sauce
1 qt	1 L	Cooking liquid from matelote
3 ½ oz	100 g	***Beurre manié***
1 ¾ oz	50 g	Butter
1 ¾ oz	50 g	Parsley, finely chopped (***haché***)
		Salt and pepper

MERLAN FRIT COLBERT

Learning Outcomes

Désossage à l'anglaise
Paner à l'anglaise
Deep-fried herbs
Beurre composé

Equipment

Knives:
Vegetable peeler (*économe*), paring knife (*office*), scissors, filet de sole

Tools:
Bowls, plastic spatula, fine chinois, araignée, whisk, cheese

Pans:
Small russe, deep fryer

Serving

4 persons

FYI...

In the preparation of fish, the term *Colbert* generally means the application of a coating (flour, beaten eggs, and breadcrumbs) and frying or deep frying as a cooking method. The Colbert butter used in this recipe as well as Colbert sauce (commonly used to accompany meat and game) are likely named after Jean-Baptiste Colbert, who was the Chief Minister to Sun King, Louis XIV.

MERLAN FRIT COLBERT

Breaded and Fried Whiting

Quantity		Ingredient
U.S.	**Metric**	
4 pcs	4 pcs	Whiting [*8 oz (250 g each)*]
1 qt	1 L	Milk
		Salt and pepper
		Paner à l'Anglaise
7 oz	200 g	Flour
4 pcs	4 pcs	Egg yolks
2 fl oz	60 mL	Oil
14 oz	400 g	Breadcrumbs
		Salt and pepper
		Oil to fry
		Beurre Colbert
7 oz	200 g	Butter, softened (***beurre pommade***)
2 pcs	2 pcs	Lemons, juice of
6 brs	6 brs	Parsley, coarsely chopped (***ciseler***)
6 brs	6 brs	Tarragon, blanched, finely chopped (***hacher***)
1 ¾ fl oz	50 mL	***Glace de viande***
		Salt and pepper
		Decoration
2 pcs	2 pcs	Lemons
2 bqs	2 bqs	Parsley
1 ¾ fl oz	50 mL	***Glace de viande***

Method

1. Preheat the deep fryer to 350°F (180°C).
2. *Clean (**habiller**) the whiting:* Using a pair of scissors, trim off the fins, the gills, and the tail of the fish. Cut out the eyes with a paring knife.
3. Lay the fish on its side and with a filleting knife make an incision from the head to the tail down the center of the spine (*the incision should expose the spine*). From this incision, cut the flesh away from both sides of the rib-cage bones. Remove the spine by cutting it at the head and tail with scissors. Repeat this procedure for the remaining 3 fish. See pages 282–285 in *Cuisine Foundations*. ***Note:** Throughout this process the skin on the belly side of the fish is kept intact.*
4. Degorge the fish in cold milk.

Beurre Colbert

1. Season the lemon juice, then beat it into the softened butter (***pommade***). Keep beating until the butter becomes lighter in color and texture.
2. Mix in the parsley and tarragon, then beat in the ***glace de viande*** until it is completely incorporated. Transfer the butter to a pastry bag and pipe it into decorative shapes onto parchment paper (***coucher***). Refrigerate until needed. Alternately, set the pastry bag aside at room temperature.

Paner à l'Anglaise

1. Remove the fish from the milk, pat dry on the inside and outside, and season with salt and white pepper. Line up 3 flat, shallow containers side by side on the counter. Pour the flour into the first container. In the second container, beat the egg yolks, oil, salt, and pepper (***anglaise***), and pour the breadcrumbs into the third. To complete the setup, place a tray lined with parchment paper beside the breadcrumbs.
2. *Treat the fish separately to the following process:* Dredge the fish completely in the flour and gently tap off any excess. Transfer the fish to the egg yolk mixture (***anglaise***) and coat completely. Finally, roll the fish in the breadcrumbs, then line them up neatly on the tray lined with parchment paper.

Fried Parsley for Decoration

1. Wash the parsley and pat it dry. Toss it in flour and gently tap off any excess. Deep fry the sprigs for a few seconds, being careful not to burn the leaves. Drain the fried parsley on a paper towel and sprinkle with salt.

Cuisson

1. One by one, gently lower the fish into the hot oil. Deep fry the whiting until the breadcrumbs are golden (*5 to 10 minutes*). Transfer them to a paper towel and sprinkle with salt.

To Serve

1. Place the hot fish on a hot plate, top with the Colbert butter, and brush around the edge of the fish with the ***glace de viande***. Decorate with fresh lemon and fried parsley.

Learning Outcomes

Ciseler
Ébarber
Réduire
Monter au beurre
Haché

Equipment

Knives:
Paring knife (*office*), slicing knife (*éminceur*)

Tools:
Fine chinois, colander, araignée, whisk

Pans:
Small russe, small marmite, bain marie

Serving

4 persons

MOULES MARINIÈRES

Steamed Mussels

Method

Clean the Mussels

1. Pull off the threads hanging from the shells, scrape off any barnacles or vegetation, and thoroughly rinse the mussels in three changes of clean, cold water.
2. Check for dead mussels by pinching them closed. Discard any that do not close shut.

Cuisson

1. Melt the butter in a large pot over low heat and sweat (***suer***) the shallots and garlic until soft. Deglaze with the white wine and add the ***bouquet garni***. Increase the heat to high and reduce the liquid by two-thirds. Add the mussels, bring the liquid back to a boil, and cover the pot to trap the steam. Cook for 5 minutes, shaking the pot to stir the mussels while cooking. When the mussels are completely cooked they will open (*discard any that do not open*). Decant the mussels from the cooking liquid and keep them hot and covered with a lid.
2. Allow the cooking liquid to rest for 5 minutes to let the grit settle to the bottom of the pot. Carefully strain the cooking liquid through a fine mesh sieve (***chinois***). ***Note:*** *Do not strain all the liquid from the pot as it contains very fine grit.*
3. Reduce the cooking liquid to taste and season if necessary. Mount the sauce with cold butter and add the chopped parsley.

To Serve

1. Arrange the mussels in a deep, heated serving dish or bowl. For presentation, completely open a few of the mussels and pour the sauce over the top. Serve immediately.

Quantity		*Ingredient*
U.S.	*Metric*	
4 lb 7 oz	2 kg	Mussels
4 ¼ oz	120 g	Butter
2 oz	60 g	Shallots, finely diced (***ciseler***)
1 pc	1 pc	Garlic clove, finely chopped (***hacher***)
1 pt	500 mL	White wine
1 pc	1 pc	***Bouquet garni***
¼ oz	10 g	Parsley, finely chopped (***hacher***)
		Salt and freshly ground pepper

FYI...

Marinière is a derivative of the French word *marinier* which, in its most specific usage, refers to a coastal or seaway mariner. This makes sense considering that mussels are most prolific in river estuaries. During the 18th century, trout and even pigeons were cooked in a marinière sauce comprised of white wine, onions, and mushrooms.

Learning Outcomes

Ciseler
Ébarber
Réduire
Monter au beurre
Haché

Equipment

Knives:
Paring knife (*office*), slicing knife (*éminceur*)

Tools:
Fine chinois, colander, araignée, whisk

Pans:
Small russe, small marmite, bain marie

Serving

4 persons

MOUCLADE

Steamed Mussels with Curry

Quantity		*Ingredient*
U.S.	***Metric***	
4 lb 7 oz	2 kg	Mussels
4 ¼ oz	120 g	Butter
Pinch	Pinch	Curry powder
1 pc	1 pc	Garlic clove. finely chopped (***hacher***)
2 oz	60 g	Shallots, finely chopped (***hacher***)
1 pt	500 mL	White wine
1 pc	1 pc	***Bouquet garni***
13 ½ fl oz	400 mL	Cream
1 br	1 br	Cilantro, finely chopped (***hacher***)
Pinch	Pinch	Salt and freshly ground pepper

Method

Clean the Mussels

1. Pull off the threads hanging from the shells, scrape off any barnacles or vegetation, and thoroughly rinse the mussels in three changes of clean, cold water. Check for dead mussels by pinching them closed. Discard any that do not close shut.

Cuisson

1. Melt the butter in a large pot over low heat, add the curry powder, and sweat (***suer***) the garlic and shallots until soft. Deglaze with the white wine and add the ***bouquet garni***. Increase the heat to high and reduce the liquid by two-thirds. Add the cream and mussels. Bring the liquid back to a boil and cover the pot to trap the steam. Cook for 5 minutes, shaking the pot to stir the mussels while cooking. When completely cooked, the mussels will open (*discard any that do not open*). Decant the mussels from the cooking liquid and keep them hot and covered with a lid.
2. Allow the cooking liquid to rest for 5 minutes to let the grit settle to the bottom of the pot. Strain the cooking liquid through a fine mesh sieve (***chinois***). ***Note:*** *Do not strain all the liquid from the pot as it contains very fine grit.*
3. Reduce the cooking liquid to taste and season, if necessary. Mount the sauce with cold butter and add the chopped cilantro.

To Serve

1. Arrange the mussels in a deep, heated serving dish or bowl. For presentation, completely open a few of the mussels and pour the sauce over the top. Serve immediately.

FYI...

A mouclade is quite similar to *moules marinières*, the main difference being that the cooking liquid contains cream and curry. This is a French recipe with international influences.

MOULES À LA CRÈME

Learning Outcomes

Ciseler
Ébarber
Réduire
Monter au beurre
Haché

Equipment

Knives:
Paring knife (***office***), slicing knife (***éminceur***)

Tools:
Fine chinois, colander, araignée, whisk

Pans:
Small russe, small marmite, bain marie

Serving

2 persons

MOULES À LA CRÈME

Steamed Mussels with Cream

Quantity		Ingredient
U.S.	Metric	
4 lb 7 oz	2 kg	Mussels
4 ¼ oz	120 g	Butter
1 pc	1 pc	Garlic clove. finely chopped (***hacher)***
2 oz	60 g	Shallots, finely chopped (***hacher***)
1 pt	500 mL	White wine
1 pc	1 pc	***Bouquet garni***
13 ½ fl oz	400 mL	Cream
¼ oz	10 g	Parsley, finely chopped (***hacher)***
		Salt and freshly ground pepper

Method

Clean the Mussels

1. Pull off the threads hanging from the shells, scrape off any barnacles or vegetation, and thoroughly rinse the mussels in three changes of clean, cold water. Check for dead mussels by pinching them closed. Discard any that do not close shut.

Cuisson

1. Melt the butter in a large pot over low heat and sweat (***suer***) the garlic and shallots until soft. Deglaze with the white wine and add the ***bouquet garni***. Increase the heat to high and reduce the liquid by two-thirds. Add the cream and mussels. Bring the liquid to a boil, cover the pot to trap the steam, and cook for 5 minutes shaking the pot to stir the mussels while cooking. When completely cooked, the mussels will open (*discard any that do not open*).
2. Decant the mussels from the cooking liquid and keep them hot and covered with a lid.
3. Alllow the cooking liquid to rest for 5 minutes to let the grit settle to the bottom of the pot. Strain the cooking liquid through a fine mesh sieve (***chinois***).

Note: *Do not strain al the liquid from the pot as it contains very fine grit.*

4. Reduce the cooking liquid to taste and season if necessary. Mount the sauce with cold butter and add the chopped parsley.

To Serve

1. Arrange the mussels in a deep, heated serving dish or bowl. For presentation, completely open a few mussels and pour the sauce over the top. Serve immediately.

FYI...

Historically, French chefs have had a penchant for showing their reverence for significant people and places through the food that they create. There are regions in France that owe their popularity to the culinary arts and the dishes that were named after them. If a dish was a specialty in one particular region, the name of the region would most likely be included in the recipe title. Moules à la crème, which is also known as *moules à la crème Normande*, is no exception.

PETIT COULIBIAC DE SAUMON FRAIS ET FUMÉ

Learning Outcomes

Making a farce
Pâte à coulibiac
Crêpes
Cooking tapioca

Equipment

Knives:
Paring knife (*office*), slicing knife (*éminceur*), scissors, filleting knife

Tools:
Bowls, cutting board, fork, wooden spatula, trussing needle, plastic spatula, rolling pin, fine chinois, colander, araignée, skimmer (*écumoire*), ladle (*louche*), whisk, tamis

Pans:
Crêpe pan, sautoir, small russe, bain marie, 8-inch nonstick pan, baking sheet, large sauté pan, medium sauce pan

Serving

8 persons

FYI...

The word *coulibiac* (sometimes spelled with a "k" instead of a "c") comes from the Russian *kulebjaka* which originally described a fish pie made with salmon or pike. The preparation was purportedly a favorite preparation among Russian Imperialists; however, during what period of the Russian empire is hard to pin point. What is known is that the word *kulebjaka* came from the German *kohlgebäck* (referring to a cabbage paté) and was introduced to Russia by German immigrants.

The presence of tapioca in modern versions of coulibiac is as a replacement for vesiga, the spinal marrow of European sturgeon, once standard in haute cuisine Russian dishes. Due to the overfishing of European sturgeon for its caviar, vesiga is nearly impossible to find.

PETIT COULIBIAC DE SAUMON FRAIS ET FUMÉ

Fresh and Smoked Salmon Baked in a Leavened Dough

Method

1. Preheat the oven to 375°F (190°C).

Pâte à Coulibiac

1. Sift the flour.
2. In a bowl, mix the fresh yeast into the warm milk and let it rest until a light foam forms on the surface (*5 to 10 minutes*). Using a rubber spatula, mix the yeasted milk into 8 oz (250 g) of flour until it forms a paste (***levain***). Cover the bowl with plastic wrap and let the ***levain*** rise in a warm place for 30 minutes.
3. Meanwhile, combine 8 oz (250 g) flour, ¼ oz (10 g) salt, and 8 oz (250 g) softened butter (***pommade***) and work this mixture with a spatula until it is homogeneous. Mix the eggs in one at a time (*incorporating each egg thoroughly before adding the next*) and combine this mixture with the ***levain***. Lightly dust the work surface with flour (***fleurer***), transfer the dough to the floured tabletop, and knead and stretch it until it stops sticking to the table. Rest (*proof*) the dough for 45 minutes in a covered bowl in a warm area.

Pâte à Crêpes

1. Whisk the eggs and flour together in a large mixing bowl until smooth.
2. Mix in the milk, then the brown butter (***beurre noisette***).
3. Let the batter rest in the refrigerator for 1 hour, then mix in the herbs and reserve the batter at room temperature in a covered bowl.

Garniture

1. Bring the fish ***fumet*** to a boil in a medium pot over medium heat and season to taste. Add the tapioca and cook it until it is soft (*stirring occasionally*). Strain out any residual liquid and reserve the tapioca at room temperature.

Riz Pilaf

1. Melt some butter in a large sauté pan over high heat until lightly colored. Add the sliced mushrooms and sauté them until lightly colored. Set aside.
2. Melt 3 ½ oz (100 g) of butter in a medium saucepan over low heat and sweat (***suer***) the onion, shallots, and garlic until translucent. Add the sautéed mushrooms and cook for 1 to 2 minutes. Add the rice and stir gently until it begins to turn opaque and shiny (***nacrer***). Wet the rice with one-and-a-half times its volume of fish ***fumet,*** then add the ***bouquet garni*** and season to taste. Turn the heat up to medium. Just before it begins to boil, remove the pan from the heat. Cover the rice with a buttered parchment paper lid (***cartouche***). Cook the rice in the oven for approximately 17 minutes. Remove the pan from the oven, discard the ***bouquet garni***, and gently separate the grains with the prongs of a carving fork (***égrener***). Gently mix in the chopped herbs and let the rice cool to room temperature.

Assembling the Farce

1. Combine the rice, smoked salmon ***brunoise***, hard cooked eggs, and tapioca. Spread it onto a large piece of plastic wrap and roll it up into a tight cylinder about 12 in. (22 cm) long. Twist the ends of the plastic and reserve the filling in the refrigerator.

Quantity		*Ingredient*
U.S.	*Metric*	*Base*
1 lb 2 oz	500 g	Fresh salmon, cut into ***escalopes***
3 ¼ oz	100 g	Smoked salmon, cut into ***brunoise***
4 pcs	4 pcs	Eggs, hard cooked, chopped (***hacher***)
		Pâte à Coulibiac
5 ¼ oz	500 g	Flour
½ oz	15 g	Yeast
4 fl oz	125 mL	Milk
¼ oz	10 g	Salt
9 oz	250 g	Butter (***pommade***)
4 pcs	4 pcs	Eggs
1 pc	1 pc	Egg for egg wash
		Pâte à Crêpes
3 ½ oz	100 g	Flour
3 pcs	3 pcs	Eggs
8 fl oz	250 mL	Milk
1 ¾ oz	50 g	Brown butter (***beurre noisette***)
¼ oz	5 g	Parsley, finely chopped (***hacher***)
¼ oz	5 g	Tarragon, finely chopped (***hacher***)
¼ oz	5 g	Chervil, finely chopped (***hacher***)
2 fl oz	60 mL	Oil or butter
		Salt and pepper

Quantity		Ingredient
U.S.	Metric	Garniture
10 fl oz	300 mL	Fish ***fumet*** *(to cook the tapioca)*
1 ¾ oz	50 g	Tapioca
Pinch	Pinch	Paprika
		Salt and pepper
		Riz Pilaf
¾ oz	20 g	Butter
10 oz	300 g	Button mushrooms, finely sliced (***émincer***)
3 ½ oz	100 g	Butter
1 pc	1 pc	Onion, finely diced (***ciseler***)
3 pcs	3 pcs	Shallots, finely diced (***ciseler***)
2 pcs	2 pcs	Garlic cloves, finely chopped (***hacher***)
10 oz	300 g	Rice, short grain
15 fl oz	450 mL	Fish ***fumet***
1 pc	1 pc	***Bouquet garni***
¾ oz	20 g	Parsley, finely chopped (***hacher***)
¾ oz	20 g	Dill, finely chopped (***hacher***)
¾ oz	20 g	Tarragon, finely chopped (***hacher***)
1 ¾ oz	50 g	Butter
		Salt and pepper
		Sauce
10 ½ fl oz	300 g	Cream
7 oz	200 g	Yogurt
		Lemon, juice of
1 bq	1 bq	Dill, finely chopped (***hacher***)

Method

Cook the Crêpes

1. Heat an 8-inch non stick pan over medium heat with a little bit of clarified butter or oil. Pour a small ladle of the batter into the pan while tilting it to make sure it covers the bottom in an even layer. Return to the heat and allow to cook for 1 minute or until the edges are browned. Flip the crêpe over to finish cooking on the other side (*approximately 30 seconds*). Transfer the crêpe to a plate and repeat until all the batter has been used.

Montage

1. Cut the salmon fillet on the bias into thin ***escalopes***. Sprinkle with paprika and set aside.
2. Divide the dough in half. Roll the first piece of dough out into a rectangle ¼ in. (½ cm) thick and large enough to wrap the cylinder. Brush with egg wash and arrange a row of crepes across the middle slightly overlapping. Brush with egg wash and cover with a layer of salmon also slightly overlapping. Remove the farce from the refrigerator, and carefully unroll it onto the center of the marinated salmon. Cover with the farce with the remaining salmon. Brush the remaining crêpes with egg wash and cover the salmon. Arrange the sides so that the farce is completely covered, first by the layer of salmon then by the crepes. Gently lift and press the dough around the wrapped farce, brushing with more egg wash as needed to help it stick. Brush completely with egg wash.
3. Roll the second piece of dough, slightly thicker than the first. Place on top of the ***coulibiac*** and press to seal the two pieces of dough together. Fold the ends and tuck underneath. Transfer to a greased baking sheet, cover with a clean cloth and set aside in a warm place until the dough is doubled in volume.

Sauce

1. Reduce the cream to a light custard consistency in a small pan over low heat. Remove from the heat, add the lemon juice and yogurt, then season to taste and reserve in a bain marie until needed.

Cuisson

1. Brush the ***coulibiac*** with another coating of egg wash and decorate scraps of unleavened dough such as pâte briseé or pate feuilletée. Brush again with egg wash if using any leftover dough. Pierce a hole in the top using a trussing needle (***cheminée***) and place the ***coulibiac*** in the oven to bake. After 10 minutes, turn the temperature down to 350°F (180°C).
2. Bake the ***coulibiac*** until the crust is golden and a trussing needle inserted into the center comes out hot to the touch. Remove the ***coulibiac*** from the oven and let it rest for 5 minutes. Add some dill to the sauce and pour it into a sauceboat.
3. Place the ***coulibiac*** on a hot serving tray and present the sauce on the side.

QUENELLES DE BROCHET, SAUCE CARDINAL

Learning Outcomes

Pocher
Tamiser
Farce à panade
Quenelles
Sauce cardinal

Equipment

Knives:
Paring knife (***office***), slicing knife (***éminceur***), scissors, filleting knife

Tools:
Bowls, wooden spatula, plastic spatula, fine chinois, cutting board, skimmer (***écumoire***), ladle (***louche***), whisk, food processor, tamis, corne, soup spoons, ice bath, rubber spatula, tweezers

Pans:
Sauce pan (***russe***), marmite, bain marie, gratin dish, roasting pan, sauté pan

Serving

4 persons

FYI...

Quenelles are shaped using two serving spoons and can be made with a combination of puréed fish, meat, or offal. Depending on the preparation, flour or breadcrumbs may be used with eggs to bind the mixture. The word *quenelle* comes from the German word *knödel*, meaning "noodle" or "dumpling." This dish is a specialty of Lyon, France.

Cardinal sauce, which figures prominently in this recipe, is one of 50 sauces that are derived from the white mother sauce, *béchamel*.

Quantity		Ingredient
U.S.	Metric	Farce à Panade
6 ¾ fl oz	200 mL	Milk
2 ¾ oz	80 g	Butter
3 ¼ oz	100 g	Flour
1 lb 5 oz	600 g	Pike fillets
1 pc	1 pc	Egg yolk
2 or 3 pcs	2 or 3 pcs	Eggs
2 ¾ oz	80 g	Butter, softened (***pommade***)
Pinch	Pinch	Nutmeg
		Salt and pepper
		Beurre de Homard
		Shells and heads from 2 lobsters
3 ¼ oz	100 g	***Mirepoix***
¾ oz	20 g	Tomato paste
2 pcs	2 pcs	Tomatoes, peeled (***émonder***), seeded (***épépiner***), and diced (***concasser***)
14 oz	400 g	Butter

QUENELLES DE BROCHET, SAUCE CARDINAL

Pike Dumplings with a Lobster Sauce

Method

1. Preheat the oven to 400°F (205°C).

Farce à Panade

1. *Make a panade:* Sift the flour (***tamiser***). Place the milk, butter, and salt in a medium saucepan and bring the mixture to a boil over medium-high heat. ***Note:*** *The butter must be melted just as the liquid reaches a boil.*
2. As soon as the mixture boils, take the pan off the heat and add the flour all at once and mix with a wooden spoon. When a dough forms, return it to the heat. Vigorously stir the dough until it forms a ball and comes cleanly off the sides of the pan. Transfer the panade to a bowl, spread it up the sides of the bowl to cool, cover and set aside.
3. *Prepare the pike fillets:* Skin the fillets. Using tweezers, remove all the bones and chop (***hacher***) the pike very fine. Using a plastic scraper, press the pulp through a drum sieve (***tamis***) into a bowl set over an ice bath.
4. Remove the bowl from the ice bath and mix the fish into the cooled panade with a rubber spatula. Add the yolk and the eggs one by one, mixing well between each addition. Mix in the softened butter (***pommade***), then add nutmeg and season the panade with salt and pepper. Mix until smooth. Cover the mixture with plastic wrap and let it set in the refrigerator for a minimum of 30 minutes.

Beurre de Homard

1. Place the lobster carcasses in a roasting pan and roast the shells in the oven until they turn red (***cardinaliser***). Remove the pan from the oven, add the ***mirepoix***, and return to the oven until the ***mirepoix*** begins to soften. Remove from the oven and add the tomato paste. Return to the oven to cook for 5 minutes (***pincer la tomate***). Remove from the oven, mix in the tomato ***concassée,*** and return to the oven for 5 to 10 minutes. Remove the roasting pan from the oven and reduce the temperature to 200°F (95°C).
2. While the temperature in the oven is lowering, add the cold butter and allow it to slowly melt. Return the roasting pan to the oven to infuse the butter for 2 hours. Remove the roasting pan from the oven and fill it with ice; this will both shock and solidify the butter and bring the impurities to the surface. When the ice has melted, skim (***écumer***) the solidified clarified butter from the surface.
3. Melt the butter in a saucepan and skim off any impurities. Pour off into a clean recipient leaving any remaining solids in the bottom of the pan.
4. Discard the liquid and reserve the fat (***beurre de homard***).

Préparation des Quenelles

1. *Shape the quenelles:* Prepare two soup spoons and a container of water. Dip one spoon and scoop up some farce à panade. Dip the second spoon in water and scoop underneath the farce, following the curve of the bowl of the spoon. Alternate between the spoons,

Method

dipping in water as you go until you have a three-sided oval shape. Scrape off any excess from the ends, then carefully place the finished quenelle in a buttered sauté pan. See page 351 in *Cuisine Foundations.*

2. Bring the fish fumet (*see ingredients under* sauce cardinal) to a boil and carefully pour it around the quenelles. Cover the pan with buttered parchment paper and poach the quenelles in the oven for 10 to 15 minutes.

Finishing

1. Brush a gratin dish with clarified butter.
2. Decant the quenelles from the fish fumet, pat them dry, and arrange them neatly in the gratin dish. Cover the dish and reserve it in a warm area.
3. Strain the fish fumet through a fine mesh sieve (***chinois***) into a clean saucepan. Reduce the fumet over low heat to 1 ¼ fl oz (50 mL) to obtain a glaze consistency (***glace de poisson***).
4. Mix the ***glace de poisson*** and ***beurre de homard*** into the béchamel sauce. Check the consistency and reduce the sauce, if needed. Taste the sauce and adjust the seasoning. Mix the truffle ***brunoise*** into the ***sauce cardinal*** and mount with cold butter. Pat (***tamponner***) the surface with a piece of butter on the end of a fork and reserve in a bain marie.

To Serve

1. In 3 tablespoons of water with a small piece of butter, heat the cooked lobster tails in a small covered saucepan over low heat for 2 to 3 minutes (*being careful not to overcook the lobster*).
2. Ladle the sauce over the quenelles to coat (***napper***) [*be careful not to completely cover the quenelles* (***napper***)].
3. Brown the dish under a salamander (***glacer***).
4. Slice the lobster tails and lay them on top of the gratin dish, alternating with truffle slices.

Tip: *Brush the lobster and truffle slices with clarified butter to give them shine* (***lustrer***).

Quantity		*Ingredient*
U.S.	*Metric*	*Sauce Cardinal*
1 ½ qt	1.5 L	Béchamel à l'ancienne (*see pages 230–231 in Cuisine Foundations*)
8 fl oz	250 mL	Lobster stock (*for poaching quenelles and later to be reduced and added to the sauce cardinal*)
3 ¼ oz	100 g	Lobster butter (***beurre de homard***)
½ pc	½ pc	Truffle, cut into ***brunoise***
		Salt and pepper
		To Serve
1 ¾ fl oz	50 g	Butter, clarified, for shining (***lustrer***)
2 pcs	2 pcs	Lobster tail, cooked and shelled
½ pc	½ pc	Truffle, sliced (***émincer***)

SAINT-JACQUES PROVENÇALES

Learning Outcomes

Émonder
Épépiner
Concasser
Hacher
Sauter
Suer

Equipment

Knives:
Paring knife (*office*), slicing knife (*éminceur*), filet de sole

Tools:
Bowls, cutting board, wooden spatula, skimmer (*écumoire*), whisk

Pans:
Poêle noir or non stick pan

Serving

4 persons

SAINT-JACQUES PROVENÇALES

Scallops with Tomato Sauce

Quantity		Ingredient
U.S.	*Metric*	
20 pcs	20 pcs	Scallops, medium sized (***noix de Saint Jacques***)
2 Tbsp	30 mL	Olive oil (*optional*)
1 oz	30 g	Shallots, finely chopped (***hacher***)
3 pcs	3 pcs	Garlic cloves, finely chopped (***hacher***)
¼ oz	10 g	Tomato paste
5 ¼ oz	150 g	Tomatoes, peeled, seeded (***émonder***), and diced (***concasser***)
1 bq	1 bq	Parsley, finely chopped (***hacher***)
		Salt and pepper
		To Serve
2 brs	2 brs	Chervil or parsley

Method

1. Remove the side muscle from the scallops. Pat the scallops dry and season them on both sides.
2. Heat just enough olive oil to coat the bottom of a black iron pan (***poêle***) over high heat.
3. When hot, place the scallops to sear (***saisir***). Once nicely colored, turn over and sear on the other side. Transfer to a wire rack, cover, and keep warm.
4. Reduce the heat to low and sweat (***suer***) the chopped shallots in the remaining oil that the scallops were cooked in. Sweat the shallots until they begin to soften. Season lightly, add the garlic, and sweat until soft. Add the tomato paste and cook for 2 to 3 minutes (***pincer la tomate***). Add the tomato ***concassée*** and cook until all the liquid evaporates. Remove the pan from the stove, stir in the chopped parsley, and season to taste.

To Serve

1. Place a bed of the tomato mixture in the center of a hot plate. Arrange the scallops on top and decorate with fresh parsley or chervil. Alternatively, a small spoonful of the tomato mixture can be placed on top of each scallop.

Tip: *Add 2 or 3 grains of coarse sea salt on top of each scallop.*

FYI...

When scallops are fished from the ocean, they are not brought directly to the market or grocers. They make a rather arduous journey from the ocean, to the port, and then to the consumer—often with wait times between each of these steps. To prolong the shelf life of a scallop, producers often immerse their catch in a preserving agent called *tripolphosphate.* While this agent maintains the product's freshness, it also causes it to absorb water. This absorption affects the overall weight of the scallop, allowing merchants to charge a slightly inflated price. Is there a way to tell the difference between a treated and a non treated scallop? Yes. A treated scallop will appear blanched, while a scallop that has not undergone this process will be beige to blush in color. Select scallops that appear moist and possess a delicate, sweet aroma.

Learning Outcomes

Marinade instantanée
Pâte à blinis
Émincer
Brunoise
Blanchir

Equipment

Knives:
Slicing knife (*éminceur*), paring knife (*office*), filleting knife

Tools:
Bowls, zester, whisk, rubber spatula, large offset spatula

Pans:
Frying pan

Serving

4 persons

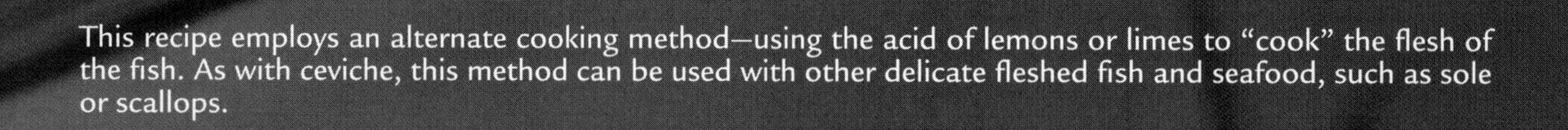

FYI...

This recipe employs an alternate cooking method—using the acid of lemons or limes to "cook" the flesh of the fish. As with ceviche, this method can be used with other delicate fleshed fish and seafood, such as sole or scallops.

SAUMON CRU À L'ANETH

Raw Marinated Salmon with Dill

Method

Blinis

1. Sift the wheat and buckwheat flour together.
2. Heat half the milk in a small pan over low heat until just warm to the touch. Mix in the yeast and let rest at room temperature until the surface is covered in a light foam (about 10 to 15 minutes). Mix in 1 ¾ oz (50 g) of flour (***levain***), cover with a damp cloth, and let rise in a warm area for 2 hours.
3. Add the remaining flour and milk, the salt, the melted butter, and the egg yolks. Mix until combined. Whip the egg whites to stiff peaks (***monter les blancs en neige***) and fold into the mixture.
4. Melt a small amount of clarified butter in a non-stick pan over high heat. Pour a few tablespoons of batter into the hot pan to create individual circles and cook until small bubbles appear on the surface. Flip over and cook the other side. Repeat until all the batter is used.
5. Reserve the blinis wrapped in a clean cloth.

Marinade

1. Mix together the lime ***brunoise***, dill, shallot, green peppercorns, capers, and lemon juice. Season. Incorporate in the olive oil in a thin stream, mixing with a whisk to emulsify.
2. Slice the salmon very thinly on the bias and lay the slices out flat in a single layer in a shallow container and cover in the marinade. Leave to marinate in the refrigerator for 15 minutes. Decant the salmon from the marinade. Mix the green and red pepper ***brunoise*** into the marinade and reserve.

Cream Sauce

1. Whip the crème fraîche to stiff peaks and fold it into the lime juice. Season to taste.

To Serve

1. Serve the salmon on a plate, pour the marinade over the top, and decorate with dill and blanched lemon or lime zest. Serve the cream sauce and blinis on the side.
2. As a classic garnish, serve chopped egg yolks and whites on the side.
3. *Optional presentation:* Serve the blinis as canapés, topped with a small amount of cream sauce, a piece of salmon, some caviar, and a sprig (***pluche***) of dill.

Quantity		*Ingredient*
U.S.	***Metric***	
14 oz	400 g	Salmon fillet, fresh and fairly thick
		Blinis
3 ½ oz	100 g	Wheat flour
2 ½ oz	75 g	Buckwheat flour
11 fl oz	325 mL	Milk
¼ oz	10 g	Fresh yeast
1 tsp	5 g	Salt
½ oz	15 g	Butter, melted
2 pcs	2 pcs	Eggs, separated
3 ½ fl oz	100 g	Butter, clarified, for cooking
		Marinade
3 pcs	3 pcs	Limes, flesh of, cut into (***brunoise***)
1 bq	1 bq	Dill, roughly chopped (***ciseler***)
1 pc	1 pc	Shallot, finely diced (***ciseler***)
1 tsp	5 mL	Green peppercorns, canned
¾ oz	20 g	Capers, small
3 ½ oz	100 g	Green bell pepper, cut into (***brunoise***)
3 ½ oz	100 g	Red bell pepper, cut into (***brunoise***)
1 pc	1 pc	Lemon, juice of
7 fl oz	200 mL	Olive oil
		Salt
		Cream Sauce
7 fl oz	200 g	Crème fraîche
½ pc	½ pc	Lime, juice of
1 ½ pcs	1 ½ pcs	Salt and pepper
		Decoration
4 brs	4 brs	Fresh dill
2 pcs	2 pcs	Lime zest, cut into ***julienne*** (*blanched*)
1 ¼ oz	40 g	Caviar (*optional*)
2 pcs	2 pcs	Eggs, hard cooked, finely chopped (***hacher***)

SOLE ET COQUILLAGES EN NAGE

Learning Outcomes

Court bouillon
Émincer
Julienne
Bouquet garni
Mijoter
Suer
Blanchir
Lever les filets

Equipment

Knives:
Cleaver, vegetable peeler (***économe***), paring knife (***office***), slicing knife (***éminceur***), scissors, filleting knife

Tools:
Bowls, cutting board, wooden spatula, metal spatula, fine chinois, araignée, skimmer (***écumoire***), ladle (***louche***), whisk

Pans:
Sauteuse, sautoir, small russe, small marmite, bain marie

Serving

4 persons

SOLE ET COQUILLAGES EN NÂGE

Sole and Shellfish in Their Cooking Broth

Method

Court Bouillon

1. Combine the carrot, leek, shallots, fennel, garlic cloves, star anis, lemon (juice and zest), and the white wine in a large saucepan. Bring to a boil over high heat, skim the surface of any foam or impurities, then reduce the heat to low and simmer gently until the vegetables are almost cooked through (*about 15 minutes*). Add the fumet to the cooking liquid and continue to simmer until the vegetables are easily pierced with a knife (*about 3 to 4 minutes*). Remove the saucepan from the heat and strain the court bouillon through a fine mesh sieve (***chinois***) into a large saucepan. Season the court bouillon to taste and reserve at room temperature. Place the strained vegetables in a bowl and remove the garlic cloves and the star anis into a separate bowl. Add some court bouillon to the garlic cloves and star anis. Cover both bowls and reserve at room temperature.

Cuisson

1. While the court bouillon is simmering, sear (***saisir***) the lobster until it changes color (***cardinaliser***). Remove from the heat, then remove the flesh from the claws and tails.
2. Remove the side muscle from the scallops, pat the scallops dry.
3. Skin and fillet the Dover sole. Tie the fillets into knots.
4. Add the sole, scallops, and lobster to the court bouillon and simmer gently over low heat until cooked, about 5 to 10 minutes. Decant the seafood from the nage, cover and keep warm
5. The term nage refers to a court bouillon after it has been used for cooking fish or seafood. It is then reduced to concentrate its flavors then served alongside.

To Serve

1. Pour 8 fl oz (250 mL) of the nage into a small saucepan and reduce it to 1 ¾ fl oz (50 mL) over medium heat. Mix in the cream, mount the sauce with cold butter, and mix in the chopped parsley.
2. Strain the remaining nage. Strain the garlic and star anise and discard. Transfer the vegetables to a warm dish, cover, and set aside. Reduce the remaining nage by two-thirds and mount with butter.
3. Arrange the fish and seafood in a heated soup plate. Add some of the vegetables from the nage and sprinkle the dish with chopped parsley. Pour nage over the contents of the plate.
4. Serve the sauce on the side.

Quantity		*Ingredient*
U.S.	*Metric*	
2 pcs	2 pcs	Lobster [1 lb (500 g) each]
8 pcs	8 pcs	Scallops
2 pcs	2 pcs	Dover sole (whole)
		Nage
1 pc	1 pc	Carrot (***canneler***, ***émincer***)
1 pc	1 pc	Leek whites (***émincer***)
2 pcs	2 pcs	Shallots (***émincer***)
1 pc	1 pc	Fennel bulb (***émincer***)
2 pcs	2 pcs	Garlic cloves
1 pc	1 pc	Star anis
1 pc	1 pc	Lemon, zest and juice of
1 pt	500 mL	White wine
1 qt	1 L	Fish ***fumet***
3 ½ oz	100 g	Butter
		Salt and white pepper
		Sauce Crème
1 ¾ fl oz	50 mL	Nage, reduced to a glaze (***glace***)
8 fl oz	250 mL	Cream
1 ¾ oz	50 g	Butter
5 brs	5 brs	Parsley (***hacher***)
		Salt and white pepper

FYI...

Sole is one of the better poaching fish; the resulting meat becomes soft and tender. To make something "en nage" is to literally serve something in its cooking liquid—in which case the sole is served in a bath of this aromatic juice.

SOLE MEUNIÈRE

Learning Outcomes

Habiller (écailler, ébarber, vider)
Turned potatoes à l'anglaise
Sauter meunière
Beurre meunière
Clarified butter

Equipment

Knives:
Filleting knife, paring knife (*office*), slicing knife (*éminceur*), turning knife, scissors

Tools:
Pastry brush, large metal spatula, sieve

Pans:
Oval fish pan (large enough to fit all four fish), medium russe

Serving

4 persons

FYI...

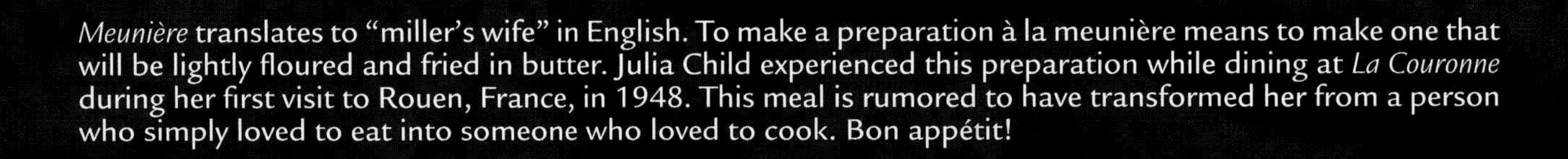
Meunière translates to "miller's wife" in English. To make a preparation à la meunière means to make one that will be lightly floured and fried in butter. Julia Child experienced this preparation while dining at *La Couronne* during her first visit to Rouen, France, in 1948. This meal is rumored to have transformed her from a person who simply loved to eat into someone who loved to cook. Bon appétit!

SOLE MEUNIÈRE

Sauteed Sole with Lemon and Parsley Butter

Quantity		Ingredient
U.S.	Metric	Sole
2 pcs	2 pcs	Dover sole [*14 oz (400 g)*]
9 oz	200 g	Butter
7 oz	250 g	Flour
		Salt and white pepper
		Beurre Meunière
7 oz	200 g	Butter
1 pc	1 pc	Lemon, juice of
		Pommes à l'Anglaise (optional)
6 pcs	6 pcs	Potatoes, turned ***à l'anglaise***
2 oz	60 g	Butter
		Salt and white pepper
		To Serve
2 pcs	2 pcs	Lemons, starred (***historié***)
2 brs	2 brs	Parsley

Method

Sole

1. Cut off the fins and trim the tail with a pair of scissors. Skin the entire fish and extract the eyes using a paring knife. Make an incision under the jaw and pull out the gills (***ouïes***) and discard. Rinse the gill cavity under running water and slice open the stomach. Remove the organs and rinse out the cavity with running water. Repeat this process with the 3 remaining fish.

Pommes à l'Anglaise (optional)

1. Turn the potatoes ***à l'Anglaise*** (*2 ¼ in. (5 to 6 cm) long and as wide as a small egg*) and place them in a large saucepan with salted cold water. Bring the water to a boil over high heat, then reduce the heat to a simmer (*to avoid the potatoes bumping and breaking against each other during cooking*) and cook the potatoes until the tip of a knife can be easily inserted.
2. Drain and lightly coat with butter. Season with salt and white pepper. Keep warm and reserve until needed.

Cuisson à la Meunière

1. *Clarify the butter:* Melt 8 oz (250 g) of butter in a small saucepan (*or in a bain marie*) over low heat without stirring or moving the pan. Keep the butter over the heat until the components separate into layers. Skim the white crust from the surface (***depouiller***) and pour off the clear, golden yellow clarified butter (leaving the milk solids residue behind).
2. *Cook the sole:* Pat the fish dry, sprinkle it with salt and white pepper, and dredge it in flour (***fariner***). Pat off any excess flour.
3. Heat the clarified butter in a large fish pan over medium-high heat. Place both fish in the hot pan presentation side down the rounded side. Sauté the fish, basting it regularly with the butter. When the presentation side is golden, carefully turn the fish over and repeat the procedure on the other side. Transfer the sole to a heated platter and keep warm.

Beurre Meunière

1. Discard the clarified butter and place the pan back on the heat. Add the 7 oz (200 g) of fresh butter and cook the butter over medium heat until it begins to brown. Add the lemon juice, which will foam, then immediately pour over the fish on the platter.

To Serve

1. Decorate with cut lemon and some sprigs of fresh parsley. Serve the potatoes on the side (***citron historié***) and some fresh parsley. Serve the potatoes on the side.

Tip: *Traditionally, the maître d'hótel in the restaurant would prepare this tableside and fillet the fish for the diner.*

TERRINE CHAUDE DE POISSON, SAUCE AMÉRICAINE

Learning Outcomes

Mousseline chaude de poisson
Sauce Américaine
Montage d'une terrine
Cuisson au bain marie

Equipment

Knives:
Filleting knife, chef knife (*couteau chef*), paring knife (*office*)

Tools:
Bowls, food processor, tamis, ice bath, plastic spatula, plastic scraper, trussing needle, colander, chinois, ladle (*louche*), terrine, pastry brush, pastry bag, piping bag, large plain tip, offset spatula, wooden spoon

Pans:
Roasting pan, sautoir, bain marie, deep fryer, baking sheet

Serving

8 persons

TERRINE CHAUDE DE POISSON, SAUCE AMÉRICAINE

Hot Fish Mousse with Shellfish and Tomato Sauce

Method

1. Preheat the deep fryer to 370°F (190°C).
2. Preheat the oven to 350°F (150°C).

Tempura Batter

1. Combine the flour, cornstarch, and baking powder and sift them into a large bowl. Using a whisk, mix in the salt, then slowly incorporate the water until smooth, cover the mixing bowl and let the batter rest for 30 minutes.

Mousseline de Poisson

1. Process the salmon and whiting fillets in a food processor using short bursts. Using a plastic scraper (***corne***), press the pulp through a drum sieve (***tamis***) into a bowl place in an ice bath. With a rubber spatula, work the egg whites into the pulp one by one mixing well after each addition. Keeping the bowl over the ice bath, beat in the cream little by little (***monter la farce***). Continue beating until the mixture is thick and pale. Mix in the softened butter (***pommade***) and season with salt, white pepper, and cayenne pepper. Cover and refrigerate the mousseline until needed.

Garniture

1. Peel the prawns and reserve the heads and shells for the ***sauce Américaine***. Make a shallow incision along the back of the prawn tails and pull out the black thread (***dénerver***). Pat the prawns dry. Season them well and dredge them lightly with flour (***fariner***), patting off any excess. One by one, dip the prawns in the tempura batter, then transfer them to the hot oil to fry until golden. Remove the cooked prawns from the deep fryer and transfer them to a tray that is lined with paper towels.
2. Tear the stems and lower ribs off the spinach leaves.
3. Bring a large pot of salted water to a boil and blanch the spinach for 30 seconds. Drain the spinach and immediately refresh it in an ice bath. Press out all the excess water, then select 12 large spinach leaves and lay them out flat on a cutting board. Brush them with egg white and roll up a tempura fried prawn in each one. Set aside. Reserve the remaining prawns for decoration.

Montage

1. Brush the inside of a terrine mold with softened butter (***beurre en pommade***) and line the bottom and sides with the blanched spinach leaves. Allow spinach to hang over the sides of the terrine. Fill a large pastry bag fitted with a large round tip with the mousseline de poisson. Fill the terrine one-third of the way up with mousseline and tap

Quantity		*Ingredient*
U.S.	*Metric*	***Tempura Batter***
1 oz	30 g	Flour
¾ oz	20 g	Cornstarch
½ tsp	2 g	Baking powder
		Salt
8 fl oz	250 mL	Water
		Mousseline de Poisson
1 lb	500 g	Salmon, fillet
2 lb	1 kg	Whiting, fillet
6 pcs	6 pcs	Egg whites
10 ¼ fl oz	300 mL	Cream
2 ¾ oz	75 g	Butter, softened (***pommade***)
Pinch	Pinch	Cayenne pepper
		Salt and white pepper
		Garniture
15 pcs	15 pcs	Prawns
1 ¾ oz	50 g	Flour
		Tempura batter
2 lb 3 oz	1 kg	Spinach, large leaves, blanched
		Salt and pepper
1 pc	1 pc	Egg white

FYI...

Mounting a sauce with butter (***monter au beurre***) is a technique that requires the chef to add small pieces of cold butter to a finished sauce prior to serving. This is done for three reasons: to add texture, to add flavor, and for aesthetics—it adds sheen to the sauce for presentation.

Quantity		Ingredient
U.S.	Metric	*Sauce Américaine*
1 ¾ fl oz	50 mL	Olive oil
		Prawn shells
1 pc	1 pc	Carrot, cut into ***mirepoix***
1 pc	1 pc	Celery, cut into ***mirepoix***
1 pc	1 pc	Onion, cut into ***mirepoix***
¼ oz	10 g	Tomato paste
¾ oz	20 g	Flour
2 pcs	2 pcs	Tomatoes (***concassée***)
1 pt	500 mL	Fish ***fumet***
4 fl oz	125 mL	Cream
1 ¾ oz	50 g	Butter, to mount
		Salt and pepper
		To Serve
3 pcs	3 pcs	Tempura-fried prawns
3 pcs	3 pcs	Sea urchin tongue
¾ oz	20 g	Butter

Method

the mold once or twice on the work surface to spread out the mixture evenly and remove any air bubbles. Arrange six of the spinach-wrapped prawns in two parallel lines along the length of the terrine (*pressing them lightly into the mousseline*).

2. Pipe out a layer of mousseline to cover the prawns. Arrange the remaining six spinach-wrapped prawns in two rows on top then fill the remainder of the terrine with the remaining mousseline. Smooth the surface of the terrine flat using a metal spatula and fold the excess spinach over. Cover any holds with additional spinach.
3. Cover the terrine mold and transfer it to a roasting pan filled with hot water (***bain marie***) (*make sure the water reaches two-thirds the way up the side of the terrine mold*). Transfer the terrine in its bain marie to the oven to cook for 40 to 45 minutes. To determine when the terrine is cooked, either use a probe [*the inside temperature should be 149°F (65°C)*] or insert the tip of a trussing needle into the center. If it comes out clean, the terrine is cooked through. Reserve the terrine in the bain marie to keep hot.

Sauce Américaine

1. Heat the olive oil in a large pot over high heat and cook the prawn shells until pink (***cardinaliser***). Add the carrot, celery, and onion ***mirepoix*** and sauté until they begin to color. Next, add the tomato paste and cook for 1 to 2 minutes (***pincer la tomate***). Add the flour and cook for 1 to 2 minutes (***singer***), stirring well. Add the tomato ***concassée*** and cook until all the liquid evaporates. Deglaze the pan with the fish fumet and bring the liquid to a boil. Reduce the heat to low and simmer for 30 minutes, skimming the surface regularly. Season to taste.
2. Reduce the cream by one-half in a small pan over low heat. When the sauce is just thick enough to coat the back of a spoon (***à la nappe***), stir in the cream. Taste and adjust the seasoning and reserve the sauce in a bain marie.

To Serve

1. Place 3 tempura-fried prawns and the sea urchin tongues on a baking sheet lined with parchment paper and warm them in the oven for about 5 minutes.
2. Unmold the terrine onto a heated serving platter.
3. *Finish the **sauce Américaine:*** Whisking continuously, add 1 ¾ oz (50 g) of cold butter, piece by piece, to the sauce (***monter au beurre***). Fill the bottom of the dish with ***sauce Américaine*** and top the terrine with the hot prawns and sea urchin tongues.

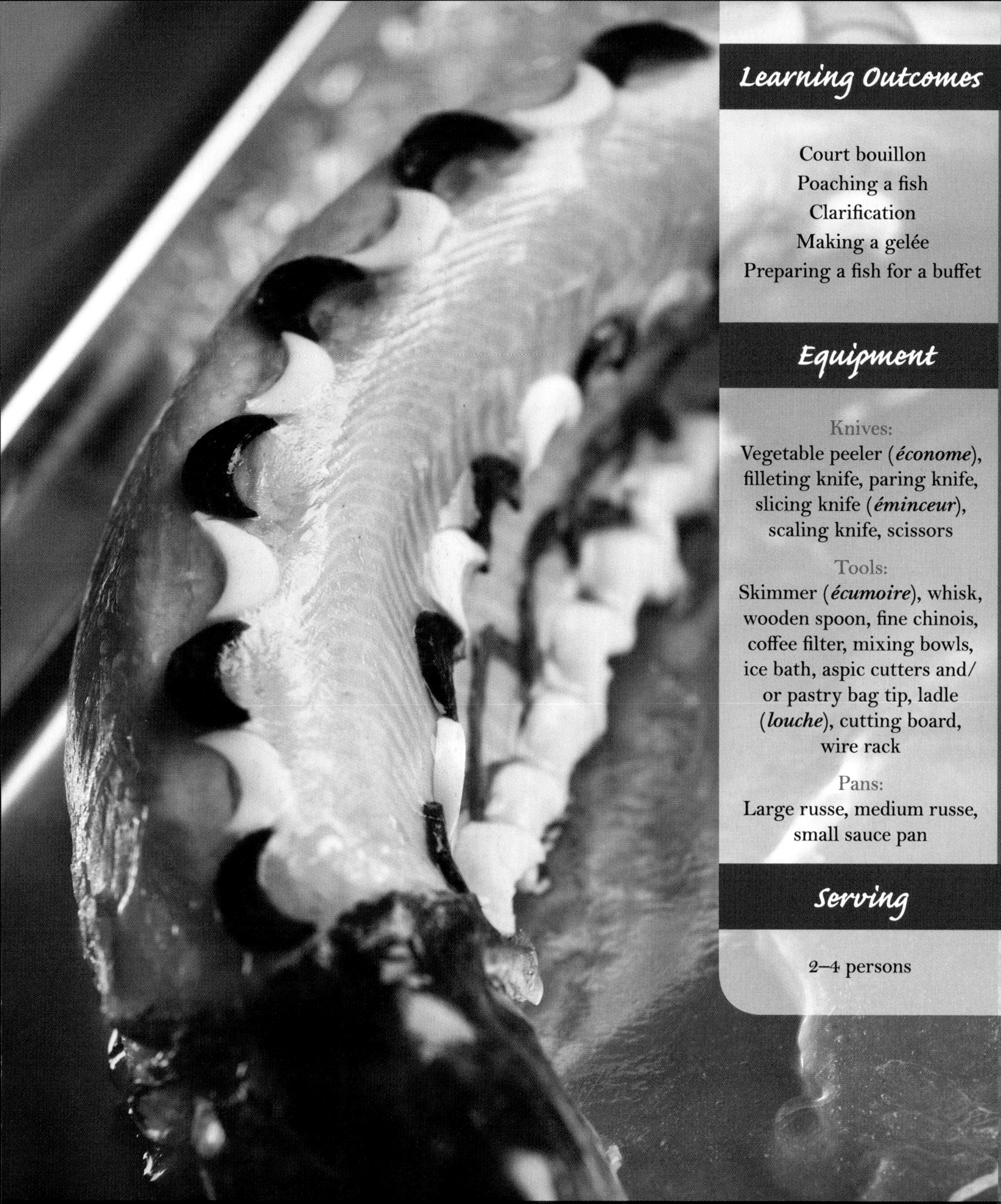

TRUITE EN BELLEVUE

Learning Outcomes

Court bouillon
Poaching a fish
Clarification
Making a gelée
Preparing a fish for a buffet

Equipment

Knives:
Vegetable peeler (***économe***), filleting knife, paring knife, slicing knife (***éminceur***), scaling knife, scissors

Tools:
Skimmer (***écumoire***), whisk, wooden spoon, fine chinois, coffee filter, mixing bowls, ice bath, aspic cutters and/or pastry bag tip, ladle (***louche***), cutting board, wire rack

Pans:
Large russe, medium russe, small sauce pan

Serving

2–4 persons

FYI...

The origin of the term *bellevue* (literally, "beautiful view") used in conjunction with food is rumored to have had its beginnings at the Château de Belleville. The owner of the chateau, Madame de Pompadour, was said to prepare tantalizing meals to arouse the appetite of King Henry XV. Madame de Pompadour was also one of the king's preferred mistresses.

TRUITE EN BELLEVUE

Cold Poached Trout in Aspic

Method

1. Place a presentation tray in the refrigerator to chill.

Court Bouillon

1. Bring all the court bouillon ingredients to a boil in a large stock pot over high heat. Once boiling, reduce the court bouillon to low heat and simmer for about 45 minutes, skimming (***écumer***) if necessary. Strain the court bouillon through a fine mesh sieve (***chinois***). Taste the liquid and adjust the seasoning. Cool until warm or at room temperature. Set it aside until needed.

Prepare the Trout

1. ***Habiller*** *the trout*: Cut off all the fins and trim the tail using scissors (***ébarber***). Scale the fish (***écailler***) and slice along the stomach to remove the organs (***vider***). Remove the gills. See page 256 in *Cuisine Foundations.*
2. *Peel the trout:* Bring a stock pot of water to a boil over high heat and remove it from the heat as soon as it comes to a boil. Using two skimmers (***écumoires***), lower the fish into the hot water and keep the fish immersed for 20 to 25 seconds. Transfer the fish to a cutting board. While the fish is still hot, peel the skin off its body using the curved edge of a spoon.
3. Immerse the peeled trout in the prepared court bouillon and place it over medium heat. Gently bring to a simmer, reduce the heat to low and leave to poach, about 15 minutes. Carefully remove the trout from the court bouillon and place it on a wire rack over a clean tray. While still hot, carefully peel off the skin. Let it cool to room temperature, then reserve it in the refrigerator. Strain the court bouillon through a fine mesh sieve (***chinois***) and refresh it over an ice bath.

Garniture/Decoration Elements

1. *Cook the carrot slices **à l'Anglaise***: Bring some salted water to a boil in a small saucepan over high heat. Add the carrot slices and cook them until soft. Refresh in an ice bath before draining.
2. *Blanch the chives:* Bring salted water to a boil in a small saucepan over high heat. Add the chives and blanch for 20 seconds. Refresh in an ice bath before draining.
3. Using a paring knife, aspic cutter, or pastry tip, cut the truffle, hard-cooked egg (*white and yolk*), blanched chives, chervil, and cooked carrot slices into small decorative shapes. Set aside until needed.

Clarification

1. Pour the court bouillon into a large saucepan. Mix the whiting meat and egg whites together and stir them into the cold court bouillon until combined. Gently heat the liquid over medium heat, stirring from time to time to keep the proteins from sticking to the bottom of the pan. When the proteins start to coagulate on the surface of the court bouillon, reduce the heat to a gentle simmer. ***Note***: *These coagulated ingredients are referred to as a raft* (***gâteau***). Make a small hole in the raft so the bubbles can escape. Let the liquid simmer gently, without stirring, for 20 minutes. Line a fine mesh sieve (***chinois***) with

Method

a coffee filter or damp cloth and carefully strain the clarified court bouillon.

Gelée

1. Soak the gelatin sheets in cold water until they are soft, then squeeze out and discard the water. Dissolve the drained sheets in 8 fl oz (250 mL) of hot court boullion and stir until the gelatin has completely dissolved. Strain this mixture through a fine mesh sieve (***chinois)*** directly into the pan of hot clarified court boullion.

Montage

1. *Temper the gelée:* Pour 1 pt (500 mL) of the liquid gelée into a bowl over an ice bath and stir it gently using a metal spoon (***tempérer***). Continue until the liquid becomes oil-like in resistance and movement.
   Tip: *Work the liquid gently to avoid creating air bubbles.*
2. Pour enough of the thickened gelée into the bottom of the presentation platter to form a thin layer. Return the platter to the refrigerator. Pour the rest of the gelée into a soup plate or deep dish to create a ½ in. to 1 in. (1 to 2 cm) thick layer and refrigerate it.
3. Temper another 8 fl oz (250 mL) of gelée and pour it over the cold trout to create a first coat (***lustrer***). Return the fish to the refrigerator for about 5 minutes.
4. Meanwhile, arrange the garniture (*the herbs, the decoratively cut truffle, carrot, and hard-cooked egg*) on a plate and spoon some tempered gelée over them.
5. Remove the fish from the refrigerator and, using the tip of a knife or tweezers, arrange the garniture in a decorative pattern along the sides and down the spine of the fish.
6. Temper more gelée if needed. Slowly pour tempered gelée over the fish to cover the decorative shapes in a second coat (***lustrer***). Refrigerate for 10 minutes to set the gelée. If needed, recoat the fish in jelly.
   Tip: *If the jelly begins to set before coating, gently warm it in a bain marie until liquid.*

To Serve

1. Remove the serving platter and soup plate from the refrigerator. Flip the contents of the soup plate out onto a chopping board and cut the gelée into medium dice.
2. Arrange the fish on the gelée-covered serving platter and make a design around it using the diced gelée.

Quantity		*Ingredient*
U.S.	***Metric***	
1 pc	1 pc	Trout [*10 ½ oz (300 g)*]
		Court Bouillon
2 qt	2 L	Water
6 ¾ fl oz	200 mL	White wine
7 oz	200 g	Onion
1 pc	1 pc	***Bouquet garni***
1 ½ tsp	5 g	Peppercorns
¾ oz	20 g	Coarse salt
		Garniture
1 pc	1 pc	Carrot, sliced lengthways
1 br	1 br	Chives
¼ oz	10 g	Truffle
1 pc	1 pc	Hard-cooked egg
1 br	1 br	Chervil
		Clarification
9 oz	250 g	Whiting meat, finely chopped (***hacher***)
2 pcs	2 pcs	Egg whites
		Gelée
20 pcs	20 pcs	Gelatin sheets

Learning Outcomes

Habiller and tronçonner a turbot
Braising a fish
Dépouiller

Equipment

Knives:
Cleaver, serrated knife, paring knife (*office*), slicing knife (*éminceur*), scissors

Tools:
Araignée, skimmer (*écumoire*), whisk

Pans:
Sauteuse, sautoir, sauce pan (*russe*), bain marie

Serving

4 persons

TRONÇONS DE TURBOTIN À LA DUGLÉRÉ

Turbot Steaks with a Tomato and White Wine Sauce

1. Preheat the oven to 350°F (175°C).
2. Using a cleaver or a serrated knife, cut the turbot in half lengthwise down the spine. Cut off the head and the fins. Slice each side into 3 equal pieces (***tronçons***) and skin the individual pieces. See pages 289–290 in *Cuisine Foundations.*

Garniture

1. Melt the butter in a large pan over low heat and sweat (***suer***) the shallot and onion until translucent. Season lightly. Add the tomatoes and parsley and continue to sweat (***suer***) for 5 minutes. Wet (***mouiller***) the pan with the white wine and reduce by half. Season to taste.

Cuisson

1. Pat the turbot pieces (***tronçons***) dry and season with salt and pepper. Place the ***tronçons*** in a large pan and spoon the tomato garnish on top. Fill the pan with just enough fish ***fumet*** to not quite cover the fish [*about 1 qt (750 mL)]*. Cover the pan with a buttered parchment lid, then bring the liquid to a gentle simmer on medium heat and transfer the pan to the oven. Braise the fish for 15 to 20 minutes, depending on the size of the pieces.

To Serve

1. Decant the fish from the pan. Strain the braising liquid through a fine mesh sieve (***chinois***), reserving the garniture intact. Reduce the braising liquid over medium heat until just thick enough to coat the back of a spoon (***à la nappe***). Mount the braising liquid with cold butter, mix in the reserved garniture and chopped parsley, and season the sauce to taste.
2. Serve the turbot covered (***napper***) with the garnisher and sauce on a hot plate. Decorate with fresh herbs.

Quantity		*Ingredient*
U.S.	*Metric*	
1 pc	1 pc	Turbot (***turbotin***), whole [*6 lb (3 kg)*]
		Garniture
1 ¾ oz	50 g	Butter
¾ oz	25 g	Shallot, finely chopped (***hacher)***
1 1¼ oz	40 g	Onion, finely chopped (***hacher)***
10 oz	300 g	Tomato, peeled (***émonder***), seeded (***épépiner***), and diced (***concasser***)
10 brs	10 brs	Parsley, finely chopped (***hacher***)
1 pt	500 mL	White wine
		Salt and pepper
		Cuisson
1 qt	1 L	Fish ***fumet*** (*reduced to 24 oz [750 mL]*)
		Salt and pepper
		To Serve
3 ½ oz	100 g	Butter
3 brs	3 brs	Parsley, finely chopped (***hacher***)
		Salt and pepper

FYI...

Chef Dugléré studied under Carême, and later went on to be one of the most celebrated chefs in France. He was chef to the Rothschild household, and the executive chef at the Café Anglais.

VOL-AU-VENT MARINIÈRE

Learning Outcomes

Pâte feuilletée
Tourer
Chiqueter
Vol-au-vent
Mousseline de poisson
Châtrer
Liaison à l'oeuf

Equipment

Knives:
Filleting knife, paring knife (*office*), slicing knife (*éminceur*), shucking knife

Tools:
Corne, baker's brush, rolling pin, docker, vol-au-vent, discs, pastry brush, bowls, rubber spatula, wooden spoon, chinois, tea spoons, skimmer (*écumoire*), small offset spatula, cooling racks, whisk, wire racks

Pans:
Baking sheet, 2 large russes, 2 medium russes, large shallow pan, gratin dish, bain marie

Serving

8 persons

VOL-AU-VENT MARINIÈRE

Seafood with White Sauce in a Puff Pastry Case

Method

1. Preheat the oven to 400°F (205°C).

Pâte Feuilletée

1. See pages 377–379 in *Cuisine Foundations,* using the quantities listed on this page.

Mousseline de Poisson

1. Fillet the sole (***lever les filets***). Remove the skin and remove any blood spots. Then transfer them to a food processor and purée them using short bursts. Press the purée through a drum sieve (***tamis***) into a bowl set in an ice bath. Using a rubber spatula, mix in the egg whites, one by one. Mix well after each addition.
2. Keeping the bowl on the ice bath, beat in the cream, little by little. Beat until the mixture is thick and pale (***monter la farce***). Season the mousseline with salt, pepper, and nutmeg.

Tip: *To check the seasoning, poach a small amount of mousseline in boiling water and taste.*

3. Cover and refrigerate until firm.

Make a Vol-au-vent Case (see pages 384–386 in Cuisine Foundations)

1. Roll out a sheet of feuilletage pastry. Using a 10-inch vol-au-vent circle as a template, mark and cut out 3 rounds of pastry. Using an 8-inch vol-au-vent circle as a template, cut out the center of one the 10-inch discs. **Note**: *Reserve only the 10-inch pastry ring for the vol-au-vent and reserve the 8-inch circle for another preparation.*
2. Place one 10-inch disc of feuilletage onto a greased baking sheet and brush a 1 ½-inch band of water around the surface of the disc (*it is important not to get water anywhere on the center of the disc as only the edges [1 ½ in. (3 cm)] are to be fused*). Lay the second 10-inch disc on top, securing it by pressing lightly around the edges. Brush a 1 ½-inch band of water around the surface of the top disc and place the 10-inch pastry ring on top. At 1-inch intervals, score the sides of the vol-au-vent with a paring knife [*this will be both decorative and will ensure that the vol-au-vent will rise evenly (**chiqueter**)*]. Brush the surface area of the pastry with egg wash. Using a sharp paring knife, score the surface inside the ring with a decorative cross-hatch pattern. Using a toothpick, deeply perforate the rim of the vol-au-vent pastry case.
3. Refrigerate for 10 to 20 minutes.

Garniture

1. *Cook the mussels à la marinière:* Melt the butter in a large pot over low heat and sweat (***suer***) the chopped shallots and onion until translucent. Wet (***mouiller***) with the

Quantity		*Ingredient*
U.S.	*Metric*	*Pâte Feuilletée*
1 lb	500 g	Flour
12 oz	350 g	Butter
8 fl oz	250 mL	Water
2 tsp	10 g	Salt
1 pc	1 pc	Egg, for egg wash
		Mousseline de Poisson
2 pcs	2 pcs	Sole [*12 oz (350 g) each*]
3 pcs	3 pcs	Egg whites
8 fl oz	250 mL	Cream
Pinch	Pinch	Nutmeg, grated
		Salt and pepper

FYI...

Vol-au-vent is a classic French pastry that is rumored to have been created by Antonin Carême. It can be described as a delicate pastry bowl that holds the savory portion of the dish. The sole is cooked with a simple white wine and herb sauce, *à la marinière*, "or mariner's style."

Quantity		Ingredient
U.S.	*Metric*	*Moules Marinières*
2 lb	1 kg	Mussels
1 ¾ oz	50 g	Butter
4 pcs	4 pcs	Shallots, finely diced (***ciseler***)
¼ pc	¼ pc	Onion, finely diced (***ciseler***)
1 pt	500 mL	White wine
1 pc	1 pc	***Bouquet garni***
		Salt and pepper
		Garniture
16 pcs	16 pcs	Shrimps
8 pcs	8 pcs	Crayfish
1 ½ pt	750 mL	Fish ***fumet***
14 oz	400 g	Mushrooms, peeled, de-stemmed, quartered
1 ¾ oz	50 g	Butter
½ pc	½ pc	Lemon, juice of
16 pcs	16 pcs	Oysters, shucked
		Salt and white pepper

VOL-AU-VENT MARINIÈRE

Method

white wine and add the ***bouquet garni***. Increase the heat to high, reduce the liquid by two-thirds, and add the mussels. Bring back to a boil and cover the pot to trap the steam. Cook the mussels for 5 minutes, shaking the pot to stir them while cooking.

2. Once the mussels are cooked, remove with a wire strainer (***araignée***). Allow the cooking liquid to settle and carefully strain through a fine mesh sieve, leaving the sediment in the pot (they contain the sand from the mussels). Remove the mussels from their shells (***décortiquer***) and pull off the mantles (***ébarber***). Bring the cooking liquid to a simmer in a small pan over medium heat and poach the mussels for 2 minutes. Reserve the mussels in 3 ½ oz (100 mL) of the cooking liquid. Strain the rest of the juice in a fine mesh sieve (***chinois***) and reserve separately in the refrigerator.
3. *Cooking the shrimp and crayfish:* Gently twist the middle tail fin of the crayfish until it comes loose (***chatrer***). Pull it out slowly and the intestine will follow.
4. Shell and devein the shrimp.
5. Bring the ***fumet*** to a gentle simmer in a medium saucepan over medium heat. Add the shrimp and crayfish and poach until they are cooked through (*5 to 10 minutes*). Drain the shrimp and crayfish and reserve them with the mussels. Add a little ***fumet*** to the container if needed to keep moist. Reserve the cooking liquid in the saucepan.
6. ***Cuire à Blanc*** *the mushrooms*: Place the mushrooms in a small saucepan with butter, lemon juice, salt, and white pepper. Cover the saucepan and cook the mushrooms over low heat until they can be easily pierced with a knife (*5 to 10 minutes*). Strain the mushrooms, set them aside with the rest of the garniture, and reserve the cooking liquid.
7. *Quenelles:* Shape 12 quenelles out of the sole mousseline using 2 teaspoons. See page 351 in *Cuisine Foundations*. Arrange the quenelles evenly in a buttered baking dish and reserve them in the refrigerator.
8. Place the saucepan with the fish ***fumet*** over high heat, bring it to a boil, and carefully pour it into the baking dish containing the quenelles. Cover the dish with foil and place it in the oven to poach for 10 to 15 minutes. Decant the quenelles onto a dish lined with paper towels. When drained, transfer the quenelles to a buttered gratin dish and refrigerate. Reserve the ***fumet*** until it is needed to heat the quenelles.
9. *Bake the vol-au-vent:* Place the vol-au-vent in the oven and bake for 10 minutes, then turn the heat down to 375°F (190°C) and continue to bake until golden. Remove the vol-au-vent from the oven.

Method

10. To make the lid for the vol-au-vent, cut around the edge of the top pastry disc with a paring knife. Carefully remove the lid and place it in the oven for 5 or 10 minutes to dry out. Reserve the vol-au-vent case and lid on a rack in a warm place.
11. *Preparing the oysters:* Shuck the oysters over a small saucepan to catch the liquid from the shell. Bring the liquid to a simmer over medium heat, add the oysters, cover them, and poach for 5 minutes. Drain the oysters (*reserving the cooking liquid*) and set them aside with the rest of the garniture.

Sauce Marinière

1. Combine all the cooking liquids [*about 1 pt (500 mL)*] in a large saucepan. Bring the liquid to a boil over medium heat and whisk in the ***beurre manié***, piece by piece, and stir until the sauce is thick enough to coat the back of a spoon (***à la nappe***). Reduce the heat to low and let the sauce simmer for 5 minutes. Adjust the seasoning to taste.
2. Mix the yolks with the cream. Add 1 or 2 ladles of the hot veloute to the yolk mixture and stir well. Off the heat, whisk the tempered yolks into the sauce. Stir over low heat until thick enough to coat the back of a spoon (***à la nappe***). Reserve the sauce in a bain marie.

Finishing

1. Bring a large pot of salted water to a boil over high heat (***chauffante***).
2. *Reheat the garniture:* Pour a small ladleful of hot fish ***fumet*** into the gratin dish containing the quenelles. Cover the dish in foil and place it in the oven for 5 minutes to reheat. Warm the vol-au-vent case in the oven at the same time. Place the bowl of garniture elements and residual cooking juice over a bain marie to reheat them.
3. Meanwhile, remove the sauce marinière from the bain marie and place it over low heat, until it reaches 158°F (70°C). Remove the sauce from the heat and mix in the chopped parsley and lemon juice. Strain the hot garniture ingredients and add them into the hot sauce.

To Serve

1. Place the vol-au-vent in the center of a heated platter.
2. Fill with the sauce and garniture.
3. Decant the quenelles onto a dish lined with paper towels; once drained, place them around the rim of the vol-au-vent case. Discard the ***fumet***.
4. Quickly reheat the crayfish in the ***chauffante*** and drain. Arrange them around the edge of the vol-au-vent case.
5. Place a sprig of fresh parsley in the center of the vol-au-vent and finish the presentation by balancing the lid on the side. Serve any leftover sauce on the side.

Tip: *To add shine, brush the vol-au-vent case and crayfish shells with clarified butter or oil (**lustrer**).*

Quantity		*Ingredient*
U.S.	*Metric*	*Sauce Marinière*
1 qt	1 L	Cooking juices
4 ¼ oz	120 g	Beurre manié
7 fl oz	200 mL	Cream
4 pcs	4 pcs	Egg yolks
2 ½ oz	75 g	Butter
¾ oz	20 g	Parsley, finely chopped (***hacher***)
1 pc	1 pc	Lemon, juice of
		Salt and white pepper
		To Serve
1 br	1 br	Parsley

Les Viandes
(Meat)

Les Viandes (Meat)

Aiguillette de bœuf bourguignonne
—Beef rump and vegetables in a red wine aspic

Blanquette de veau à l'ancienne, riz pilaf
—Traditional veal blanquette with pilaf rice

Canon et carré d'agneau rôtis, polenta
—Roasted saddle and rack of lamb

Cœur de filet Henri IV, sauce béarnaise
—Tenderloin steaks with Béarnaise sauce

Contre-filet rôti, pommes boulangères
—Roasted beef sirloin, potato and onion casserole

Côte de boeuf grillée, beurre marchand de vin, tomates provençales
—Grilled beef prime rib with shallot and red wine butter

Côte de porc charcutière, pommes purée
—Pork chops with mashed potatoes and demi-glace sauce and gherkins

Escalope de veau à la crème, petits pois à la française
—Veal escalope with a mushroom cream sauce

Escalope de veau à la Viennoise
—Breaded veal escalope with traditional garnish

Navarin d'agneau printanier
—Lamb stew with spring vegetables

Osso-Bucco Piédmontaise, risotto
—Braised slices of veal knuckle with saffron risotto

Paupiettes de veau, carottes vichy
—Stuffed veal escalopes with carrots

Pot-au-feu, raviolis d'agneau
—Poached meats and vegetables with lamb ravioli

Selle d'agneau en croûte de sel
—Lamb saddle baked in a salt crust
Steak au poivre
—Sirloin steak with pepper

AIGUILLETTE DE BOEUF BOURGUIGNONNE

Learning Outcomes

Braiser à brun
Larder
Marinade crue
Étuver
Légumes glacés
Clarification
Aspic

Knives:
Fillet knife, chef knife (*couteau chef*), vegetable peeler (*économe*), turning knife

Tools:
Bowls, wooden spatula, larding needle, colander, skimmer (*écumoire*), wire rack, chinois, ice bath, whisk, ladle (*louche*), cheesecloth or coffee filter, mold

Pans:
Sauce pans (*russe*), sauté pans, large Dutch oven (*cocotte*), stock pot with lid (*marmite*)

Serving

6–8 persons

HISTORY

"Bourguignonne" in a recipe refers both to the region of France and more specifically to preparations cooked with red wine. As a region, Burgundy (*Bourgogne*) is known principally for its vineyards and fine wine but also for its cuisine. Dijon mustard and pain d'épices, for instance, are produced in this central region of France. This recipe is a variation of boeuf bourguignonne, which is served cold in a red wine aspic with the same garnish of carrot, pearl onions, and lardons. *Aiguilette* refers to the fact that the finished piece of meat is cut into thin slices before being molded in the aspic.

AIGUILLETTE DE BOEUF BOURGUIGNONNE

Beef Rump and Vegetables in a Red Wine Aspic

Quantity		Ingredient
U.S.	**Metric**	
6 lb	3 kg	Beef (*point rump*)
13 oz	400 g	Lard strips (bard)
5 fl oz	150 mL	Vegetable oil
2 ½ qt	2.5 L	Veal stock
		Salt and pepper
		Marinade Sèche (for lard)
1 ¾ oz	50 g	Butter
2 pcs	2 pcs	Garlic cloves, finely chopped (***hacher***)
4 pcs	4 pcs	Shallots, finely chopped (***hacher***)
1 bq	1 bq	Parsley, finely chopped (***hacher***)
¾ fl oz	25 mL	Cognac
		Salt and pepper
		Marinade for Beef
2 pcs	2 pcs	Carrots, cut into ***mirepoix***
2 pcs	2 pcs	Onions, cut into ***mirepoix***
1 pc	1 pc	Leek, cut into ***mirepoix***
1 pc	1 pc	Celery stalk, cut into ***mirepoix***
4 pcs	4 pcs	Garlic cloves
1 pc	1 pc	***Bouquet garni***
3 ½ fl oz	100 mL	Cognac
3 ½ fl oz	100 mL	Red wine vinegar
5 fl oz	150 mL	Vegetable oil

Method

1. Preheat the oven to 300°F (150°C).

Marinade Sèche

1. Melt the butter in a medium sauté pan over medium heat and sauté the chopped garlic and shallots until they just begin to color. Remove the pan from the heat and mix in the chopped parsley and cognac. Transfer the marinade to a clean bowl and let it cool to room temperature.
2. Cut the lard into ¼ in. (1 cm) thick strips and stir them around in the marinade to coat them. Lay the strips of lard on a parchment paper-lined tray. Transfer the lard strips to the freezer for 15 minutes to firm them up; this will make them easier to insert into the meat.

Marinade for Beef

1. Combine all the marinade ingredients in a container that is large enough to contain both the marinade and the beef rump. Remove the strips of lard from the freezer and, working in the direction of the meat grain, insert them at regular intervals into the rump (***larder***) using a larding needle. Place the rump in the marinade and cover the container. Transfer to the refrigerator to marinate for 24 hours.

Cuisson

1. Decant the rump from the marinade and pat it dry. Strain the marinade through a strainer directly into a large saucepan. Drain the vegetables well and set them aside. Bring the marinade to a boil over high heat and skim (***écumer***) the surface of any impurities. Lower the heat to a simmer and reduce the liquid by one-third.
2. Season the meat with salt and pepper.
3. Heat the vegetable oil in a large Dutch oven (***cocotte***) over high heat. Place the rump in the pot and sear it on all sides until the meat has colored evenly. Transfer the rump to a wire rack. Place the vegetables in the pot and sear (***saisir***) them until golden. Deglaze the cocotte with the marinade liquid and add the veal stock. Return the rump to the cocotte, cover, and transfer it to the oven. Braise the meat gently for 6 hours.
4. When the meat is cooked and very tender, turn off the oven and let it cool (*overnight, if possible*) in the oven with the door closed.
5. Once cool, decant the meat from the cocotte and reserve it in the refrigerator.
6. Skim the fat (***dégraisser***) off the braising liquid and strain the contents through a fine mesh sieve (***chinois***), discarding the vegetables. If the liquid tastes bland, reduce it in a large saucepan over low to medium heat until it becomes more flavorful. Adjust the seasoning and let the braising juice (***jus de braisage***) cool to room temperature (*if needed, stir over an ice bath to speed up the process*).

Garniture

1. *Blanch the lardons:* Place the lardons in a small saucepan of cold water and bring to a boil over high heat. Let the water boil for 1 minute, then refresh the lardons under cold running water and drain them well.
2. Heat the oil in a small sauté pan over high heat and sauté the lardons until golden. Drain them on a plate that has been lined with paper towels.
3. ***Glacer à blanc*** *the pearl onions:* Place the onions in a pan that is large enough to hold them in a single layer. Pour in enough cold water so the onions are two-thirds immersed. Add the sugar and butter and season with salt. Cover with a buttered parchment paper lid (***cartouche***) and cook the onions over low heat until all the water has evaporated (***étuver***). Roll the onions in the resulting syrup to glaze them. Reserve in a clean bowl at room temperature.

Quantity		Ingredient
U.S.	Metric	Marinade for Beef, continued
3 qt	3 L	Red wine (*4 bottles*)
1 tsp	5 g	Peppercorns
¾ oz	20 g	Coarse salt
		Garniture
24 pcs	24 pcs	Lardons
4 tsp	20 mL	Vegetable oil
24 pcs	24 pcs	Pearl onions
2 tsp	10 g	Sugar, granulated
1 oz	30 g	Butter
24 pcs	24 pcs	Carrots, turned
1 tsp	5 g	Sugar, granulated
1 oz	30 g	Butter
		Salt and pepper
		Clarification
2 ½ qt	2 ½ L	Braising juice (***jus de braisage***)
⅓ pc	⅓ pc	Onion, finely chopped (***hacher***)
1 pc	1 pc	Carrot, finely chopped (***hacher***)
⅓ pc	⅓ pc	Leek white, finely chopped (***hacher***)
½ pc	½ pc	Celery stalk, finely chopped (***hacher***)
1 pc	1 pc	Tomato, finely chopped (***hacher***)
4 pcs	4 pcs	Egg whites
8 oz	250 g	Beef, finely ground (***hacher***)
1 ½ oz	40 g	Gelatin leaves
		Salt and pepper
		Sauce Gribiche
4 pcs	4 pcs	Egg yolks, hard-cooked
1 Tbsp	15 g	Dijon mustard
8 fl oz	250 mL	Vegetable oil
1 pc	1 pc	Lemon, juice of
10 pcs	10 pcs	Gherkins, finely chopped (***hacher***)
5 oz	150 g	Capers, finely chopped, (***hacher***)
2 pcs	2 pcs	Shallots, finely chopped, (***hacher***)
½ pc	½ pc	Parsley, finely chopped, (***hacher***)
		Salt and pepper

Method

4. ***Glacer à blanc** the turned carrots:* Place the carrots in a pan that is large enough to hold them in a single layer. Pour in enough cold water so the carrots are two-thirds immersed. Add the sugar and butter and season with salt. Cover with a buttered parchment paper lid (***cartouche***) and cook the carrots over low heat until all the water has evaporated (***étuver***). Roll the carrots in the resulting syrup to glaze them. Reserve with the onions.

Clarification

1. Combine the chopped onion, carrot, leek, celery, tomato, egg whites, and ground beef in a large saucepan. Season with salt and pepper. Measure out 2.5 qt (2.5 L) of the cold braising juice and slowly pour it into the saucepan, stirring well to combine all the ingredients.
2. Bring the mixture to a gentle simmer over low heat, stirring constantly to prevent the proteins from sticking to the bottom of the pan. When the liquid begins to simmer and the clarification ingredients float to the surface, stop stirring. Once a crust (*raft*) has formed on the surface of the liquid, make a hole in the center using a ladle. Let the clarification simmer gently over low heat for 1 hour, occasionally ladling the liquid (*from the hole in the raft*) to prevent raft from drying out.
3. Line a fine mesh sieve (***chinois***) with a wet cheesecloth or coffee filter. Once the clarification is complete and the liquid is clear, carefully strain the liquid through the prepared sieve into a clean recipient.

Gelée

1. Soak the gelatin leaves in cold water until they are soft, then squeeze out and discard the excess water. Dissolve the drained leaves in the hot clarified liquid (***consommé***) and stir carefully to combine. Reserve at room temperature.

Montage

1. Place a mold in the freezer to chill (***sangler***) for 10 to 15 minutes.
2. Cut the cooled beef rump into thin slices, about 1 cm thick.
3. *Temper 1 pt (500 mL) of gelée*: Pour the liquid gelée into a bowl over an ice bath and stir it gently using a metal spoon (***tempérer***). Continue until the liquid becomes oil-like in resistance and movement.

Tip: *Stir the liquid gently to avoid creating air bubbles.*

4. Remove the mold from the freezer and pour in the gelée. Gently tilt the mold to line the bottom and sides with a layer of gelée. When the gelée is set, decoratively arrange a portion of the beef slices, lardons, carrots, and onions in the bottom of the mold. Temper more gelée if needed and pour it into the mold to cover the ingredients. Let it set in the refrigerator, then create a new layer with the beef and other ingredients. Repeat this process until the mold is full. Reserve in the refrigerator for 1 to 2 hours before serving.

Sauce Gribiche

1. Press the hard-cooked egg yolks through a drum sieve (***tamis***) using a plastic scraper. Whisk the yolks and mustard together in a large round-bottomed bowl and season with salt and pepper. Continue whisking while pouring in the oil in a thin stream until the mixture is thick and glossy. Mix in the lemon juice. Taste and adjust the seasoning. Finish the sauce by stirring in the chopped gherkins, capers, shallots, and parsley.

To Serve

1. Dip the mold in hot water for 2 or 3 seconds to loosen the gelée. Unmold onto a serving dish. Serve the ***sauce gribiche*** on the side.

Tip: *If the surface of the gelée shows any irregularities or bubbles, melt them with a blowtorch; this will also give the preparation shine.*

BLANQUETTE DE VEAU À L'ANCIENNE, RIZ PILAF

Learning Outcomes

Pocher
Fond blanc de veau
Making of a velouté
Cuire à blanc
Glacer à blanc
Roux
Liaison with yolks
Cooking rice pilaf

Equipment

Knives:
Chef knife (***couteau chef***) boning knife, vegetable peeler (***économe***), paring knife (***office***)

Tools:
Wooden spatula, carving fork, kitchen twine, slotted spoon, bowls, wooden spatula, whisk, chinois

Pans:
Stock pot (***marmite***), medium saucepans, large saucepans, medium sauté pans

Serving

4 persons

FYI...

À l'ancienne means to cook "in the old style." What makes this particular recipe *à l'ancienne* is the preparation of the garniture; the pearl onions and mushrooms are prepared *glacer à blanc* and *à blanc*, respectively.

Quantity		Ingredient
U.S.	Metric	
1 pc	1 pc	Veal shoulder, deboned [2 lbs (1 kg)]
1 pc	1 pc	Veal flat rib (***plat de côte***) [2 lbs (1 kg)]
		Cuisson
1 pc	1 pc	Carrot, large
1 pc	1 pc	Onion
1 pc	1 pc	Clove
1 pc	1 pc	Celery stalk
1 pc	1 pc	Leek white
1 pc	1 pc	***Bouquet garni***
		Mushroom skins and stems (*from garniture*)
¾ oz	20 g	Coarse salt
½ tsp	5 g	White peppercorns
4 qt	4 L	Water
		Riz pilaf
		Garniture
1 ½ oz	40 g	Butter
3 ½ oz	100 g	Onion, finely chopped (***hacher***)
8 oz	250 g	Long-grain rice
12 ½ fl oz	375 mL	Water
1 pc	1 pc	***Bouquet garni***

BLANQUETTE DE VEAU À L'ANCIENNE, RIZ PILAF

Traditional Veal Blanquette with Pilaf Rice

Method

1. Preheat the oven to 350°F (180°C).
2. Cut the veal into 1-inch cubes and degorge overnight in a mixture of half milk and half water (*keep refrigerated*). Drain the meat and discard the liquid.
3. *Blanch the meat:* Place the meat in a large stock pot, cover with cold water, and bring to a boil over high heat. Let the water boil for 10 minutes and skim (***écumer***) the surface of any fat (***dégraisser***) or impurities. Drain the meat, then refresh it in an ice bath.

Cuisson

1. Wash and peel the vegetables and stud the onion with a clove (***clouter***). Place the veal, all the vegetables, the ***bouquet garni***, the coarse salt, and the whole peppercorns in a large stock pot. Add enough cold water 2 or 3 in. (5 to 8 cm) above the top of the ingredients and bring to a boil over high heat. Reduce the heat to a gentle simmer, cover the pot, and poach the veal until tender (*40 to 45 minutes*), skimming (***écumer***) regularly.

Garniture

1. *Prepare the* ***riz pilaf****:* Melt the butter in a medium saucepan over low heat and sweat (***suer***) the chopped onions until they are transparent.
2. Add the rice and stir it gently, until it begins to turn opaque and shiny (***nacrer***).
3. Wet (***mouiller***) with cold water (*1.5 times the volume of the rice*) and bring almost to a boil over medium-high heat. Remove the pan from the heat and add the ***bouquet garni***.
4. Cover the pan with a buttered parchment paper lid (***cartouche***). Place the pan in the oven for 17 minutes.
5. Remove the pan from the oven, lift the cartouche and dot the surface with butter. Replace the cartouche until the butter has melted and gently separate the grains of rice with the prongs of a carving fork (***égrener***). Set aside near the stove top to keep it warm.
6. ***Glacer à blanc*** *the pearl onions:* Place the onions in a pan that is large enough to hold them in a single layer. Pour in enough cold water so that the onions are two-thirds immersed. Add the sugar and butter and season with salt. Cover with a buttered parchment paper lid (***cartouche***) and cook the onions over low heat until all the water has evaporated (***étuver***). Roll the onions in the resulting syrup to glaze them. Reserve.

Method

7. *Cook the mushrooms **à blanc***: Peel the mushrooms and remove the stems. Add both the peelings and the stems to the veal poaching liquid. If the mushrooms are large, cut them into quarters.
8. Melt the butter in a medium saucepan over medium heat and add the lemon juice with a few tablespoons of water. Add the mushroom caps to the pan, season with salt and white pepper, and stir to coat the mushrooms in the melted butter. Cover the pan and cook the mushrooms over medium heat for 5 minutes. Drain them and set aside in a covered container.

Sauce

1. *Prepare a roux blanc:* Melt the butter in a medium pan over medium heat. When the butter begins to foam, add the flour and stir well until the mixture begins to bubble. Cook for 1 or 2 minutes to obtain a white roux (*be careful not to let it color*). Transfer the roux to a small dish and set aside to cool.
2. When the meat is cooked and tender, decant it into a bowl.
3. Strain the resulting white veal stock through a fine mesh sieve (***chinois***) and measure 1 ½ qt (1.5 L) into a large saucepan. Bring the stock to a boil over high heat. Whisk the cold roux, piece by piece, into the boiling stock until it is completely incorporated and the sauce begins to thicken. Reduce the heat to low and simmer the sauce for 10 minutes. Add three-quarters of the cream and simmer gently until the sauce is thick enough to coat the back of a spoon (***à la nappe***). Taste and adjust the seasoning.
4. Combine the egg yolks with the remaining cream. Remove the sauce from the heat and whisk 1 or 2 ladles of the hot liquid into the yolks to temper them, then whisking continuously, pour in the cream and yolk mixture in a thin stream (this produces a ***liaison***). Return the pan to low heat and stir in a figure 8 motion until the sauce coats the back of a spoon (***à la nappe***).
5. Mount the sauce with cold butter.

To Serve

1. Reheat the mushrooms, pearl onions, and meat in the sauce for 5 minutes. Then decant these ingredients into a hot serving dish. Strain the sauce through a fine mesh sieve (***chinois***) and ladle it over the meat to coat (***napper***). Sprinkle the dish with chopped parsley and serve the rice pilaf on the side.

Quantity		*Ingredient*
U.S.	*Metric*	*Garniture*
2 ½ oz	80 g	Butter, cold
20 pcs	20 pcs	Pearl onions
1 ½ oz	40 g	Butter
1 ½ oz	40 g	Sugar, granulated
10 oz	300 g	Button mushrooms (*peeled and stemmed*)
¾ oz	20 g	Butter
0.5 pc	0.5 pc	Lemon, juice of
		Salt and pepper
		Sauce
2 oz	60 g	Butter
2 oz	60 g	Flour
1 ½ qt	1.5 L	Veal cooking liquid (***fonds de cuisson***)
2 pcs	2 pcs	Egg yolks
½ pt	250 mL	Cream
1 ½ oz	40 g	Butter, cold
2 brs	2 brs	Parsley, finely chopped (***hacher***)
		Salt and pepper

Learning Outcomes

Rôtir
Désosser a carré and selle d'agneau
Manchonner
Parer
Ficeler
Farcir
Jus
Polenta

Equipment

Knives:
Cleaver (*couteau à batte*), boning knife (*désosseur*), vegetable peeler (*économe*), paring knife (*office*), slicing knife (*éminceur*), scissors (*ciseaux*), wire rack

Tools:
Bowls, cutting board, wire rack, wooden spatula, plastic spatula, metal spatula, fine chinois, colander, skimmer (*écumoire*), ladle (*louche*), whisk, kitchen twine (*ficelle*)

Pans:
Sauteuse, sautoir, sauce pan (*russe*), roasting pan (*plaque à rôtir*), bain marie, baking sheet

Serving

8 persons

CANON ET CARRÉ D'AGNEAU RÔTIS, POLENTA

Roasted Saddle and Rack of Lamb

Quantity		Ingredient
U.S.	*Metric*	
2 pcs	2 pcs	Rack of lamb with ribs
½ pc	½ pc	Lamb saddle, with flaps (***panoufle***)
1 pc	1 pc	Caul fat
4 pcs	4 pcs	Bay leaves
6 fl oz	175 mL	Butter, clarified
		Salt and freshly ground pepper
		Farce Simple
10 oz	300 g	Lean ground veal
5 oz	150 g	Lean ground pork
1 fl oz	30 mL	Vegetable oil
½ pc	½ pc	Onion, finely chopped (***ciseler***)
2 pcs	2 pcs	Garlic cloves, finely chopped (***ciseler***)
1 pc	1 pc	Shallot, finely chopped (***ciseler***)
5 oz	150 g	Mushrooms, wild
1 ½ oz	50 g	Lard
1 pc	1 pc	Egg
1 Tbsp	15 g	Parsley, finely chopped (***hacher***)
1 tsp	2.5 g	Cayenne pepper
		Salt and white ground pepper

Method

1. Preheat the oven to 375°F (190°C).
2. Clean the rack of lamb (***parer/manchonner***), truss it (***ficeler***), and cover the bones in foil to prevent them from coloring during cooking.
3. Debone (***désosser***) and clean (***parer***) the lamb saddle.

Farce Simple

1. Heat the vegetable oil in a medium sauté pan over medium-high heat and sweat (***suer***) the chopped onions until translucent. Add the garlic and shallot and sweat until soft. Add the mushrooms and sweat until cooked through, then remove the pan from the heat and let the contents cool to room temperature.
2. In a round-bottomed bowl, combine the mushroom mixture, the ground meat, the lard, and a raw egg to bind the farce (***lier***). Mix in the chopped parsley.
3. Spread out the caul fat on the work surface and arrange the bay leaves in a line down the middle. Place the saddle on top, flaps open, and spread the farce down the center of the saddle. Tightly fold the flaps (***panoufles***) over the farce to create a tight parcel. Wrap this securely in the caul fat and truss it with kitchen twine (***ficeler***).
4. Roughly chop the leftover bones with a cleaver.

Polenta

1. Bring the chicken stock to a boil in a large saucepan over high heat. Slowly sprinkle in the cornmeal while stirring with a whisk. Turn the heat down to low and cook for 25 minutes, stirring occasionally to prevent the polenta from sticking to the bottom of the pan. Season to taste and stir in the olive oil. Grease a deep dish with the butter and pour in the polenta. Let the polenta cool to room temperature, then cover and refrigerate for 1 hour. Once set, cut the polenta into even shapes using ring molds as a template. Evenly space the polenta circles on a baking sheet lined with parchment paper.

Ratatouille

1. Heat some olive oil in a large shallow pan over high heat and sear the zucchini until colored. Add one-third of the thyme, cayenne pepper, and garlic. Reduce the heat to medium and cook for 1 to 2 minutes. Season to taste, transfer to a bowl, and set aside.
2. Heat some olive oil in the same pan over high heat and sear the eggplant until colored. Add another one-third of the thyme, cayenne pepper, and garlic. Reduce the heat to medium and cook for 1 to 2 minutes. Season to taste, transfer to a bowl, and set aside.
3. Heat some olive oil in a large shallow pan over low heat and sweat the chopped onions until soft. Add the remaining garlic, cayenne pepper, and thyme. Season to taste. Add the tomato paste and cook 1 to 2 minutes (***pincer la tomate***). Add the red and green peppers, the tomatoes, and the ***bouquet garni***. Season to taste and simmer for 25 to 30 minutes. Remove the ***bouquet garni*** and toss in the shredded (chiffonade) basil.

Quantity		Ingredient
U.S.	**Metric**	*Polenta*
1 pt	½ L	Chicken stock
4 ¼ oz	125 g	Cornmeal
2 ½ fl oz	75 mL	Olive oil
¾ oz	25 g	Butter (*to grease dish*)
		Salt and pepper
		Ratatouille
4 fl oz	120 mL	Olive oil
2 pcs	2 pcs	Zucchini, medium, cut into ***bâtonnets***
1 pc	1 pc	Eggplant medium, degorged, cut into ***bâtonnets***
2 pcs	2 pcs	Garlic cloves, finely chopped
1 pc	1 pc	Onions, finely chopped (***hacher***)
2 pcs	2 pcs	Shallot, finely chopped (***hacher***)
½ oz	15 g	Tomato paste
½ pc	½ pc	Red pepper, cut into ***bâtonnets***
½ pc	½ pc	Green pepper, cut into ***bâtonnets***
2 pcs	2 pcs	Tomatoes large, peeled (***émonder***), and seeded (***épépiner***)
1 pc	1 pc	***Bouquet garni***
3 brs	3 brs	Thyme, rubbed (***hacher***)
½ bq	½ bq	Basil, shredded (***chiffonade***)
3 ½ oz	100 g	Breadcrumbs
		Salt and pepper
		Jus
1 pc	1 pc	Carrot, cut into ***mirepoix***
1 pc	1 pc	Onion, cut into ***mirepoix***
¼ pc	¼ pc	Leek white, cut into ***mirepoix***
1 pc	1 pc	Celery stalk, cut into ***mirepoix***
3 pcs	3 pcs	Garlic cloves, peeled
1 pc	1 pc	***Bouquet garni***
		Salt and pepper
		To Serve
5 oz	150 g	Parmesan cheese, grated
3 ½ oz	100 g	Butter
1 bq	1 bq	Parsley

Method

Cuisson

1. Season the lamb (*both the rack and the saddle*) with salt and pepper. Heat the oil in a large roasting pan over high heat and sear the saddle on all sides until golden. Transfer the saddle to a wire rack. Repeat all the same steps with the rack of lamb. Add the chopped bones to the bottom of the pan and cover them with a wire rack. Place the saddle of lamb on top. In a separate roasting pan, place the rack of lamb on a wire rack. Transfer the pans to the oven and roast the meat for 20 minutes (*basting well*). When the rack of lamb is cooked to medium-rare [***rosé***: *internal temperature of 130°F (54°C)*], remove it from the oven and let it rest on a wire rack. Let the lamb saddle roast until it is medium-rare (*about 10 more minutes*), then transfer it to a wire rack to rest.

Jus

1. Remove the excess fat from the roasting pan (***dégraisser***), then add the carrot, onion, leek, celery, and garlic to the bones and sauté over medium-high heat until colored. Add the ***bouquet garni*** and deglaze with 1 ½ qt (1.5 L) of water. Bring the water to a boil, then reduce the heat to a simmer. Scrape the bottom of the pan to loosen the cooking residues (***sucs***). Simmer gently until thick and rich tasting, skimming the surface (***écumer***) regularly. Taste and adjust the seasoning, then strain the ***jus*** through a fine mesh sieve (***chinois***) into a clean saucepan. Mount the ***jus*** with cold butter and reserve it in a bain marie.

To Serve

1. Sprinkle the polenta with grated parmesan cheese. Use a round cutter to cut rounds from the ratatouille. Transfer to a parchment-lined baking sheet. Place the polenta and the ratatouille in the oven for 5 to 10 minutes, until hot and lightly colored.
2. Carve the meat and arrange it on a hot serving dish. Arrange the polenta and ratatouille on either side. Pour some ***jus*** onto the dish, being careful not to cover the meat or garniture. Decorate with fresh parsley and serve the rest of the ***jus*** on the side.

Learning Outcomes

Sauter
Turning and cooking artichokes (***blanc de cuisson***)
Preparing a sauce Béarnaise
Sabayon
Cooking and cutting pommes Pont-Neuf (***friture en 2 bains***)
Tournedos

Equipment

Knives:
Chef knife (***couteau chef***), turning knife, vegetable peeler (***économe***), paring knife (***office***)

Tools:
Bowls, kitchen twine (***ficelle***), whisk, chinois, wire rack

Pans:
Medium saucepan, bain marie, small saucepan, large saucepan, medium sauté pan, deep fryer

Serving

4 persons

FYI...

Coeur de filet is a thick cut of meat taken from the center muscle (tenderloin) that follows the length of the back and rests just below the spine. It is also known as *chateaubriand*.

CŒUR DE FILET HENRI IV, SAUCE BÉARNAISE

Tenderloin Steaks with Béarnaise Sauce

Quantity		*Ingredient*
U.S.	***Metric***	
4 pcs	4 pcs	Center cut tenderloin steaks *[6 oz each (180 g)]*
½ lb	250 g	Bard
1 ½ fl oz	50 mL	Vegetable oil
		Salt and pepper
		Artichoke Hearts
4 pcs	4 pcs	Large artichokes, turned
¾ oz	20 g	Flour
1 pc	1 pc	Lemon, juice of
		Salt
		Pommes Pont-Neuf
2 ½ lb	1.2 kg	Potatoes, cut into ***Pont-Neuf***
		Oil for deep frying
		Salt

Method

1. Preheat the oven to 400°F (205°C).
2. Preheat the deep fryer to 285°F (140°C).
3. Wrap the steaks in bard (***barder***) and tie them securely with kitchen twine (***ficeler***). Set aside.

Artichoke Hearts

1. Turn the artichokes (*see pages* 160–165 in *Cuisine Foundations*) and rub them with the cut end of a lemon half to prevent discoloring. Reserve the artichoke hearts in a large bowl of cold water with the juice of 1 lemon.
2. *Prepare a* ***blanc de cuisson***: Mix the flour into a medium saucepan half full of cold water. Stir in the lemon juice and season with salt.
3. Add the artichoke hearts. If needed, add more water to cover them. Cover the saucepan with a parchment paper lid and cook the artichokes in the ***blanc de cuisson*** over medium heat until they are easily pierced with the tip of a knife (*30 to 35 minutes*). Remove the saucepan from the heat and let the artichokes cool in the ***blanc de cuisson***. When cool enough to handle, remove the choke with a spoon and return the hearts to the ***blanc de cuisson***.

Pommes Pont-Neuf

1. Peel and wash the potatoes and cut them into rectangles, ¼ in. × ¼ in. × 2 ½ to 3 in. (1 cm × 1 cm × 6 to 8 cm) (***Pont-Neuf***) (see page 145 in *Cuisine Foundations*). Rinse them well in clean cold water and pat them dry.
2. Immerse the potatoes into the hot oil and cook them until soft enough to be easily pierced with a knife but not long enough that they color. Drain the ***pommes Pont-Neuf*** and transfer them to a paper towel–lined plate. Turn the deep fryer up to 355°F (180°C).

Sauce Béarnaise

1. Place the peppercorns, chopped shallot, and half the chopped chervil and tarragon in a large saucepan. Add the white wine vinegar and water and season lightly with salt. Bring the liquid to a boil over medium-high heat and reduce it by two-thirds.
2. Remove the pan from the heat and mix the egg yolks into the reduction with a large whisk until thick. Place the pan on a cloth to prevent it from slipping and whisk in one-third of the butter in small pieces. When completely incorporated, gradually mix in the rest of the butter. The finished sauce should have the texture of a mayonnaise and be light in color.

Method

3. Pressing well with a ladle (***fouler***), strain the mixture through a fine mesh sieve (***chinois***) into a clean pan.
4. Stir in the second half of the chopped chervil and tarragon, taste the sauce, and adjust the seasoning. Reserve the sauce in a gentle bain marie.

Cuisson

1. Pat the steaks dry and season well. Heat some oil in a large pan over high heat and sauté the steaks on both sides until they reach the desired doneness. Untie the meat, remove the bard, and sear the sides of the steaks until lightly colored. Transfer the steaks to a wire rack to rest. Reserve the pan.

Finishing

1. Melt some butter in a medium saucepan over medium heat. Place the artichoke hearts upside down in the bottom of the saucepan and baste them (***arroser***) with the melted butter for 1 to 2 minutes. Cover the saucepan and transfer it to the oven for 5 minutes to reheat the artichoke hearts. Remove the pan from the oven and set aside (*still covered*).
2. Immerse the ***pommes Pont-Neuf*** into the hot oil, cook them until golden, then drain them on a dish lined with paper towels. Season immediately.
3. In the reserved pan (*used to sauté the steak*), roll the steaks in melted butter to reheat them and give them shine before serving.

To Serve

1. Place a steak on a hot plate. Stack eight ***pommes Pont-Neuf*** in a neat tower on the side of the plate. Baste (***arroser***) the artichoke hearts in melted butter (*to add shine*) and place two on the hot plate. Fill the cavities with ***sauce Béarnaise***.
2. Sprinkle the steak with ***fleur de sel*** and decorate the plate with fresh chervil.

Quantity		*Ingredient*
U.S.	***Metric***	*Sauce Béarnaise*
10 pcs	10 pcs	Peppercorns, crushed (***mignonette***)
1 pc	1 pc	Shallot, finely chopped (***hacher***)
2 fl oz	60 mL	White wine vinegar
4 pcs	4 pcs	Egg yolks
1 ¾ fl oz	50 mL	Water
10 oz	300 g	Butter
3 brs	3 brs	Fresh chervil, finely chopped (***hacher***)
3 brs	3 brs	Fresh tarragon, finely chopped (***hacher***)
		Salt and pepper
		Finishing
1 ¾ oz	50 g	Butter
		To Serve
1 br	1 br	Fresh tarragon
		fleur de sel

CONTRE-FILET RÔTI, POMMES BOULANGÈRES

Learning Outcomes

Rôtir
Ficeler
Saisir
Jus
Monter au beurre

Equipment

Knives:
Boning knife (***désosseur***), vegetable peeler (***économe***), paring knife (***office***), slicing knife (***éminceur***), scissors (***ciseaux***)

Tools:
Bowls, cutting board, fine chinois, colander, skimmer (***écumoire***), ladle (***louche***), wire rack, kitchen twine (***ficelle***)

Pans:
Small russe, gratin dish, roasting pan (***plaque à rôtir***), sauté pan, bain marie

Serving

8 persons

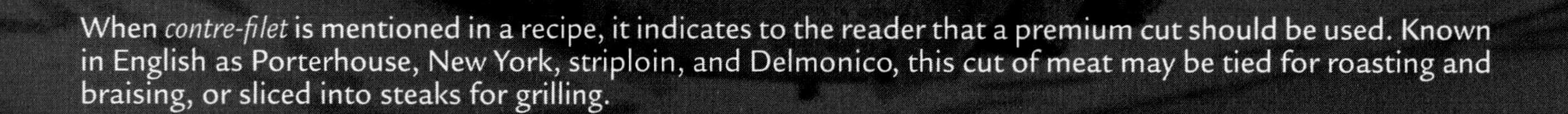

FYI...

When *contre-filet* is mentioned in a recipe, it indicates to the reader that a premium cut should be used. Known in English as Porterhouse, New York, striploin, and Delmonico, this cut of meat may be tied for roasting and braising, or sliced into steaks for grilling.

CONTRE-FILET RÔTI, POMMES BOULANGÈRES

Roasted Beef Sirloin, Potato and Onion Casserole

Quantity		Ingredient
U.S.	*Metric*	
5 lb	2.5 kg	Beef sirloin
2 pcs	2 pcs	Bard
6 oz	180 g	Butter
		Salt and pepper
		Jus
1 pc	1 pc	Carrot, cut into ***mirepoix***
1 pc	1 pc	Onion, cut into ***mirepoix***
1 head	1 head	Garlic, halved
1 pc	1 pc	***Bouquet garni***, large
3 ½ oz	100 g	Butter
		Salt and pepper
		Pommes Boulangères
6 pcs	6 pcs	Potatoes, large, sliced (***émincer***)
7 oz	200 g	Butter
3 pcs	3 pcs	Onions, large, sliced (***émincer***)
1 qt	1 L	White chicken stock or water
1 pc	1 pc	***Bouquet garni***
		Salt and pepper

Method

1. Preheat the oven to 400°F (200°C).

Contre-filet Rôti

1. Clean and trim the sirloin of silver skins, tendons, and fatty tissue. Roughly chop the trimmings and set them aside. Wrap the sirloin in bard (***barder***), truss (***ficeler***), and pat it dry. Season well.
2. Choose a roasting pan that is large enough to accommodate the sirloin. Melt the butter in the pan over medium-high heat and sear the sirloin on all sides until golden on the bard and dark on the flesh. Transfer the meat to a wire rack and add the trimmings to the pan. Place the wire rack and sirloin on top of the trimmings. Transfer the pan to the oven to roast until the sirloin is cooked medium rare [*to an internal temperature of 130°F to 140°F (55°C to 60°C)*]. Throughout the roasting process, baste (***arroser***) the sirloin with its own cooking juices.
3. Transfer the sirloin to a wire rack to rest. Reduce the oven temperature to 370°F (185°C).

Jus

1. Pour any excess fat out of the roasting pan (***dégraisser***) and place it over medium-high heat to darken the cooking residues (***pincer les sucs***). Add the carrot and onion ***mirepoix*** and sauté until colored. Add the garlic and ***bouquet garni,*** then deglaze the pan with enough water (*or chicken stock*) to cover the ingredients. Bring the water to a simmer and reduce the heat to medium. Let the ***jus*** reduce over medium heat until it is thick enough to just coat the back of a spoon (***à la nappe***). Taste and adjust the seasoning. Strain the ***jus*** through a fine mesh sieve (***chinois***) into a clean saucepan and mount the ***jus*** with cold butter. Reserve the ***jus*** in a bain marie.

Pommes Boulangères

1. Melt the butter in a large sauté pan over medium heat and sauté the onions until golden. Remove the pan from the stove.
2. Butter a large gratin dish and alternately layer the potato, then the onion slices, seasoning each layer with salt and pepper and finishing with a neatly arranged layer of potatoes. Pour in the water or chicken stock, add the bouquet garni and cover the potatoes with a circle of buttered parchment paper. Cover the dish in foil, transfer it to the oven, and bake until all the liquid is absorbed and the potatoes are easily pierced with the tip of a knife (*about 1 hour*).
3. Remove the foil and parchment paper and return the potatoes to the oven for 5 to 10 minutes to brown the surface (***gratiner***).

Note: *To preserve the starch, the potato slices should not be rinsed.*

To Serve

1. Remove the kitchen twine and bard from the sirloin. If necessary, reheat the sirloin in the oven for 5 minutes, basting it with butter to give the meat shine. For presentation, cut 2 or 3 slices of sirloin to show the doneness of the meat. Transfer the sirloin to a hot serving dish and serve the ***jus*** and ***pommes boulangères*** on the side.

Optional: *Sprinkle the roast with **fleur de sel** and fine sea salt.*

CÔTE DE BOEUF GRILLÉE, BEURRE MARCHAND DE VIN, TOMATES PROVENÇALES

Learning Outcomes

Griller
Cleaning a rib
Manchonner
Ficeler
Barder
Making a beurre composé
Reduction
Pommes Macaire
Tomates Provençales

Equipment

Knives:
Cleaver (***couteau à batte***), paring knife (***office***), slicing knife (***éminceur***), boning knife (***désosseur***), scissors (***ciseaux***)

Tools:
Whisk, rubber spatula, metal spatula, chinois, cheese grater, wire rack, metal ring, kitchen twine (***ficelle***),

Pans:
1 large pan, 1 baking sheet, small saucepan, grill, 9-by 13-inch gratin dish, medium russe

Serving

4 persons

FYI...

Beurre marchand de vin translates loosely as wine merchant's butter, and makes reference to the wine in this compound butter.

CÔTE DE BŒUF GRILLÉE, BEURRE MARCHAND DE VIN, TOMATES PROVENÇALES

Grilled Beef Prime Rib with Shallot and Red Wine Butter

Method

1. Heat the oven to 400°F (205°C).
2. Clean, oil, and heat the grill.
3. *Prepare the prime rib:* Trim off the excess fat and gristle (***parer***).
4. Remove the chine bone (***désosser***). French the bone (***manchonner***). Wrap the flesh in bard (***barder***) and tie it (***ficeler***) to maintain a regular shape. Wrap the bone in foil to prevent it from burning during cooking.
5. *Prepare the tomatoes:* Cut the tomatoes in half and remove their contents with a spoon (***évider***). Sprinkle the cavities with salt and place them upside down on a wire rack to degorge. Set aside.

Beurre Marchand de Vin

1. Reduce the red wine (*by two-thirds*) with the shallots in a small saucepan over low heat. Let the reduction cool to room temperature and season to taste.
2. Pour the reduction into the softened butter in a thin stream while whisking to create an emulsion. When all the reduction is incorporated into the butter, mix in the chopped parsley and the lemon juice.
3. Tightly roll the butter in plastic wrap to create a long cylinder about 1 ½" in. (3 cm) diameter and twist the ends to seal it. Reserve it in the refrigerator.

Note: *The butter should be soft but not be melted as into the reduction is incorporated.*

Tomates Provençales

1. Mix the breadcrumbs with the garlic and parsley (*this mixture is known as* ***persillade***). Season lightly.
2. Rinse out the tomatoes, pat them dry, and cut a small piece off the bottom to give them a stable base.
3. Season the inside of the tomato halves and fill the cavity with the ***persillade***. Set aside on a parchment paper–lined baking sheet.

Gratin Dauphinois

1. Coat a deep, ovenproof dish with softened butter and reserve in the refrigerator.
2. Combine the milk, freshly grated nutmeg, ***bouquet garni***, and garlic in a medium saucepan over medium heat. Bring the liquid to a boil, remove it from the heat, and leave it to infuse for 5 minutes.
3. Meanwhile, wash and peel the potatoes, and slice (***émincer***) them thinly [about ⅛" in. (3 mm)]. Do not rinse the potatoes as the starch is needed. Layer the potato slices tightly in the buttered dish and season each new layer with salt and pepper.
4. Stir the cream into the hot milk, then pour through a fine mesh sieve (***chinois***) onto the potatoes until the dish is full. Let rest for 1 to 2 minutes. The potatoes will absorb the liquid and the level will decrease. Top off with the remaining liquid. Season the top and sprinkle generously with grated cheese. Cover the dish with buttered parchment paper and foil. Place it in the oven to cook for 1 hour or until the potatoes are easily pierced with a knife.

Finishing

1. Remove the foil and parchment paper and return the dish to the oven to brown (***gratiner***) until the surface is golden.
2. Remove from the oven and let rest for 5 to 10 minutes before serving.
3. Once cooked, allow the gratin dauphinois to rest. Once cooled enough, cut out portions using a deep ring mold.

Cuisson

1. Season the prime rib and lightly brush them with oil. Grill to the desired doneness, giving each side two quarter turns. Let the prime rib rest on a wire rack in a warm area before serving.
2. Meanwhile, drizzle the tomatoes with olive oil and transfer them to the oven to bake until soft (*10 minutes*).

To Serve

1. If needed, reheat the ***gratin dauphinois***. Top them each with a small piece of cold butter and transfer them to the oven to reheat for 5 minutes.
2. Untie the beef ribs (*remove the bard and the foil on the bone*) and place them on a hot serving dish with the ***tomates provençales*** and the ***gratin dauphinois.*** Unwrap the ***beurre marchand de vin,*** cut it into thick slices, and arrange four slices on each piece of meat. Decorate the dish with branches of fresh watercress and serve immediately.

Quantity		*Ingredient*
U.S.	*Metric*	
2 pcs	2 pcs	Prime rib *[2 lb (1 kg each)]*
2 pcs	2 pcs	Bard
1 ¾ fl oz	50 mL	Oil
		Salt and pepper
		Beurre Marchand de Vin
13 ½ fl oz	400 mL	Red wine, good quality
2 pcs	2 pcs	Shallots, finely chopped (***hacher***)
7 oz	200 g	Butter, softened (***pommade***)
¼ bq	¼ bq	Parsley, finely chopped (***hacher***)
1 pc	1 pc	Lemon, juice of
		Salt and pepper
		Tomates Provençales
1 ½ oz	40 g	White fresh breadcrumbs
2 pcs	2 pcs	Garlic cloves, finely chopped (***hacher***)
¼ bq	¼ bq	Parsley, finely chopped (***hacher***)
4 pcs	4 pcs	Tomatoes
1 ¾ fl oz	50 mL	Olive oil
		Salt and pepper
		Gratin Dauphinois
2 ½ lb	1.2 kg	Potatoes, firm-fleshed
1 ½ qt	1 ½ L	Milk
		Freshly grated nutmeg
1 pc	1 pc	***Bouquet garni***
2 pcs	2 pcs	Cloves garlic
5 ½ fl oz	165 mL	Heavy cream
4 oz	120 g	Grated Gruyère cheese
		Unsalted butter, softened, for gratin dish
		Salt and pepper
		To Serve
1 bq	1 bq	Watercress

CÔTES DE PORC CHARCUTIÈRE, POMMES PURÉE

Learning Outcomes

Sauter
Parer
Désosser
Manchonner
Déglacer
Pincer les sucs
Monter au beurre
Sauce charcutière

Equipment

Knives:
Cleaver (*couteau à batte*), boning knife (*désosseur*), vegetable peeler (*économe*), paring knife (*office*), slicing knife (*éminceur*)

Tools:
Bowls, cutting board, wooden spatula, fine chinois, colander, skimmer (*écumoire*), ladle (*louche*), whisk, tamis, food mill, wire rack, kitchen twine (*ficelle*)

Pans:
Sauteuse, sautoir, sauce pan (*russe*), bain marie

4 persons

The etymology of the word *charcutière* can be traced back to the moniker *cuiseur de chair*, meaning "cooker of meat."

There is no English equivalent to *charcutier*. The term describes someone who specializes in the preparation of pork and pork products. Thus "charcutière" means in the style of the pork butcher's wife and is served with pork.

Sauce charcutière is a derivative of sauce Robert, one of the oldest sauces within classic French cuisine. Sauce Robert is made from onions and demi-glace that is then flavored with mustard, which also serves as a liaison. The addition of julienned cornichons turns it into a sauce charcutière.

CÔTES DE PORC CHARCUTIÈRE, POMMES PURÉE

Pork Chops with Mashed Potatoes and Demi-Glace Sauce and Gherkins

Method

1. Preheat the oven to 300°F (150°C).
2. *Prepare the pork chops*: Trim off excess fat and gristle (***parer***). Remove the chine bone (***désosser***). French the bone (***manchonner***). Tie the flesh (***ficeler***) to create a regular shape. Wrap the bone in foil to prevent it from burning during cooking.

Pommes Purée

1. Peel the potatoes and cut them into quarters. Place them in a large pot of cold salted water. Bring the water to a boil over high heat and cook the potatoes until they are easily pierced with the tip of a knife. Drain the potatoes and transfer them to a roasting pan. Place the pan in the oven for 5 to 10 minutes to dry the potatoes.
2. Press the potatoes through a drum sieve (***tamis***) or purée them through a food mill. Taste the purée and adjust the seasoning. Using a rubber spatula, work the butter into the purée until it is completely melted. Pour in the reduced cream in a thin stream, beating the purée until it reaches the desired texture. Transfer the purée to a clean container and reserve it covered on a bain marie.

Sauce Charcutière

1. Heat the oil in a large sauté pan over high heat and sauté the pork bones and trimmings until they are deeply colored. Pour any excess fat out of the pan (***dégraisser***) and sauté the chopped onion until golden. Deglaze with the white wine vinegar and reduce it until dry. Wet (***mouiller***) with the white wine and add the peppercorns and ***bouquet garni***. Reduce by half. Add the demi-glace and reduce the sauce over low heat until thick. Add gherkin (***cornichon***) juice to taste. Simmer gently for 10 minutes and set aside.

Cuisson

1. Pat the chops dry and season well. Heat the clarified butter in a large ***sauté*** pan over high heat. Place the chops in the pan. Basting them often with the butter, sauté the meat on both sides until cooked and golden colored. Transfer the pork chops to a wire rack to rest.

Finishing

1. Pour the fat out of the pan (***dégraisser***) and return it to medium heat to brown the cooking residues (***pincer les sucs***). Deglaze with the ***sauce charcutière***. Check the seasoning and consistency and adjust if necessary. Whisk in the mustard to create a ***liaison*** and strain the sauce through a fine mesh sieve (***chinois***) into a clean saucepan.

Note: *Once the mustard has been added to the sauce, do not let it boil or it will separate.*

2. Mount the sauce with cold butter and mix in the gherkin (***cornichon***) ***julienne***.

To Serve

1. Untie the chops and remove the foil from the bones.
2. Mound the ***pommes purée*** in the middle of a hot serving dish and arrange the chops on either side. Coat (***napper***) the chops in the ***sauce charcutière*** and decorate with fresh parsley.

Quantity		*Ingredient*
U.S.	***Metric***	
4 pcs	4 pcs	Pork chops *[5 oz each (160 g)]*
5 fl oz	150 mL	Butter, clarified
		Salt and pepper
		Pommes Purée
4 pcs	4 pcs	Potatoes, medium, peeled and quartered
5 oz	150 g	Butter
8 fl oz	250 g	Cream, boiled and reduced by ⅓
		Salt and pepper
		Sauce Charcutière
4 tsp	20 mL	Vegetable oil
		Pork chop trimmings and bones
2 pcs	2 pcs	Onions, medium, finely chopped (***hacher***)
1 ¾ fl oz	50 mL	White wine vinegar
7 fl oz	200 mL	White wine
1 pc	1 pc	***Bouquet garni***
1 tsp	5 g	Peppercorns, crushed (***mignonette***)
3 ½ fl oz	100 mL	Demi-glace
½ fl oz	15 mL	Gherkin (***cornichons***) juice
4 tsp	20 g	Dijon mustard
2 ½ oz	75 g	Butter
10 pcs	10 pcs	Gherkins (***cornichons***), cut into ***julienne***
		To Serve
2 brs	2 brs	Parsley (***hacher***)

ESCALOPE DE VEAU À LA CRÈME, PETITS POIS À LA FRANÇAISE

Learning Outcomes

Sauter
Rafraîchir
Escaloper
Cut lardons
Suer
Braiser à blanc
Fariner
Saisir
Débarrasser
Pincer les sucs
Dégraisser
Déglacer
Mouiller
Réduire
Chiffonade

Equipment

Knives:
Vegetable peeler (***économe***), paring knife (***office***), slicing knife (***éminceur***), meat hammer (***marteau à viande***)

Tools:
Bowls, cutting board, fork, wooden spatula, fine chinois, colander, skimmer (***écumoire***), ladle (***louche***), whisk, wire rack

Pans:
Sauteuse, sautoir, small russe, bain marie, large non stick pan, baking sheet

Serving

4 persons

FYI...

The name *petit pois* has been associated with French cuisine since the 1500s. When Catherine de Medici married King Henry II of France, she brought with her many traditional Italian dishes, including *piselli novella* (a dish featuring petit pois). Prior to this time period, peas were eaten mostly by peasants because they dried well and could be eaten through the long winter months.

ESCALOPE DE VEAU À LA CRÈME, PETITS POIS À LA FRANÇAISE

Veal Escalope with a Mushroom Cream Sauce

Quantity		*Ingredient*
U.S.	*Metric*	
4 pcs	4 pcs	Veal escalopes *[5 oz (160 g each)]*
3 ½ oz	100 g	Flour
3 ½ oz	100 g	Butter, clarified
		Salt and white pepper
		Pommes Macaire
1 lb	500 g	Potatoes
3 ½ oz	100 g	Butter
3 ½ oz	100 g	Flour
3 ½ oz	100 g	Breadcrumbs, dried
2 oz	60 g	Butter
		Salt and white pepper
		Petits Pois à la Française
2 ½ oz	75 g	Smoked pork belly, cut into small lardons
½ oz	15 g	Butter
8 pcs	8 pcs	Pearl onions
1 tsp	5 g	Sugar, granulated
1 ¾ oz	50 g	Butter
7 oz	200 g	Peas
8 pcs	8 pcs	Boston lettuce leaves, shredded (***chiffonade***)
1 ¾ oz	50 g	Butter
		Salt and pepper
		Sauce Crème
1 ¾ oz	50 g	Butter
5 oz	150 g	Button mushrooms, peeled, sliced (***émincer***)
1 ½ fl oz	40 mL	Cognac
8 fl oz	250 mL	Brown veal stock
7 fl oz	200 mL	Cream
		Salt and white pepper
		To Serve
5 brs	5 brs	Parsley, finely chopped (***hacher***)

Method

1. Preheat the oven to 370°F (190°C).

Pommes Macaire

1. Bake the potatoes in their skins (***en robe des champs***) on a baking sheet until they can be easily pierced with the tip of a knife. When they are cooked through, scoop out the flesh and season it with salt and white pepper. Using a fork, work the cold butter into the flesh of the potatoes. Press the potato mixture into small 2 cm (1 ½ in.) deep ring molds on a parchment paper-lined tray, scraping off any excess with a metal spatula. Let the molded potatoes (***pommes macaire***) cool to room temperature, then transfer the tray to the refrigerator to set for 10 minutes. Combine the breadcrumbs and flour in a large bowl. Unmold the ***pommes macaire*** and carefully roll them in the breadcrumb and flour mixture to coat them completely. Return them to the tray and press down on their centers to create depressions. Reserve the ***pommes macaire*** in the refrigerator until needed.

Tip: *To release the maximum amount of humidity from the potato flesh, pierce the skins before baking. Also, bake the potatoes on a bed of rock salt; the salt will draw out the humidity.*

Petits Pois à la Française

1. *Blanch the lardons:* Place the lardons in a small saucepan of cold water and bring to a boil over high heat. Let the water boil for 1 minute, then refresh the lardons under cold running water and drain them well.
2. ***Glacer à blanc*** *the pearl onions:* Place the onions in a sauté pan large enough to hold them in a single layer. Pour in enough cold water so the onions are two-thirds immersed. Add the sugar and butter and season with salt. Cover with a buttered parchment paper lid (***cartouche***) and cook the onions over low heat until all the water has evaporated. Roll the onions in the resulting syrup to glaze them. Reserve the onions in a clean bowl at room temperature.
3. *Cook the peas* ***à l'Anglaise***: Bring a medium saucepan of salted water to a boil over high heat. Add the peas, return the water to a boil, and cook until just tender and a vibrant shade of green (*1 to 2 minutes*). Refresh the peas immediately in an ice water bath and drain them well.
4. Melt the butter in a large sauté pan over medium-high heat and sauté the lardons until they just begin to color. Reduce the heat to low, add the lettuce, and cook it until it wilts, then add onions. Add the peas to the lettuce and lardons, then taste and adjust the seasoning. Add a pinch of sugar to taste. Once the peas are hot, add a piece of cold butter to the pan and toss its contents until the butter melts and all the ingredients are coated (***lier au beurre***). Cover the pan and set aside.

Sauce Crème

1. Melt the butter in a large sauté pan over medium-high heat and sauté the mushrooms until they are lightly colored. Season the mushrooms lightly, pour in the cognac, and flambé. Pour in the veal stock and reduce by half over medium heat. Add the cream and reduce the sauce until it is thick enough to coat the back of a spoon (***à la nappe***). Taste and adjust the seasoning. Cover the pan and set aside.

Cuisson

1. *Prepare the veal escalopes:* Lightly pound the veal ***escalopes*** with a meat hammer and pat them dry. Season them with salt and pepper and dust them very lightly with flour.
2. *Sauté the veal escalopes:* Melt the butter in a large sauté pan over medium-high heat until golden. Add the veal escalopes and sauté them on each side [baste (***arroser***) with the hot butter as they cook]. Once cooked, transfer the escalopes to a wire rack to rest. Leave the pan on the heat to concentrate the cooking residues (***pincer les sucs***) until brown, then deglaze the pan with 7 fl oz (200 mL) of water. Scrape the bottom of the pan to dissolve the cooking residue (***sucs***) and reduce (***jus***) by half. Strain the ***jus*** through a fine mesh sieve (***chinois***) and mix it into the ***sauce crème***.
3. *Cook the* ***pommes macaire***: Melt the butter in a large non stick pan over medium-high heat. Place the ***pommes macaire*** in the pan (*flat side down*) and baste them (***arroser***) with the hot butter until they are a golden color. Set aside.

To Serve

1. Arrange the veal escalopes on a heated serving dish and surround them with ***pommes macaire***. Coat (***napper***) the escalopes with ***sauce crème***. Fill the depressions in the ***pommes macaire*** with ***petits pois à la française*** and scatter the rest of the garniture around the meat.

Optional: *Sprinkle chopped parsley over the finished dish or mix it into the sauce crème just before serving.*

ESCALOPE DE VEAU À LA VIENNOISE

Learning Outcomes

Frire
Paner à l'anglaise
Flattening a piece of meat
Beurre noisette
Peler à vif
Garniture viennoise

Equipment

Knives:
Cleaver (***couteau à batte***), paring knife (***office***), slicing knife (***éminceur***)

Tools:
Spoon, shallow containers, tray, cutting board, bowls

Pans:
Large frying pan

Serving

4 persons

This is a traditional Austrian (more precisely, Viennese) preparation that is also known as *Schnitzel.* There is some discrepancy, however, regarding the origins of this dish. Some historians believe that it originated in Austria; others believe it was imported from Italy by Marshal Radetzky in 1848.

ESCALOPE DE VEAU À LA VIENNOISE

Breaded Veal Escalope with Traditional Garnish

Method

Escalope de Veau Panée à l'Anglaise

1. Line up 3 flat, shallow containers side by side on the counter. Pour the flour into the first container. In the second container, beat the egg yolks, oil, salt, and pepper (***anglaise***), and pour the breadcrumbs into the third. To complete the setup, place a tray lined with parchment paper beside the breadcrumbs.
2. *Treat the **escalopes** separately to the following process*: One at a time, lay out the ***escalopes*** between sheets of plastic wrap on a cutting board and pound them flat with the side of a cleaver (*to a thickness of ¼ in. (0.5 cm)*).
3. Roll the ***escalopes*** in the flour and gently tap off any excess. Transfer the ***escalopes*** to the egg yolk mixture (***anglaise***) and coat them completely. Finally, roll the ***escalopes*** in the breadcrumbs, then place them neatly on the tray lined with parchment paper.

Cuisson

1. Heat the oil and clarified butter in a large sauté pan over medium-high heat and sauté the ***escalopes*** until they are a golden color. Turn the ***escalopes*** and sauté them until they are golden. Drain the veal escalopes on a dish that is lined with paper towel.

Beurre Noisette

1. Melt the butter in a small sauté pan over medium heat and cook it until it begins to brown. Remove the butter from the heat and pour in the lemon juice to stop the cooking process. Swirl the contents of the pan around and set aside.

To Serve

1. Place the ***escalope Viennoise*** on a heated plate decorated with lines of egg white, yolk, parsley, and caper. Drizzle with the ***beurre noisette***. Top the ***escalope*** with a slice of lemon, an olive wrapped in an anchovy fillet, and finish the presentation with a sprig of fresh parsley.

Quantity		*Ingredient*
U.S.	*Metric*	
4 pcs	4 pcs	Veal ***escalopes*** [*5 oz (150 g) each*]
1 ¾ fl oz	50 mL	Oil
7 fl oz	200 g	Butter, clarified
		Panure à l'Anglaise
7 oz	200 g	Flour
2 pcs	2 pcs	Egg yolks
1 ¾ fl oz	50 mL	Oil
10 oz	300 g	Breadcrumbs, white
		Salt and pepper
		Beurre Noisette
7 oz	200 g	Butter
2 pcs	2 pcs	Lemons, juice of
		Garniture
1 pc	1 pc	Lemon, peeled with a knife (***peler à vif***), sliced (***trancher***)
4 pcs	4 pcs	Green olives, pitted
5 oz	150 g	Capers, finely chopped (***hacher***)
2 pcs	2 pcs	Eggs, hard-cooked, separated, finely chopped (***hacher***)
1 ¾ oz	50 g	Parsley, finely chopped (***hacher***)
4 pcs	4 pcs	Anchovy fillets (*canned*)

NAVARIN D'AGNEAU PRINTANIER

Learning Outcomes

Braiser à brun
Désosser a lamb shoulder
Parer and détailler
Saisir
Singer
Pincer
Tourner vegetables
Glacer à blanc
Glacer à brun

Equipment

Knives:
Cleaver (***couteau à batte***), boning knife (***désosseur***), vegetable peeler (***économe***), paring knife (***office***), slicing knife (***éminceur***), turning knife

Tools:
Bowls, cutting board, wooden spatula, fine chinois, ice bath colander, ladle (***louche***), wire rack

Pans:
Cocotte, sauteuse, small russes

Serving

8 persons

FYI...

Spring lamb is a designation for a lamb that is born in early spring and is slaughtered at an age no older than 8 months. Young lamb possesses a meat that is both mild and tender.

NAVARIN D'AGNEAU PRINTANIER

Lamb Stew with Spring Vegetables

Method

1. Preheat the oven to 350°F (175°C).
2. Bone (***désosser***) the lamb shoulders and trim off (***parer***) the largest pieces of fat and gristle. Cut the meat into 1 to 2 in. (2.5 to 5 cm) cubes (*cutting against the grain*). Pat the meat dry and season it.

Cuisson

1. Heat the peanut oil in a large Dutch oven (***cocotte***) over high heat. Add just enough lamb to cover the bottom of the pot and sear it until dark but not burnt (***saisir à brun***). Repeat this process until all the meat is seared and then set aside. Lower the heat to medium and sear the shallots, onions, and garlic until they pick up the colors from the residue (***sucs***) left by the meat. Add the tomato paste and cook for 1 to 2 minutes (***pincer la tomate***), stirring well. Add the flour and cook (***singer***) for 1 to 2 minutes, also stirring well. Add the tomato quarters and cook until the juices evaporate. Return the lamb to the cocotte, turn the heat up to high, then add the water and the ***bouquet garni***. Bring the water to a boil, cover, and transfer it to the oven to braise gently for approximately 1 ½ to 2 hours.

***Note:** Be sure to stir the ingredients thoroughly before putting the cocotte in the oven.*

Garniture

1. ***Glacer à blanc** the turned carrots:* Place the carrots in a pan that is large enough to hold them in a single layer. Pour in enough cold water so the carrots are two-thirds immersed. Add half of the sugar and half of the butter and season with salt. Cover the carrots with a buttered parchment paper lid (***cartouche***) and cook over low heat until all the water has evaporated. Roll the carrots in the resulting syrup to glaze them. Reserve in a covered bowl at room temperature.
2. ***Glacer à blanc** the turned turnips:* Place the turnips in a pan that is large enough to hold them in a single layer. Pour in enough cold water so that the turnips are two-thirds immersed. Add the remaining sugar and butter and season with salt. Cover the turnips with a buttered parchment paper lid (***cartouche***) and cook over low heat until all the water has evaporated. Roll the turnips in the resulting syrup to glaze them. Reserve the turnips with the carrots.

***Note:** Depending on the type of turnip and the season, a degree of bitterness may be present. Taste a piece of raw turnip and, if it is bitter, blanch the turnips before cooking them.*

Quantity		*Ingredient*
U.S.	*Metric*	
2 pcs	2 pcs	Lamb shoulder *[5 lbs (2.5 kg) each]*
½ pt	150 mL	Peanut oil
		Salt and pepper
		Garniture Aromatique de Cuisson
2 pcs	2 pcs	Shallots, cut into ***mirepoix***
2 pcs	2 pcs	Onions, cut into ***mirepoix***
½ pc	½ pc	Garlic head
2 oz	60 g	Tomato paste
2 ½ oz	80 g	Flour
2 pcs	2 pcs	Tomatoes, large, quartered
1 pc	1 pc	***Bouquet garni***
2 qt	2 L	Water
		Salt and pepper
		Garniture
40 pcs	40 pcs	Carrots, turned
40 pcs	40 pcs	Turnips, turned
3 ½ oz	100 g	Butter
1 oz	30 g	Sugar
24 pcs	24 pcs	Pearl onions, turned
1 oz	30 g	Butter
½ oz	15 g	Sugar
7 oz	200 g	Green beans, topped, tailed
7 oz	200 g	Peas
		Salt and pepper

Method

3. ***Glacer à brun** the pearl onions:* Place the onions in a sauté pan that is large enough to hold them in a single layer. Pour in enough cold water so the onions are two-thirds immersed. Add the sugar and butter and season with salt. Cover with a buttered parchment paper lid (***cartouche***). Cook the onions over low heat until all the water has evaporated and the sugar begins to caramelize and turn a golden color (***à brun***). Roll the onions in the resulting syrup to glaze them and reserve with the carrots and onions.
4. *Cook the green beans **à l'anglaise***: Bring a medium saucepan of salted water to a boil over high heat. Add the beans, return the water to a boil, and cook until just tender and a vibrant shade of green (*5 minutes*). Refresh the beans immediately in an ice water bath and drain them well. Set aside.
5. *Cook the peas **à l'anglaise***: Bring a medium saucepan of salted water to a boil over high heat. Add the peas, return the water to a boil, and cook until just tender and a vibrant shade of green (*1 to 2 minutes*). Refresh the peas immediately in an ice water bath and drain them well. Reserve them with the green beans.

Finishing

1. Bring a large saucepan of salted water to a simmer over medium heat (***chauffante***).
2. Decant the lamb into a clean bowl. Strain the braising liquid (***jus de braisage***) through a fine mesh sieve (***chinois***) into a large saucepan. If necessary, reduce the liquid over medium heat until it is thick enough to just coat the back of a spoon (***à la nappe***). Taste and adjust the seasoning, then add the braised lamb and three-fourths of the garniture vegetables.

To Serve

1. Place the braised lamb and garniture in a heated serving dish.
2. Reheat the green beans and peas by placing them in some melted butter for 30 seconds. Combine the green beans and peas with the one-quarter remaining garniture and serve them on top of the lamb.
3. The reserving of some of the garniture to be placed on top is a modern presentation meant to bring out the color of the spring vegetables. The navarin is a very traditional dish, so normally its presentation would be fairly rustic with the vegetables mixed in with the meat and sauce.

Learning Outcomes

Braising
Preparing veal knuckles
Brunoisette
Cooking a risotto
Nacrer
Lier au beurre

Equipment

Knives:
Boning knife (***désosseur***), vegetable peeler (***économe***), paring knife (***office***), slicing knife (***éminceur***)

Tools:
Bowls, cutting board, fork, wooden spatula, fine chinois, araignée, skimmer (***écumoire***), ladle (***louche***), cheese grater, wire rack

Pans:
Cocotte pot, sauce pan (***russe***), small sauté pan

Serving

4 persons

FYI...

À la piémontaise is used to describe a dish that is influenced by the produce of the Piedmont region of northern Italy. This could include Piedmont truffles, risotto, polenta, ravioli, or macaroni.

Osso-Bucco is Italian in its origins and translates as "bone with a hole."

Quantity		*Ingredient*
U.S.	*Metric*	
4 pcs	4 pcs	Veal knuckles (***jarret***) *[8 oz (250 g) each]*
3 ½ oz	100 g	Flour
		Salt and pepper
		Cuisson
1 pc	1 pc	Onion, cut into ***mirepoix***
1 pc	1 pc	Carrot, cut into ***mirepoix***
1 pc	1 pc	Celery stalk, cut into ***mirepoix***
½ pc	½ pc	Garlic head
4 tsp	20 g	Tomato paste
2 ½ oz	75 g	Flour
2 pcs	2 pcs	Tomatoes, seeded (***épépiner***), cut into quarters or eighths
1 pc	1 pc	Orange, zest of
10 fl oz	300 mL	White wine
1 ½ qt	1 ½ L	Brown veal stock
1 pc	1 pc	***Bouquet garni***
1 ¾ fl oz	50 g	Olive oil
¾ oz	50 g	Butter
		Salt and pepper

OSSO-BUCCO PIÉMONTAISE, RISOTTO

Braised Slices of Veal Knuckle with Saffron Risotto

Method

1. Preheat the oven to 350°F (175°C).
2. Remove the marrow from the veal knuckles. Degorge the knuckles and marrow in milk (*overnight in the refrigerator*).
3. Rinse off the knuckles and marrow and pat them dry. Set aside the marrow for the risotto.
4. Score the silver skins that surround the knuckles to prevent them from retracting during cooking. Season the veal knuckles with salt and pepper and lightly dust them with flour on both sides.

Cuisson

1. Heat the olive oil and butter together in a large Dutch oven (***cocotte***) over high heat. Add the veal knuckles to the pot, sear (***saisir***) them until golden, then transfer them to a wire rack. In the same pot, sauté the onion, carrot, and celery ***mirepoix*** until they begin to color. Add the garlic and sauté for 2 to 3 minutes, then add the tomato paste and cook for another 1 to 2 minutes (***pincer la tomate***), stirring well. Add the flour and cook 1 to 2 minutes (***singer***), stirring well. Add the tomatoes and cook until the liquid evaporates. Stir in the orange zest and deglaze the pan with the white wine. Reduce the wine by half, add the ***bouquet garni***, and return the veal knuckles to the pot. Wet (***mouiller***) the cocotte with the cold brown veal stock. Taste and adjust the seasoning. Cover and transfer to the oven to braise until the meat is soft (*1 ½ to 2 hours*).

Note: *The level of the braising liquid should be slightly lower than the height of the veal knuckles.*

Garniture

1. Melt the butter in a small sauté pan over low heat. Add the onion and carrot, season with a pinch of salt, then cover the pan and allow the vegetables to cook until soft [*without allowing them to color (**étuver**)*]. Add the tomato and cook for 1 minute. Toss in the parsley and lemon and orange zest and remove the pan from the heat. Adjust the seasoning and set the garniture aside until needed.

Risotto

1. Heat the olive oil in a large sauté pan over low heat. Add the bone marrow and wait until it is completely melted, then sweat (***suer***) the onion until it is soft. Add the rice and cook it until it is shiny and opaque (***nacrer***). Wet (***mouiller***) the pan with the white wine and add the saffron. Turn the heat up to medium and gently stir the contents of the pan with a wooden spatula until all the wine is absorbed. Season lightly, add some of the chicken

Method

stock, and stir the rice continuously until the liquid is completely absorbed; add more stock if needed. Traditionally, the desired consistency of risotto is *al-dente,* which means it is easily bitten but still offers some resistance. Mix the cold butter into the rice until it is melted (***lier au beurre***), then add the grated parmesan cheese. The consistency should be creamy but not sticky. Remove the pan from the heat and reserve.

Note: *Risotto cannot be left too long after cooking. It is recommended to pre cook the rice a little firm and finish the risotto just before serving it (**à la minute**). Also, the quality of the rice can have an influence on how much or how little liquid it absorbs.*

Finishing

1. Remove the cocotte from the oven and decant the veal knuckles onto a wire rack. Strain the cooking liquid (***jus de braisage***) through a fine mesh sieve (***chinois***) and discard the ***garniture de braisage***. If needed, reduce the liquid in a medium saucepan over medium heat until it is just thick enough to coat the back of a spoon (***à la nappe***). Taste and adjust the seasoning. To keep it moist, reserve the veal knuckles in the finished sauce.
2. Gently reheat the garniture in a small sauté pan over low heat and add to the sauce.
3. If the risotto is a little sticky, add a small ladle of chicken stock and briefly stir it over low heat. This will return the risotto to its creamy consistency.

To Serve

1. Arrange the veal knuckles on a heated serving dish. Coat them with the ***jus de braisage*** (***napper***).
2. Serve the risotto on the side.

Quantity		*Ingredient*
U.S.	*Metric*	*Garniture*
1 ¾ oz	50 g	Butter
3 ½ oz	100 g	Onion, finely chopped (***hacher***)
3 ½ oz	100 g	Carrot, cut into ***brunoisette***
3 ½ oz	100 g	Tomato, peeled (***émonder***), seeded (***épépiner***), cut into ***brunoisette***
10 brs	10 brs	Parsley, finely chopped (***hacher***)
1 pc	1 pc	Lemon, zest of
1 pc	1 pc	Orange, zest of
		Salt and pepper
		Risotto
1 ¾ fl oz	50 g	Olive oil
1 ¾ oz	50 g	Bone marrow
1 pc	1 pc	Onion, finely chopped (***hacher***)
7 oz	200 g	Arborio rice
½ pt	250 mL	White wine
¼ tsp	2 g	Saffron
1 pt	500 mL	Chicken stock
3 ½ oz	100 g	Butter
2 ½ oz	75 g	Parmesan cheese
		Salt and pepper

Learning Outcomes

Braiser
Farce simple
Barder
Paupiettes
Carottes vichy
Blanching zest

Equipment

Knives:
Cleaver (***couteau à batte***), vegetable peeler (***économe***), paring knife (***office***), slicing knife (***éminceur***), scissors (***ciseaux***)

Tools:
Bowls, cutting board, fork, wooden spatula, fine chinois, skimmer (***écumoire***), ladle (***louche***), wire rack, kitchen twine (***ficelle***)

Pans:
Cocotte, sautoir, small russe, bain marie

Serving

4 persons

HISTORY FYI...

Blanching citrus fruits is an effective means to remove bitterness from the rind.

Vichy is a spa town in France that has been popular for its healing baths since Roman times. Its naturally sparkling and salty water is known not only for its culinary uses but also for its healing properties. The first appearance of *carottes Vichy* in print shows up in a book titled *365 Manières d'Accommoder les Oeufs et 100 Potages* (1883). The recipe was called *oeufs polignac aux carottes Vichy*. In *La Cuisinière Cordon Bleu* (1896) it says they are called *carottes Vichy* because the carrots are either blanched in Vichy water or Vichy salt is used in the cooking process.

PAUPIETTES DE VEAU, CAROTTES VICHY

Stuffed Veal Escalopes with Carrots

Quantity		Ingredient
U.S.	Metric	
8 pcs	8 pcs	Veal ***escalopes*** *[3½ oz (100 g) each]*
8 pcs	8 pcs	Strips of bard
3 ½ fl oz	100 g	Butter, clarified
		Salt and pepper
		Farce Simple
7 oz	200 g	Veal, finely chopped (***hacher***)
2 ¾ oz	80 g	Bard, finely chopped (***hacher***)
1 ¾ oz	50 g	Butter
2 pcs	2 pcs	Garlic cloves, finely chopped (***hacher***)
2 pcs	2 pcs	Shallots, finely chopped (***hacher***)
1 pc	1 pc	Egg
7 oz	20 g	Parsley, finely chopped (***hacher***)
		Salt and pepper
		Garniture Aromatique
3 ½ oz	100 g	Onions, cut into ***brunoise***
3 ½ oz	100 g	Carrots, cut into ***brunoise***
1 ¾ oz	50 g	Leek whites, cut into ***brunoise***
1 ¾ oz	50 g	Celery stalk, cut into ***brunoise***
1 ¾ oz	50 g	Shallots, finely chopped (***hacher***)
2 pcs	2 pcs	Garlic cloves, finely chopped (***hacher***)
1 oz	30 g	Tomato paste
2 pcs	2 pcs	Tomatoes, peeled (***émonder***), seeded (***épépiner***), and diced (***concasser***)
1 pc	1 pc	***Bouquet garni***
1 pc	1 pc	Orange, juice of
1 pc	1 pc	Lemon, juice of
3 ½ fl oz	100 mL	White wine
1 qt	1 L	Brown veal stock
		Salt and pepper
		Carottes Vichy
4 pcs	4 pcs	Carrots, sliced (***émincer***)
3 ½ oz	100 g	Butter
1 pt	500 mL	Vichy water
¾ oz	20 g	Sugar, granulated (*optional*)
1 br	1 br	Parsley, finely chopped (***hacher***)
		Salt and pepper
		To Serve
1 ½ oz	50 g	Butter
2 brs	2 brs	Parsley, finely chopped (***hacher***)
1 pc	1 pc	Lemon zest, cut into ***julienne***
1 pc	1 pc	Orange zest, cut into ***julienne***

Method

1. Preheat the oven to 350°F (175°C).

Farce Simple

1. Melt the butter in a small sauté pan over low heat and sweat (***suer***) the garlic and shallots until soft. Remove the pan from the heat and let the contents cool to room temperature. Combine the cooled shallots, garlic, bard, and ground veal. Mix in the egg (*to bind the farce*) and the chopped parsley, then season with salt and pepper. Reserve in the refrigerator until needed.

Paupiettes

1. One at a time, lay out the ***escalopes*** between sheets of plastic wrap and pound them flat with a meat pounder or the bottom of a saucepan. Arrange 1 oz (35 g) of the farce down the center of each ***escalope*** and roll them into cylinders. Wrap a strip of bard around each paupiette (***barder***) to close the ends and hide the overlap of meat. Tie the paupiettes securely in kitchen twine (***ficeler***).

Cuisson

1. Pat the paupiettes dry and season well.
2. Heat the clarified butter in a medium Dutch oven (***cocotte***) over medium-high heat. Place the paupiettes in the Dutch oven, sear (***saisir***) them until golden, then transfer to a wire rack. Add the ***brunoises*** of onion, carrot, leek whites, and celery, and sauté until they just begin to color. Season lightly. Add the chopped shallots and garlic, reduce the heat to medium, and cook for 1 to 2 minutes, stirring well. Add the tomato paste and cook for 1 to 2 minutes (***pincer la tomate***), stirring well. Add the tomato dice and the ***bouquet garni***, then deglaze with the lemon and orange juice. Wet (***mouiller***) the cocotte with white wine and reduce by half. Add the veal stock and return the paupiettes to the cocotte. Bring the liquid to a boil over high heat, cover, and transfer to the oven to braise for 25 to 30 minutes.

Carottes Vichy

1. Melt the butter in a pan, add the carrots and stir to coat. Add the Vichy water or other sparkling mineral water, season with salt and pepper. Bring to a simmer, cover with a paper cover (***cartouche***) and cook gently until the liquid is reduced to a glaze. Set aside and keep warm.

Finishing

1. *Blanch the orange and lemon zests:* Place the zests in a small saucepan of cold water. Bring the water to a boil over high heat, then immediately drain and refresh the zests under cold running water. Repeat this process two more times.
2. Remove the paupiettes from the oven and decant them onto a wire rack. Cover the paupiettes to keep them from drying out.
3. Strain the contents of the cocotte through a fine mesh sieve (***chinois***). If needed, reduce the liquid (***jus de braisage***) in a medium saucepan over medium heat until it is thick enough to coat the back of a spoon (***à la nappe***). Skim (***écumer***) off any impurities from the surface of the liquid.
4. Mount the ***jus de braisage*** with cold butter.

To Serve

1. If necessary, reheat the carottes Vichy over medium heat.
2. Remove the string and the bard from the paupiettes and arrange them on a heated serving dish. Spoon on the ***jus de braisage*** to coat the paupiettes. Pour the rest of the ***jus*** into the bottom of the dish and top each paupiette with zest and chopped parsley.
3. Toss the chopped parsley into the carottes Vichy and serve them on the side.

Learning Outcomes

Pocher
Tourner
Habiller
Brider
Bouquet garni
Mijoter
Écumer
Pâte à pâtes
Hacher
Sauter
Pincer
Mouiller
Réduire
Abaisser

Equipment

Knives:
Vegetable peeler (***économe***), paring knife (***office***), slicing knife (***émninceur***), scissors (***ciseaux***), tourner knife

Tools:
Bowls, cutting board, fork, wooden spatula, plastic spatula, rolling pin, fine chinois, colander, araignée, skimmer (***écumoire***), ladle, cheese grater, wire rack

Pans:
Cocotte pot, russe, stock pot (***marmite***), bain marie

Serving

8 persons

POT-AU-FEU, RAVIOLIS D'AGNEAU

Poached Meats and Vegetables with Lamb Ravioli

Quantity		Ingredient
U.S.	*Metric*	
3 lb	1 ½ kg	Beef leg (***gite***)
		Or
		Beef shoulder (***paleron***)
1 lb	500 g	Chicken
		Salt and white pepper
		Garniture de Cuisson
1 pc	1 pc	Onion, stuck with a clove (***clouté***)
2 pcs	2 pcs	Carrots
2 pcs	2 pcs	Leeks
2 pcs	2 pcs	Celery
½ pc	½ pc	Head of garlic
1 pc	1 pc	Onion, halved, grilled for color
1 pc	1 pc	***Bouquet garni***
		Salt
		Pâte à Pâtes
8 oz	250 g	Flour
1 tsp	5 g	Salt
2 pcs	2 pcs	Eggs
½ fl oz	10 mL	Olive oil
½ fl oz	10 mL	Water (*optional*)

Method

1. Remove the excess fat and silver skins from the beef (***parer***).
2. Degorge the beef, bones, and chicken in cold water in the refrigerator for 24 hours (*changing the water every few hours*).
3. *Blanch the meat:* Place the beef (*with bones*) and chicken in a large pot and pour in enough water to cover the ingredients by 2 inches. Bring the water to a boil over high heat and let it boil for 10 minutes, periodically skimming off the impurities (***écumer, dépouiller***) and fat (***dégraisser***). Refresh the meat under cold running water and transfer it to a clean pot with the studded onion (***oignon clouté***), grilled onion, carrots, leeks, celery, garlic, and the ***bouquet garni.*** Pour in enough cold water to cover the ingredients by 3 or 4 inches. Season with salt and bring the water to a boil over high heat. Skim the surface of impurities (***écumer, dépouiller***) and fat (***dégraisser***). Reduce the heat to a gentle simmer and let it cook (***mijoter***) for a minimum of 4 hours, but remove the chicken after 1 ½ hours. Skim (***écumer***) the surface regularly. Taste the broth and adjust the seasoning.

Pâte à Pâtes

1. Sift the flour onto the work surface. Using a plastic scraper (***corné***), make a well (***fontaine***) in the center of the pile of flour. Beat the salt, egg, and oil together with a whisk and pour them into the well. Using a ***corne***, incorporate the flour into the egg mixture a little at a time. Add water if the dough is too dry. Work the dough until it is homogeneous and roll it into a ball. Wrap it in plastic and let it relax in the refrigerator for a minimum of 20 minutes.

Farce Simple

1. Melt the butter in a small sauté pan and sweat (***suer***) the shallot and garlic until soft. Season to taste, remove the pan from the heat, and allow its contents to cool to room temperature. Once cool, mix the shallot and garlic into the chopped meat and fat. Whisk the egg and ***glace d'agneau*** together and mix them into the farce to bind it. Finish the farce by mixing in the chopped parsley and mint. Season with salt and white pepper and reserve in the refrigerator until needed.

Ravioli

1. Using a rolling pin or a pasta roller, roll out (***abaisser***) the dough until it is 1–2 millimeters thick. Cut the sheet of pasta into 2 equal parts. On one sheet, space small mounds of farce (*approximately the size of a pearl onion*) 2 inches apart. Brush egg wash on the dough in between the farce and lay the second sheet of dough on top. Being careful not to create any air bubbles, seal the dough around each mound of farce by pressing with your fingers. Cut out the ravioli (***détailler***) using a 1 ½ inch round cookie cutter. Reserve the ravioli on a tray lined with parchment paper and dust with flour to prevent them from sticking.

Quantity		Ingredient
U.S.	Metric	Farce Simple
¾ oz	25 g	Butter
1 pc	1 pc	Shallot, finely chopped (***hacher***)
1 pc	1 pc	Garlic clove, finely chopped (***hacher***)
5 oz	150 g	Lamb shoulder, finely chopped (***hacher***)
1 ¾ oz	50 g	Pork fat, finely chopped (***hacher***)
½ pc	½ pc	Egg
2 ½ fl oz	75 mL	Lamb stock, reduced to a glaze (***glace d'agneau***)
3 brs	3 brs	Parsley, finely chopped (***hacher***)
1 br	1 br	Mint, finely chopped (***hacher***)
		Salt and white pepper
		Garniture
4 pcs	4 pcs	Leek whites (*tied in a bundle*)
1 pc	1 pc	Celery heart
10 oz	300 g	Carrots, turned
10 oz	300 g	Turnips, turned
		Salt and white pepper
		To Serve
½ bq	½ bq	Parsley, finely chopped (***hacher***)
32 pcs	32 pcs	Pickled pearl onions
32 pcs	32 pcs	Gherkins (***cornichons***)
1 ¼ oz	35 g	Dijon mustard
1 ¼ oz	35 g	Coarse salt
1 ¼ oz	35 g	Gruyère cheese, grated
8 pcs	8 pcs	Baguette slices, toasted
2 ½ fl oz	75 g	Butter, clarified

Method

Tip: *Because farce can spoil quite easily, ravioli are best served on the day they are prepared. Ravioli can be kept in an airtight container (dusted with semolina flour) for a maximum of 2 days. Other options include flash-freezing the ravioli or pre cooking and reserving them in the refrigerator.*

Cooking the Garniture

1. Wrap the vegetables individually in a clean cloth and tie them with string. Place the bundles in the pot with the simmering meat starting with the longer cooking vegetables first (leeks, carrots, celery, and finally the turnips). Cook until soft. Drain the vegetables, unwrap them, and set them aside. This keeps them separated from the aromates while adding more flavor to the bouillon.

To Serve

1. Decant the beef from the broth (***bouillon***) and carve the beef and chicken into serving portions. Reserve both the carved meat and the resulting marrow bones.
2. Strain the broth (***bouillon***) through a fine mesh sieve (***chinois***) into a clean saucepan over low heat. Reheat the meat, marrow bones, and vegetables in the hot broth (***bouillon***), then divide them evenly between 8 heated bowls.
3. *Cook the ravioli:* Heat some bouillon in a pan and poach the ravioli 2 to 3 minutes (***until al-dente***). Drain the ravioli and divide them between the 8 bowls. Strain the ravioli liquid back into the bouillon.
4. Pour a small ladle of hot broth (***bouillon***) into each bowl and sprinkle with chopped parsley.
5. On the side, serve little dishes of pickled pearl onions, gherkins, Dijon mustard, coarse salt, and grated Gruyère cheese to season the pot-au-feu. Also serve with slices of toasted baguette brushed with clarified butter to spread the marrow on.

Notes: *Leftover broth (**bouillon**) can be used to make a **consommé double**, or to make vermicelli soup (**soupe du lendemain**).*

SELLE D'AGNEAU EN CROÛTE DE SEL

Learning Outcomes

Preparing sweetbreads
Braising, boning, and stuffing a lamb saddle
Roasting in a salt crust
Making pommes fondantes

Equipment

Knives:
Boning knife (***désosseur***), paring knife (***office***), chef knife (***couteau chef***), turning knife, cleaver (***couteau à batte***)

Tools:
Colander, ice bath, wooden spatula, kitchen twine, wire rack, slotted spoon, rolling pin, pastry brush

Pans:
Medium saucepan, small and large sauté pans, large roasting pan, bain marie, baking sheet

Serving

4 persons

FYI...

The salt crust in this preparation creates a sealed environment and ensures that the meat stays moist.

Quantity		Ingredient
U.S.	Metric	
1 pc	1 pc	Lamb saddle *[6 lbs (3 kg)]*
1 pc	1 pc	Strip of bard, large
10 fl oz	300 g	Butter, clarified
		Salt and pepper
		Braised Sweetbread
1 pc	1 pc	Veal sweetbread
1 ¾ fl oz	50 g	Butter, clarified
1 ¾ fl oz	50 mL	Veal stock
		Salt and pepper
		Braised Chestnuts
1 ¾ fl oz	50 g	Butter
5 oz	150 g	Chestnuts
½ pc	½ pc	Onion, cut into ***mirepoix***
½ pc	½ pc	Carrot, cut into ***mirepoix***
½ pc	½ pc	Celery stalk, cut into ***mirepoix***
1 pc	1 pc	Garlic clove
1 pc	1 pc	***Bouquet garni***
12 fl oz	350 mL	Veal stock
		Salt and pepper

SELLE D'AGNEAU EN CROÛTE DE SEL

Lamb Saddle Baked in a Salt Crust

Method

1. Preheat the oven to 400°F (205°C).

Braised Sweetbreads

1. Prepare the sweetbreads (***ris de veau***) in advance. See pages 321-322 in *Cuisine Foundations*.
2. Pat the sweetbreads dry and season with salt and pepper. Heat the clarified butter in a small sauté pan over medium-high heat and sear (***saisir***) the sweetbreads on all sides until it is a golden color. Deglaze the pan with the veal stock, cover the pan, and transfer it to the oven to braise for 20 minutes. Remove the pan from the oven and decant the sweetbreads onto a wire rack to cool to room temperature. Reserve the braising liquid for the sauce.

Braised Chestnuts

1. Melt the butter in a large sauté pan over medium-high heat and sauté the chestnuts until they are lightly colored. Add the onion, carrot, and celery ***mirepoix*** and sauté for 1 to 2 minutes. Season with salt and pepper, add the garlic clove and ***bouquet garni***, then deglaze with veal stock. Cover the pan and transfer it to the oven to braise for 15 minutes.
2. Remove the pan from the oven and decant the chestnuts to a clean container to cool to room temperature. Reserve the braising liquid for the sauce.

Farce Simple

1. Melt the butter in a small sauté pan over low heat and sweat (***suer***) the garlic, onion, and shallots until soft. Remove the pan from the heat and let its contents cool to room temperature. Once cooled, mix the onion, shallots, and garlic with the chopped meat and fat. Cut the sweetbreads into a ***brunoise*** and mix it, along with the whole chestnuts, into the farce. Mix in the chopped sage and reserve the farce in the refrigerator until needed.

Preparing the Selle d'agneau

1. Debone the lamb saddle. Roughly chop the bones (***concasser***) with a cleaver and set them aside. Lay the saddle out flat on the work surface and lay the farce along its center (*lengthwise*). Roll the saddle up around the farce to form a cylinder and wrap it from end to end in the bard. Make sure the overlapping seam as well as the open ends of the meat are covered. Tie the stuffed saddle with kitchen twine (***ficeler***). Reserve it in the refrigerator until needed.

Pâte à Sel

1. Combine the salt, water, flour, and egg whites to make a dough. Mix in the peppercorns and various herbs. Cover with a damp cloth and reserve at room temperature.

Cuisson

1. Pat the saddle of lamb dry and season it well. Heat the clarified butter in a roasting pan over high heat. Add the saddle of lamb and sear (***saisir***) it until it is well colored on all sides (*5 to 10 minutes*). Transfer the saddle to a wire rack to cool completely.

Method

Jus

1. Meanwhile, pour off and reserve the cooking juices from the lamb. Place the roasting pan with the bones on high heat, add the ***mirepoix*** of carrot, onion, leek, and celery, and sauté them until they begin to color. Add the garlic cloves and the ***bouquet garni***. Deglaze the roasting pan with the braising liquid from the chestnuts and sweetbreads as well as with the roasting juice from the lamb. Add some veal stock, if necessary. Scrape the bottom of the roasting pan with a wooden spoon [*to dissolve all the cooking residues* (***sucs***)] and pour the liquid (***jus***) into a large saucepan. Let the ***jus*** simmer over medium-low heat for 20 to 25 minutes, skimming (***écumer***) the surface often. Strain the ***jus*** through a fine mesh sieve (***chinois***) into a clean saucepan and reserve in a bain marie.

Cuisson en Croûte de Sel

1. Remove the kitchen twine (***déficeler***) and bard (***débarder***) from the lamb saddle. Lightly flour (***fleurer***) the work surface and, with a rolling pin, roll out (***abaisser***) the salt dough (***pâte à sel***) until it is large enough to wrap around the saddle. Brush the salt dough with egg wash and place the cold roast on top of the dough. Roll the roast up in the dough so that it is completely sealed and place it, seam side down, on a baking sheet that is lined with parchment paper. Brush the salt dough with egg wash and transfer it to the hot oven. Reduce the oven temperature to 180 and roast for at least 45 minutes, or until a trussing needle pierced through the center comes out hot.
2. Remove the baking sheet from the oven and let the meat rest in the closed salt crust (*5 to 10 minutes*) before serving.

Tip: *If letting the meat rest longer, cut the crust open to allow steam to escape and prevent the meat from overcooking.*

Pommes Fondantes

1. Turn 16 potatoes (***fondant*** shape: *90 g*, see pages 149–150 in *Cuisine Foundations*). Heat the clarified butter in a large sauté pan over medium-high heat and sauté the potatoes, turning them often until evenly colored. Deglaze the pan with white chicken stock and season with salt and pepper. Cover with a buttered parchment paper lid (***cartouche***). Place the pan in the oven and cook the potatoes until the liquid is completely reduced (*45 minutes*). Remove the pan from the oven and add fresh butter to it. Delicately roll the potatoes in the butter to glaze them.

To Serve

1. Cut the salt crust off the lamb saddle. Remove the string and barding fat. Transfer the saddle to a heated serving platter and surround it with ***pommes fondantes***. Decorate the dish using the cut salt crust. Serve the ***jus*** on the side.

Note: *A preparation of this sort would be served in the dining room by the maître d'hôtel. The roast would be presented to the client in its salt crust, and it would be cut and carved tableside.*

Quantity		*Ingredient*
U.S.	*Metric*	*Farce Simple*
8 oz	250 g	Pork, finely chopped (***hacher***)
1 oz	30 g	Butter
½ pc	½ pc	Onion, finely chopped (***hacher***)
2 pcs	2 pcs	Shallots, finely chopped (***hacher***)
2 pcs	2 pcs	Garlic cloves, finely chopped, (***hacher***)
1 ¾ oz	50 g	Pork fat, finely chopped (***hacher***)
10 pcs	10 pcs	Sage leaves, finely chopped (***hacher***)
		Salt and pepper
		Pâte à Sel
12 ½ oz	375 g	Coarse salt
6 fl oz	180 mL	Water
1 lb 4 oz	600 g	Flour
6 pcs	6 pcs	Egg whites
1 tsp	5 g	Peppercorns, crushed (***mignonette***)
8 brs	8 brs	Thyme, rubbed (***émietter***)
8 pcs	8 pcs	Bay leaves, coarsely chopped (***concasser***)
8 brs	8 brs	Rosemary, coarsely chopped (***concasser***)
8 brs	8 brs	Basil, coarsely chopped (***concasser***)
1 pc	1 pc	Egg for egg wash
		Jus
1 pc	1 pc	Carrot, cut into ***mirepoix***
1 pc	1 pc	Onion, cut into ***mirepoix***
¼ pc	¼ pc	Leek white, cut into ***mirepoix***
1 pc	1 pc	Celery stalk, cut into ***mirepoix***
3 pcs	3 pcs	Garlic cloves
1 pc	1 pc	***Bouquet garni***
		Salt and pepper
		Pommes Fondantes
16 pcs	16 pcs	Potatoes, turned, ***fondante*** size
7 fl oz	200 g	Butter, clarified
1 pt	500 mL	White chicken stock
1 ¾ oz	50 g	Butter
		Salt and pepper

STEAK AU POIVRE

Learning Outcomes

Sauter
Pommes sautées à cru
Sauce minute

Equipment

Knives:
Vegetable peeler (***économe***), office, slicing knife (***émincer***)

Tools:
Bowls, wooden spatula, fine chinois, whisk, wire rack

Pans:
Large pan

Serving

4 persons

FYI...

This dish was popularized in the early 1900s and several chefs have since claimed it as their own creation. However, information on when and where the preparation was first cooked is mostly speculative. The *New York Times Food Encyclopedia* hazards that *steak au poivre* may be traced back to Germany in the late 1700s.

STEAK AU POIVRE
Sirloin Steak with Pepper

Method

Pommes Sautée à Cru

1. Wash and peel the potatoes. Cut them into ***coin de la rue***-sized dice (see page 148 in *Cuisine Foundations*). Heat the duck fat in a large sauté pan over high heat and sauté the potatoes on all sides until soft inside. Reduce the heat to low, add the chopped garlic and shallots, and cook them until they are soft. Pour the fat out of the pan and add the cold butter. Toss the potatoes in the pan until the butter is completely melted. Set aside and keep warm.

Cuisson

1. Pat the steaks dry and season them with salt. Spread the crushed peppercorns onto a plate. Cover the steaks in peppercorns by pressing them down on the plate (*cover both sides*). Heat the butter in a large sauté pan over high heat. Gently place the steaks in the pan and sauté them on both sides until they are cooked to the desired doneness. Transfer them to a wire rack and keep warm.

Sauce au Poivre

1. Pour the cooking fat out of the pan. Place the pan over medium heat to concentrate the cooking residues (***pincer les sucs***) and deglaze with cognac. Wet (***mouiller***) the pan with the ***demi-glace*** and add the cream. Reduce the sauce over medium-low heat until it is just thick enough to coat the back of a spoon (***à la nappe***). Taste and season the sauce, then strain it through a fine mesh sieve (***chinois***). Mount the sauce with cold butter.

To Serve

1. Place the steaks on a hot plate. Serve the ***sauce au poivre*** and ***pommes sautées à cru*** on the side.

Quantity		*Ingredient*
U.S.	*Metric*	
4 pc	4 pc	New York strip *[6 oz (170 g) each]*
1 oz	30 g	Black peppercorns (***mignonette***)
		Salt and pepper
¾ fl oz	20 mL	Butter, clarified
		Sautées à Cru
2 pcs	2 pcs	Potatoes, large, cut into ***coin de la rue***
3 ½ oz	100 g	Duck fat
3 pcs	3 pcs	Garlic (***hacher***)
2 pcs	2 pcs	Shallots (***ciseler***)
½ bq	½ bq	Parsley (***hacher***)
1 ¾ oz	50 g	Butter
		Salt
		Sauce au Poivre
4 fl oz	125 mL	Cognac
7 oz	200 g	Demi-glace
4 fl oz	125 mL	Cream
1 ¾ oz	50 g	Butter
		Salt

Les abats et les gibiers
(Offal and Game)

Recipes

Les abats et les gibiers (Offal and Game)

Pigeon rôti sur canapé, pommes soufflées
—Roasted pigeon on a croûton, soufflé potatoes

Cervelles à la grenobloise
—Sautéed calves brains with capers, lemon, and croûtons

Foie de veau au vinaigre et aux pommes
—Veal liver with vinegar served with apples

Foie gras chaud
—Hot foie gras

Foie gras en terrine
—Foie gras terrine

Pigeon aux choux
—Pigeon braised with cabbage

Ris de veau braisés Demidoff
—Braised sweetbreads with vegetables and truffles

Rognons de veau au genièvre, lasagne de veau et bolets
—Veal kidneys with gin juniper berries, veal and porcini lasagna

Tripes (à la mode de Caen)
—Stewed tripe with carrots and potatoes

PIGEON RÔTI SUR CANAPÉ, POMMES SOUFFLÉES

Learning Outcomes

Rôtir
Roasting small birds
Making pommes soufflées
Making a farce à gratin
Making canapés

Equipment

Knives:
Chef knife (***couteau chef***), vegetable peeler (***économe***), serrated knife

Tools:
Scissors, kitchen twine, tamis, plastic scraper, bowls, pastry brush, wire rack, wooden spatula, chinois, araignée

Pans:
Medium sauté pan, medium roasting pan, deep fryer, baking sheet, bain marie

Serving

4 persons

PIGEON RÔTI SUR CANAPÉ, POMMES SOUFFLÉES

Roasted Pigeon on a Croûton, Soufflé Potatoes

Quantity		Ingredient
U.S.	Metric	
4 pcs	4 pcs	Pigeons
4 pcs	4 pcs	Bard strips
3 ½ oz	200 g	Butter
		Salt and pepper
		Farce à Gratin
1 oz	30 g	Butter
2 ¾ oz	80 g	Bacon, cut into lardons, blanched
4 oz	120 g	Chicken livers
1 pc	1 pc	Shallot, finely chopped (***hacher***)
¾ fl oz	25 mL	Cognac
3 ½ fl oz	100 mL	Cream
		Canapés
1 pc	1 pc	Country bread (***pain de campagne***)
1 oz	30 g	Butter, clarified
		Jus
1 pc	1 pc	Shallot, cut into small ***mirepoix***
½ pc	½ pc	Onion, medium, cut into small ***mirepoix***
½ pc	½ pc	Carrot, cut into small ***mirepoix***
¼ pc	¼ pc	Celery branch, cut into small ***mirepoix***
1 ¾ fl oz	50 mL	Madeira
1 pt	500 mL	Veal stock
1 ¾ oz	50 g	Butter
		Pommes Soufflées
18 oz	500 g	Potatoes
		Salt to taste

Method

1. Preheat the oven to 350°F (175°C).
2. Preheat the deep fryer to 285°F (140°C).
3. *Prepare the pigeons:* Cut the extremities off the wings and legs, remove any extra skin, and remove the wish bone. Season the birds with salt and pepper and wrap them in bard (***barder***), making sure the breast meat is covered. Tie them in kitchen twine (***ficeler***). Set the pigeons aside in the refrigerator.
4. *Prepare the chicken livers:* Pat the livers dry, denerve and slice them on the bias (***escaloper***), and season with salt and pepper.

***Note:** When cleaning the livers, if any are found with green spots (**fiel**), discard them as they will be bitter.*

Farce à Gratin

1. Melt the butter in a medium sauté pan over high heat and sauté the lardons until they begin to color. Add the chicken livers and sauté for 2 to 3 minutes. Stir in the shallot and briefly sauté it (*being careful not to burn it*). Add the cognac and flambé, then pour in the cream and reduce the heat to a simmer. Cook the cream until it reduces by half. Press the resulting mixture through a drum sieve (***tamis***) into a clean bowl. Cover and reserve at room temperature.

Canapés

1. Cut four ¾ in. (1 cm) thick slices of the bread. Trim in rectangulars, make a small cavity in the center of each rectangle. Arrange on a parchment paper–lined baking sheet, brush generously with melted, clarified butter. Place the baking sheet to the oven and cook until golden. Turn the canapés (*adding more butter, if needed*) and return them to the oven to brown on the second side.
2. When toasted, drain the canapés on a paper towel. Turn the oven up to 400°F (205°C).

Cuisson

1. Place the pigeons on a wire rack set in a medium roasting pan. Top the birds with pieces of cold butter and place in the oven to roast for 20 minutes. Every 5 minutes, remove the pigeons from the oven to baste (***arroser***) with melted butter. When cooked, remove the rack with the pigeons onto a plate, cover and set it aside to rest. Pour off all but 1 to 2 tablespoons of fat from the pan. Place the pan over medium-high heat and sauté the shallot, onion, carrot, and celery ***mirepoix*** until just colored. Deglaze the pan with the Madeira and reduce it until almost dry. Wet (***mouiller***) the pan with the veal stock and reduce by half. Mount the sauce with butter. Strain through a fine mesh sieve (***chinois***) into a clean saucepan and reserve covered in a bain marie.

Pommes Soufflées

1. Preheat the oil to 257 to 285°F (135 to 140°C).
2. Peel and cut the potatoes into blocks and trim the corners (see page 41 in *Cuisine Foundations*). Cut the trimmed potatoes into ⅛ in. (3 mm) slices. Rinse the slices well in cold water, until the water runs clear. Drain and pat the potatoes dry. Arrange them on a paper towel–lined tray and allow to dry in a warm place. A white skin of starch may form on the surface.
3. *Premier soufflée:* Once the oil is hot, place 10–15 potatoes into the hot oil and cook them until they expand slightly without any coloring. Be sure to stir them so they cook evenly and do not stick to each other. Remove and drain and repeat with the remaining potatoes. Arrange the potatoes on a cloth and set aside.
4. *Deuxième soufflée:* Increase the heat of the oil to 360°F (180°C). Remove the pan from the heat and add 10–15 potato slices. Delicately stir and flip them with a wire skimmer (***araignée***) to ensure even cooking and coloring. As soon as they puff up and turn golden, remove them to drain on paper towel. Season well.

To Serve

1. Remove the twine (***déficeler***) and bard (***débarder***) from the birds. Fill each canapé with the ***farce à gratin*** and place them in the oven to warm up on a baking sheet. Place each canapé in the center of a hot plate and rest a pigeon on top. Lightly brush the pigeons with clarified butter to add shine (***lustrer***). Drizzle the plate with a little sauce and serve the rest on the side. Serve the ***pommes soufflées*** on the side.

CERVELLES À LA GRENOBLOISE

Learning Outcomes

Sauter
Preparing brains
Cuisson meunière
Garniture grenobloise
Beurre noisette

Equipment

Knives:
Slicing knife (***éminceur***), paring knife (***office***), serrated knife

Tools:
Ladle (***louche***), fork, wire rack, mixing bowl

Pans:
1 stock pot,
1 frying pan,
1 large sauté pan

Serving

4 persons

CERVELLES À LA GRENOBLOISE

Sautéd Calves Brains with Capers, Lemon, and Croûtons

Method

Prepare the Brains

1. Prepare the brains in advance. See pages 323–325 in *Cuisine Foundations*.

Cuisson à la Meunière

1. Season the brains and dredge them in flour (*tapping off the excess*). Heat the clarified butter in a large sauté pan over medium-high heat. Add the brains and sauté until golden, basting them regularly. Once they are cooked, set them aside on a wire rack. Remove the remaining fat from the pan (***dégraisser***). Add butter to the same pan and cook until the butter reaches a light brown color (***beurre noisette***). Add lemon juice and set aside.

Garniture à la Grenobloise

1. In another pan, melt fresh butter in the pan and add the diced bread. Toss in the butter until evenly colored and crisp. Add the capers, then the diced lemon and parsley. Toss together. Season to taste.

To Serve

1. Transfer the brains to the pan of ***garniture à la grenobloise*** to rest for a few minutes. Spoon the ***garniture*** onto a plate and place the brains on top.
2. Drizzle with the browned butter. Decorate with lemon slices (***rondelles***) and sprinkle with chopped parsley.

Quantity		*Ingredient*
U.S.	*Metric*	
4 pcs	4 pcs	Calf brains
		Dégorger
4 qt	4 L	Water
5 fl oz	150 mL	White vinegar
1 pc	1 pc	Lemon
		Blanchir
2 brs	2 brs	Thyme
1 pc	1 pc	Bay leaf
½ oz	10 g	Coarse salt
½ oz	10 g	Peppercorns
½ pc	½ pc	Lemon
		Cuisson Meunière et Garniture Grenobloise
5 oz	150 g	Flour
4 oz	125 g	Butter, clarified
1 pc	1 pc	Lemon
2 slices	2 slices	Bread, white, cut into dice
3 ½ oz	100 g	Capers
2 pcs	2 pcs	Lemon, peeled with a knife (***peler à vif***), cut into dice
1 oz	30 g	Parsley, finely chopped (***hacher***)
		Salt and pepper
		To Serve
1 pc	1 pc	Lemon, sliced (***rondelles***)
½ oz	10 g	Parsley, finely chopped (***hacher***)

FOIE DE VEAU AU VINAIGRE ET AUX DEUX POMMES

Learning Outcomes

Sauter
Potatoes cooked "*en robe des champs*"
Make a pâte à choux
Frire
Suer
Réduire
Monter au beurre

Equipment

Tools:
Bowls, cutting board, fork, wooden spatula, plastic spatula, fine chinois, pastry bag, araignée, skimmer (***écumoire***), ladle (***louche***), tamis, food mill, wire rack

Pans:
Sauteuse, small russe, small marmite, bain marie, deep fryer

Serving

4 persons

FOIE DE VEAU AU VINAIGRE ET AUX DEUX POMMES

Veal Liver with Vinegar Served with Apples

Quantity		*Ingredient*
U.S.	*Metric*	
4 pcs	4 pcs	Calf liver slices [*5 oz (150 g) each*]
3 ½ oz	100 g	Butter
		Pommes Dauphines
4 pcs	4 pcs	Waxy potatoes, large, in their jackets (***en robe des champs***)
7 to 10 oz	200 to 300 g	Rock salt
		Pâte à Choux
4 fl oz	125 mL	Water
1 oz	30 g	Butter
2 oz	60 g	Flour
2 pcs	2 pcs	Eggs
		Salt
		Pommes Fruit
4 pcs	4 pcs	Golden Delicious apples
3 ½ oz	100 g	Butter
		Salt and pepper
		Sauce
2 pcs	2 pcs	Shallots, sliced (***émincer***)
7 fl oz	200 mL	Red wine vinegar
1 ¾ oz	50 g	Butter
¼ oz	10 g	Parsley, finely chopped (***hacher***)

Method

1. Preheat the oven to 350°F (175°C).
2. Preheat the deep fryer to 320°F (160°C).

Pommes Dauphines

1. *Potato purée:* Wash the potatoes in cold water.
2. Prick their skins with the tip of a knife and transfer the potatoes onto a baking sheet lined with a layer of rock salt. Place in the oven to bake with the skins on (***en robe des champs***) until the tip of a knife can easily be inserted into their centers (*30 to 40 minutes*). Peel the potatoes while still hot and purée their flesh through a food mill or a drum sieve (***tamis***). Season the purée to taste and set aside.

Note: *The addition of rock salt will help to extract humidity from the potatoes, adding a better texture to the pommes dauphines mix.*

Prepare a Pâte à Choux

1. *Prepare a* ***panade***: Combine the water, butter, flour, and salt in a large pan and bring to a boil over high heat. Once the butter has completely melted, remove from the heat and add the flour. Stir with a wooden spatula until combined; then, over medium heat, stir until the mixture forms a smooth ball and comes cleanly away from the edge of the pan. Transfer the ***panade*** to a clean bowl and cool until warm. Mix the eggs into the ***panade*** one by one using a rubber spatula, making sure each one is completely incorporated before adding the next. The dough should be elastic and slightly sticky. Measure out 3 ½ oz (100 grams) of dough and refrigerate the remainder for other use.
2. Combine 8 oz (250 g) of potato purée with 3 ½ oz (100 g) of ***pâte à choux*** and mix with a rubber spatula until smooth. Transfer the mixture to a pastry bag fitted with a large plain tip. Onto a baking sheet lined with parchment paper, pipe out small balls of the mixture (***choux***) 1 to 1 ½ in. (2 ½ to 4 cm) in diameter. Set aside in the refrigerator.

Pommes Fruit

1. Peel, quarter, and core the apples, and cut each quarter into 2 or 3. Melt the butter in a large sauté pan over medium-high heat. Arrange the apple slices in the pan and sauté until the undersides are golden. Turn the apples over and color the other side. Reserve and keep warm.

Foie de Veau au Vinaigre

1. Pat the liver dry and season them. Melt the butter in a large sauté pan over medium-high heat and sauté the liver on both sides until the desired doneness is reached. Transfer the liver to a wire rack and pour most of the fat out of the pan, leaving just enough fat to cook the shallots. Return the pan to medium-high heat and add the shallots. Sauté the shallots until colored. Deglaze the pan with the vinegar and reduce it to a syrupy consistency (***glace***). Mount the sauce with cold butter. Season to taste, then mix in the parsley. Remove the pan from the heat.

Finishing

1. Drop the ***pommes dauphines*** into the deep fryer, cook them until puffed up and golden, then drain them on a dish lined with a paper towel. Season the ***pommes dauphines*** immediately.

To Serve

1. On a hot serving dish, arrange the apples and ***pommes dauphines*** around the liver slices. Top the preparation with the shallot mixture.

Learning Outcomes

Sauter
Étuver
Compoter
Gastrique-based reduction sauce

Equipment

Knives:
Chef knife (*couteau chef*), paring knife (*office*)

Tools:
Wooden spatula, wire rack, teaspoons

Pans:
Large saucepan, large non stick sauté pan

Serving

5 to 6 persons

FOIE GRAS CHAUD

Hot Foie Gras

Method

1. Preheat the oven to 350°F (175°C).

Compote d'oignons (confiture)

1. Heat the olive oil in a large saucepan over low heat and add the onions. Season the onions with salt and stir until well coated. Cover the saucepan and leave the onions to cook (***étuver***) for 10 minutes.
2. Add the citrus zests, cardamom, raisins, and honey and stir them into the onions. Cover the saucepan again and leave to cook (***étuver***) for an additional 10 minutes. Deglaze with balsamic vinegar and place the coriander on top. Cover the saucepan once more and transfer it to the oven to cook (***étuver***) for 35 to 40 minutes.
3. Remove the pan from the oven and take off the lid. Place the pan over low heat and cook its contents until the liquid evaporates. Taste and adjust the seasoning. Remove the coriander and transfer the compote to a clean bowl. Set aside.

Tip: *The compote is ready when it can be easily formed into quenelles.*

Cuisson

1. Slice the foie gras on the bias into thick pieces (***escalopes***) weighing 3 oz (90 g) each. Season the ***escalopes*** on both sides with salt and pepper. Heat a large non stick pan over high heat.
2. Place the ***escalopes*** (*1 or 2 at a time*) in the hot pan and sauté for 2 to 3 minutes until the undersides are a caramel color (*baste regularly in the cooking fat*). Turn the foie gras over and repeat the same cooking process.

Tip: *If the foie gras leaks too much fat, pour it out of the pan to ensure proper cooking.*

3. Once the foie gras is golden on both sides, transfer it to a wire rack to rest. Repeat the cooking process with the remaining ***escalopes***.

Sauce

1. Pour the fat out of the pan, then place the pan over medium heat. Add the honey to the pan and cook the honey until it begins to caramelize. Deglaze with the balsamic vinegar and reduce the mixture to a syrup (***glace***). Mount the reduction with cold butter. If the cooking fat from the foie gras is clear, a teaspoon of it can be added to the sauce for flavor. Taste the sauce and adjust seasoning.

Optional: *If the sauce is too sweet, add a tablespoon of meat glaze (**glace de viande**) to balance out the flavor, or a squeeze of lemon juice.*

To Serve

1. Using two teaspoons, shape one ***quenelle*** (*per serving*) with the onion compote. Place a ***quenelle*** on a heated plate with a ***suprême*** of orange and ***suprême*** of lemon. Spoon a line of sauce over these ingredients. Place one piece of foie gras in the center of the plate and decorate the preparation with a sprig of fresh coriander.

Tip: *If needed, reheat the foie gras on a wire rack in the oven for 1 to 2 minutes just before serving.*

Quantity		*Ingredient*
U.S.	*Metric*	
1 lb	500 g	Foie gras, catégorie 1
		Salt and pepper
		Compote d'oignons (confiture)
3 pcs	3 pcs	Onions, large, sliced (***émincer***)
3 ½ fl oz	100 mL	Olive oil
2 pcs	2 pcs	Oranges, zest of
2 pcs	2 pcs	Lemons, zest of
3 ½ oz	100 g	Raisins
5 pcs	5 pcs	Cardamom pods
1 ¾ oz	50 g	Honey
1 ¾ fl oz	50 mL	Balsamic vinegar
¼ bq	¼ bq	Fresh coriander
		Salt and pepper
		Sauce
1 oz	30 g	Honey
3 ½ fl oz	100 mL	Balsamic vinegar
1 ¾ oz	50 g	Butter
1 Tbsp	1 Tbsp	***Glace de viande*** (*optional*)
		Salt and pepper
		To Serve
1 pc	1 pc	Orange, cut into ***suprêmes***
1 pc	1 pc	Lemon, cut into ***suprêmes***
2 brs	2 brs	Fresh coriander

FOIE GRAS EN TERRINE

Learning Outcomes

Pocher
Preparation of a foie gras for terrines (***dénerver, dégorger, mariner***)
Making chutney
Glacer à terrine with a gelée
Clarification

Equipment

Knives:
Boning knife (***désosseur***), paring knife (***office***), slicing knife (***éminceur***)

Tools:
Bowls, wooden spatula, chinois, skimmer (***écumoire***), ladle (***louche***), ice bath, whisk, probe, coffee filter, wire rack, tray

Pans:
Terrine dish, casserole dish, medium saucepans, bain marie

Serving

8 persons

FOIE GRAS EN TERRINE

Foie Gras Terrine

Method

1. Preheat the oven to 350°F (175°C).

Preparing the Foie Gras

1. Using a paring knife (*and being careful to keep the foie gras intact*), cut the veins and nerves out of the foie gras lobes (***déveiner***, ***dénerver***).
2. Place the foie gras in a recipient and cover it in cold milk. Leave it to degorge (***dégorger***) in the refrigerator for 12 hours.

Marinade

1. Remove the foie gras from the milk, pat dry, and transfer it to a clean recipient. Combine the salt, pepper, allspice, cognac, white port wine, and Sauterne wine. Mix these ingredients into the foie gras, being careful not to damage the lobes. Cover and let it marinate in the refrigerator for 6 hours.

Cuisson

1. Drain the foie gras and pat it dry. Use the largest and smoothest pieces to line the inside of a terrine dish. Press the smaller pieces into the center as tightly as possible without crushing them. Cover the terrine dish with foil and place it in a large roasting pan. Fill the pan with hot water two-thirds of the way up the sides of the terrine. Transfer the terrine and bain marie to the oven to cook for 30 minutes. When cooked to an internal temperature of 133°F (56°C), remove the terrine from the bain marie and let it rest for 10 to 15 minutes. Next, press it down using the bottom of an identical terrine. If needed, fill the second terrine with weights on top to increase the weight. The fat should drain out of the sides of the dish without pushing out any foie gras. Let the foie gras terrine cool to room temperature, then refrigerate it overnight (*still covered with the weight*).

Fig Chutney

1. Quarter the figs, then cut each quarter in half lengthwise. Combine the figs, sherry vinegar, cardamom, curry, and anis powder in a medium saucepan. Season lightly and place the pan over low heat. Once the ingredients are warm, cover the pan and transfer it to the oven to cook (***compoter***) for 30 minutes. If the compote is still too wet when removed from the oven, transfer the pan to low heat to evaporate the excess liquid. Pick out the cardamom pods, then taste and adjust the seasoning. Transfer the compote to a clean recipient and let it cool to room temperature. Cover and refrigerate until needed.

Clarification

1. Pour the chicken stock into a medium saucepan. Mix in the chopped chicken meat, egg whites, ***mirepoix***, and port wine. Season with salt and pepper and stir until all the ingredients are combined. Gently heat the liquid over medium heat, stirring from time to time to prevent the proteins from sticking to the bottom of the pan. When the proteins begin to coagulate and create a crust on the surface of the liquid, reduce the heat to a gentle simmer. Pierce a small hole in the crust to allow the simmering bubbles to escape. Let the liquid simmer gently for 20 minutes, occasionally wetting (***arroser***) the crust by ladling some stock onto it from the hole in the crust. Carefully strain the resulting ***consommé*** into a bowl through a fine mesh sieve (***chinois***) lined with a coffee filter or wet cloth.

Gelée

1. Soak the gelatin leaves in cold water until soft. Squeeze out the excess water and add the leaves to the hot ***consommé***. Gently stir the mixture with a wooden spatula to avoid creating bubbles.

Glaçage

1. Unmold the foie gras from the terrine dish and place it upside down on a wire rack set over a clean tray. Carefully remove any excess cooking fat with a spatula and ensure that the terrine keeps a rectangular shape. Reserve in the refrigerator.
2. *Temper the gelée:* Place the *gelée* in a bowl over an ice bath. Stir it gently with a metal spoon (***tempérer***) until the liquid is oil-like in resistance and movement.
3. Ladle the *gelée* onto the foie gras to coat it in a thin even layer. Return the foie gras to the refrigerator for 5 minutes to chill, then repeat the coating process. Continue until all the gelée is used up. Reserve the foie gras and the tray that collected the *gelée* in the refrigerator.

To Serve

1. Place the terrine on a chilled serving dish. Press the fig compote into a small mold and unmold it onto the dish.
2. Finely chop (***hacher***) the gelée that gathered in the tray and use it for decoration.
3. Serve the foie gras with either French ginger bread (***pain d'épices***) or toasted rustic French bread (***pain de campagne***).

Quantity		*Ingredient*
U.S.	***Metric***	
1 pc	1 pc	Duck foie gras [1 lb (*500 g*)]
		Marinade
¼ oz	6 g	Salt
Pinch	Pinch	Pepper
Pinch	Pinch	Allspice
1 ¾ fl oz	50 mL	Cognac
1 ¾ fl oz	50 mL	White port wine
Splash	Splash	Sauterne or a sweet white wine
		Fig Chutney
8 pcs	8 pcs	Ripe black figs
1 ¾ fl oz	50 mL	Sherry vinegar
3 pcs	3 pcs	Cardamom pods
Pinch	Pinch	Curry powder
Pinch	Pinch	Anis powder
		Salt and pepper
		Clarification
1 pt	500 mL	Chicken stock
3 ½ oz	100 g	Chicken meat, finely chopped (***hacher***)
2 pcs	2 pcs	Egg whites
2 oz	60 g	Small ***mirepoix***
1 fl oz	30 mL	Port wine
		Salt and pepper
		Gelée
10 fl oz	300 mL	Consommé (*from clarification*)
5 pcs	5 pcs	Gelatin leaves

PIGEON AUX CHOUX

Learning Outcomes

Braiser
Trussing and barding a small bird
Making croûtons
Making a jus

Equipment

Knives:
Chef knife (*couteau chef*), serrated knife, paring knife (*office*)

Tools:
Trussing needle, kitchen twine, colander, skimmer (*écumoire*), ice bath, wire rack, wooden spatula, chinois, bowls, cheese grater

Pans:
Oven dish, Dutch oven (*cocotte*), large stock pot (*marmite*), roasting pan, baking sheet

Serving

4 persons

PIGEON AUX CHOUX

Pigeon Braised with Cabbage

Method

1. Preheat the oven to 400°F (205°C).
2. Bring a stock pot of salted water to a boil over high heat. Cut the cabbage in 4 and blanch it in the boiling water for 2 to 3 minutes. Refresh the cabbage in an ice bath, drain it, and press out the excess water. Cut out the core and thinly slice the leaves (***ciseler***). Set aside (see pages 82–83 in *Cuisine Foundations*).
3. Truss the pigeons (***brider***) and season them with salt and pepper, then wrap them in bard (***barder***) and tie the bard with kitchen twine (***ficeler***).

Jus

1. Place the pigeons on a wire rack in a roasting pan, top with a piece of butter and roast in the oven for 10 to 15 minutes basting once the butter melts. Remove the pigeons from the pan and set aside on the wire rack. Pour off the fat and place the pan on medium-high heat. Add the oil in the pan and sauté the ***mirepoix*** until it is lightly colored (***blond***). Pour out the fat and add the thyme and bay leaves. Deglaze the pan with 4 fl oz (120 mL) of cold water and stir well to loosen and dissolve any cooking juices. Wet (***mouiller***) with the veal stock and bring to a simmer. Reduce the heat to low and simmer gently for 15 to 20 minutes. Strain the ***jus*** through a fine mesh sieve (***chinois***) and set aside.

Cuisson en Cocotte

1. Melt the butter in a Dutch oven (***cocotte***) over high heat. Add the ***lardons*** and the whole sausage and sauté until golden. Add the sliced carrot and onion and sauté until they are lightly colored (***blond***). Stir in the tomato paste and cook for 1 to 2 minutes (***pincer la tomate***). Add the chopped cabbage, sprinkle it with flour, and cook for 2 to 3 minutes (***singer***), stirring well. Deglaze with white chicken stock, add the ***bouquet garni***, cover the ***cocotte***, and transfer it to the oven to cook (***étuver***) for 20 to 25 minutes.

Croûtons

1. Slice the bread ¾ in. (1 cm) thick and cut the slices in half. Arrange them on a baking sheet, brush with clarified butter and toast on both sides in a hot oven or under a salamander. Set aside.

Finishing

1. Remove the cocotte from the oven. Check that the liquid has been mostly absorbed and that the cabbage is cooked through. If not, return the cocotte to the oven to cook for a further 5 to 10 minutes.
2. Remove the bard from the pigeons and remove the string. Pour the ***jus*** into the sausage/cabbage mixture and arrange the pigeons on top. Cover the cocotte and transfer it to the oven to braise for another 10 minutes. Remove the cocotte from the oven and decant the pigeons and sausage to a wire rack. Place the cocotte over low heat and cook until the ***jus*** is thick enough to thinly coat the back of a spoon (***à la nappe***). Slice the sausage and return it to the cocotte.
3. Rub the croûtons with garlic and sprinkle with the cheese. Place in the oven to brown.

To Serve

1. Create a bed of cabbage in a heated serving dish. Arrange the pigeons on top. Decorate with fresh parsley. Serve the croûtons on the side.

Quantity		*Ingredient*
U.S.	*Metric*	
4 pcs	4 pcs	Pigeons
4 pcs	4 pcs	Bard strips
1 pc	1 pc	Savoy cabbage
5 oz	150 g	Butter
		Salt and pepper
		Jus
1 ¾ oz	50 g	Olive oil
5 oz	150 g	***Mirepoix***
5 brs	5 brs	Thyme
2 pcs	2 pcs	Bay leaves
7 fl oz	200 mL	Brown veal stock
		Salt and pepper
		Cuisson en Cocotte
1 ¾ oz	50 g	Butter
4 oz	125 g	Bacon, cut into lardons
1 pc	1 pc	Toulouse sausage
2 pcs	2 pcs	Carrots, channeled (***canneler***), cored (***évider***), and sliced (***émincer***)
1 pc	1 pc	Onion, sliced (***émincer***)
½ oz	10 g	Tomato paste
1 oz	25 g	Flour
1 pt	500 mL	White chicken stock
1 pc	1 pc	***Bouquet garni***
		Salt and pepper
		Croûtons
1 pc	1 pc	French rustic bread (***pain de campagne***)
5 oz	150 g	Clarified butter
2 pcs	2 pcs	Garlic cloves
4 oz	120 g	Gruyère cheese, grated
		To Serve
1 br	1 br	Fresh parsley

Learning Outcomes

Braiser
Preparing sweetbreads
Piquer
Larder
Truffer
Ficeler
Suer
Étuver

Equipment

Knives:
Paring knife (*office*), turning knife, chef knife (*couteau chef*), channeling knife

Tools:
Bowls, skimmer (*écumoire*), colander, larding needle, kitchen twine, wooden spatula, wire rack, chinois

Pans:
Dutch ovens

Serving

4 persons

FYI...

Named after the Russian prince Anatole Demidoff who married the niece of Napoleon I, *ris de veau Demidoff,* with its truffles and braised veal sweetbreads (*ris de veau*), today may seem food fit for royalty, but truffles were used liberally during this time period.

RIS DE VEAU BRAISÉS DEMIDOFF

Braised Sweetbread with Vegetables and Truffles

Method

1. Preheat the oven to 350°F (175°C).
2. Prepare the sweetbreads in advance. See pages 321–322 in *Cuisine Foundations*. Using a small larding needle, insert strips of bard into the sweetbreads (***larder***). Stud the sweetbreads with truffles (***piquer, truffer***) and tie them (***ficeler***) into even shapes.

Braisage

1. Pat the sweetbreads dry, then season and lightly coat them in flour patting off the excess. Heat the clarified butter in a Dutch oven (***cocotte***) over medium-high heat and sear (***saisir***) the sweetbreads until golden on all sides. Transfer them to a wire rack and pour the butter out of the ***cocotte***. Replace the ***cocotte*** on medium-high heat and sauté the ***mirepoix*** until colored. Add the ***bouquet garni***, deglaze with Madeira wine, and reduce it by half. Wet (***mouiller***) the mixture with veal stock, season it lightly, and return the sweetbreads to the cocotte. Cover and place it in the oven to braise for 30 minutes.

Garniture Demidoff

1. Heat the clarified butter in a cocotte over low heat and sweat the carrots, turnips, celery, and pearl onions for 5 minutes. Season lightly and add 4 fl oz (120 mL) of water. Cover the pot and cook on low heat (***étuver***) until all the vegetables are soft. Mix the chopped parsley into the garnish. Decant the sweetbreads from their braising liquid (***jus de braisage***) and rest them on the garnish. Strain the braising liquid (***jus de braisage***) through a fine mesh sieve (***chinois***) into the cocotte containing the sweetbreads and garnish. Add the truffle, cover, and cook over low heat (***étuver***) for 10 minutes. Taste and adjust the seasoning.

To Serve

1. Decant the truffle and slice it thinly. Decant the sweetbreads, remove the twine (***déficeler***), and cut into slices ½ to ¾ in. (1 cm) thick. Strain the garnish and spoon it into the center of a hot plate. Arrange alternating slices of sweetbread and truffle on top of the garnish. Mount the braising liquid (***jus de braisage***) with cold butter in a medium saucepan over high heat. Pour it over the sweetbreads to give them shine (***lustrer***).

Quantity		*Ingredient*
U.S.	*Metric*	
4 pcs	4 pcs	Veal sweetbreads (***noix***)
1 ¾ oz	50 g	Bard, cut into strips
1 ¾ oz	50 g	Truffles, cut into ***bâtonnets***
5 oz	150 g	Flour
		Salt and pepper
		Fond de Braisage
3 ½ oz	100 g	Butter, clarified
7 oz	200 g	***Mirepoix***
1 pc	1 pc	***Bouquet garni***
3 ½ fl oz	100 mL	Madeira wine
10 fl oz	300 mL	Brown veal stock
		Salt and pepper
		Garniture Demidoff
3 ½ oz	100 g	Butter, clarified
7 oz	200 g	Carrots, channeled (***canneler***), cored (***évider***), sliced on the bias (***sifflet***)
7 oz	200 g	Turnips, cut into bouchons, channeled (***canneler***), cored (***évider***), sliced on the bias (***sifflet***)
7 oz	200 g	Celery branch, channeled (***canneler***), cored (***évider***), sliced on the bias (***sifflet***)
7 oz	200 g	White pearl onions, thickly sliced (***rouelle***)
5 brs	5 brs	Parsley, finely chopped (***hacher***)
1 ¾ oz	50 g	Truffle
		Salt and pepper
		To Serve
1 ¾ oz	50 g	Butter

Learning Outcomes

Rôtir
Cleaning kidneys
Roasting kidneys
Making pasta

Equipment

Knives:
Boning knife (***désosseur***), paring knife (***office***), chef knife (***couteau chef***)

Tools:
Wire rack, kitchen twine, wooden spatula, chinois, plastic scraper, rolling pin or pasta roller, bowls, round cutters, ring molds, cheese grater, whisk, meat pounder

Pans:
Medium saucepan, oven dish, bain marie, medium sauté pan, large pot, baking sheet, roasting pan

Serving

4 persons

ROGNONS DE VEAU AU GENIÈVRE, LASAGNE DE VEAU ET BOLETS

Veal Kidneys with Gin Juniper Berries, Veal and Porcini Lasagna

Quantity		Ingredient
U.S.	Metric	
2 pcs	2 pcs	Veal kidneys with their own fat [13 oz (400 g)]
¾ fl oz	20 mL	Gin
¾ oz	20 g	Fresh juniper berries
		Salt and pepper
		Sauce
1 fl oz	30 mL	Gin
8 fl oz	250 mL	Veal stock
5 fl oz	150 mL	Cream
1 oz	30 g	Fresh juniper berries
		Salt and pepper
		Pâte à Pâtes
8 oz	250 g	Flour
2 pcs	2 pcs	Eggs
¼ oz	5 g	Salt
½ fl oz	10 mL	Olive oil
½ fl oz	10 mL	Water (*optional*)
		Farce Simple
1 oz	30 g	Butter
2 pcs	2 pcs	Shallots, finely chopped (***hacher***)
3 pcs	3 pcs	Garlic cloves, finely chopped (***hacher***)
4 oz	120 g	Porcini mushrooms, finely chopped (***hacher***)
¾ fl oz	20 mL	Cognac
5 brs	5 brs	Parsley, finely chopped (***hacher***)
8 oz	250 g	Veal, finely chopped (***hacher***)
3 ½ oz	100 g	Bacon, finely chopped (***hacher***)
		Salt and pepper
		To Serve
2 oz	60 g	Parmesan cheese, grated
1 ¾ oz	50 g	Butter
5 brs	5 brs	Parsley

Method

1. Preheat the oven to 350°F (175°C).

Cuisson

1. Cut the kidneys out of their fat, delicately open them, and cut out their nerves (***dénerver***).
2. Place the fat between two sheets of plastic wrap and pound it with a meat pounder or the bottom of a saucepan to flatten. Season the kidneys with salt and pepper and place them in the center of the fat. Sprinkle with gin and juniper berries, and wrap the fat around them. Tie (***ficeler***) the fat closed with kitchen twine.
3. Place the kidneys on a wire rack in a roasting pan and transfer them to the oven to roast for 25 minutes. Remove the kidneys from the oven and set aside.

Sauce

1. Pour off the fat, place the roasting pan on medium-high heat, and deglaze with gin. Reduce the gin until almost dry and wet (***mouiller***) with veal stock. Reduce by half and add the cream. Reduce the heat to low and simmer the sauce until it is thick enough to coat the back of a spoon (***à la nappe***). Taste and adjust the seasoning. Strain the sauce through a fine mesh sieve (***chinois***) into a clean saucepan. Add the juniper berries and reserve the sauce in a bain marie.

Pâte à Pâtes

1. Sift the flour onto the work surface. Using a plastic scraper, make a well (***fontaine***) in the center of the pile of flour. Beat the salt, egg, and oil together with a whisk and pour them into the well. Incorporate the flour into the egg mixture, bit by bit, with a plastic scraper. If the dough is too dry, add water. Work the dough until it is homogeneous, then roll it into a ball. Wrap the dough in plastic and let it relax for a minimum of 20 minutes in the refrigerator.

Farce Simple

1. Melt the butter in a medium sauté pan over low heat. Add the shallots and garlic and sweat (***suer***) until soft. Season lightly. Add the ***brunoise*** of porcini mushrooms and sweat until cooked. Increase the heat to high. Deglaze with cognac and flambé. Remove the pan from the heat and mix in the chopped parsley, veal, and bacon. Reserve covered in the refrigerator.

Lasagna

1. Bring a large pot of salted water to a boil over high heat. Roll out the pasta dough (***pâte à pâtes***) to a thickness of ¹⁄₁₆ in. (1 mm). Cut the sheet into large rectangles and cook them one at a time in the boiling water until *al-dente* (*1 to 2 minutes*). Refresh the pasta immediately in ice water and pat them dry. Using a round dough cutter, cut the cooked pasta into discs to fit into the ring molds for the lasagna. Reserve the pasta discs on a baking sheet between two sheets of lightly oiled parchment paper.
2. Butter the inside of the ring molds and arrange them on a parchment paper–lined baking sheet. Place a pasta disc in the bottom of each ring mold, add a teaspoon of farce, and press it down level. Place another pasta disc on top and continue layering the pasta and farce until the ring molds are full (*finishing with a pasta disc*). Sprinkle the finished lasagna molds with grated parmesan, dot with butter and transfer them to the oven to bake for 15 minutes.

To Serve

1. Remove the fat casing from the kidneys. Slice the kidneys and arrange them on a heated plate. Unmold the hot lasagna and place next to the kidneys. Mount the sauce with butter and spoon it onto the plate (*being careful not to get any on the meat*). Decorate with fresh parsley.

TRIPES (À LA MODE DE CAEN)

Learning Outcomes

Preparation of tripes
Dégorger
Blanchir
Canneler vegetables
Tourner potatoes
(*anglaise size*)

Equipment

Knives:
Paring knife (***office***), vegetable peeler (***économe***), chef knife (***couteau chef***), channeling knife, cleaver (***couteau à batte***)

Tools:
Bowls, skimmer (***écumoire***), ladle (***louche***), colander, kitchen twine

Pans:
Dutch oven (***cocotte***), medium stock pot (***marmite***)

Serving

4 persons

FYI...

Caen is the capital of the Basse-Normandie region of northwestern France, including the region of Calvados. In terms of this recipe, Calvados is important for its apple orchards and more precisely for its production of the distilled apple cider, known as Calvados. In *tripes à la mode de Caen*, the tripe is simmered for four or five hours with Calvados, white wine, and hard cider, not only tenderizing it but also adding a distinctive flavor to the preparation.

TRIPES (À LA MODE DE CAEN)

Stewed Tripe with Carrots and Potatoes

Quantity		*Ingredient*
U.S.	***Metric***	
2 lb 4 oz	1.2 kg	Veal tripe (*whole: 4 parts*)
½ pc	½ pc	Veal hoof (***pied de veau***)
		Cuisson
3 ½ oz	100 g	Butter
3 pcs	3 pcs	Onions, sliced (***émincer***)
2 pcs	2 pcs	Carrots, channeled (***canneler***), sliced (***émincer)***
1 pc	1 pc	***Bouquet garni***, strong
¼ oz	5 g	Black peppercorns
1 pc	1 pc	Cloves
3 ½ fl oz	100 mL	Calvados
10 fl oz	300 mL	White wine
1 ½ pt	750 mL	Cider
1 qt	1 L	Water
		Salt and pepper
		Garniture
12 pc	12 pc	Potatoes, turned ***anglaise***
5 br	5 br	Parsley, finely chopped (***hacher***)

Method

Preparation of Tripe

1. Soak the tripe and veal hoof in cold water overnight (***dégorger***). Before cooking, rinse once more under running water.
2. Place the hoof and the tripe in a medium stock pot (***marmite***) and fill with water until covered. Add salt and bring to a boil over high heat. When the water has reached the boiling point, lower the heat and allow to blanch for 10 minutes. While blanching, remove impurities by skimming the surface (***écumer***).
3. Remove the tripe and hoof to a recipient. Run cold water over the tripe to cool it down and to remove any remaining impurities before the cooking process.
4. Cut the tripe into strips (***laniers***) 2 in. by ½ in. (5 cm by 1 cm). Set aside.

Cuisson

1. Peel, channel, and slice the carrots to a ⅛ in. (3 mm) thickness. Peel and cut the onions into slices (***emincer***).
2. Melt the butter in a cocotte over medium heat, add the vegetables, and sweat (***suer***) without coloring. Add the tripe to the cocotte. Season with salt and pepper.
3. Deglaze (***deglacer***) the tripe with Calvados and reduce until dry (***réduire à sec***). Add white wine, reduce by half, and finish with cider.
4. Add the ***bouquet garni***, black peppercorns, and the studded onion to the cocotte and bring to a boil. Skim the surface (***écumer***), cover, and gently simmer for 4 to 5 hours. If the liquid evaporates too quickly, reduce the heat and add some water.
5. When the tripe is cooked and tender, remove the studded onion and ***bouquet garni***. Reduce the liquid, if necessary. Adjust seasoning.

Garniture

1. While the tripe is cooking, turn potatoes *anglais* (egg size 1 ¾ oz to 2 oz (50 to 60 g), with 7 sides). See page 155 in *Cuisine Foundations*. Place in cold salted water and bring to a boil. Reduce the heat to a low boil and cook 20 to 25 minutes or until the tip of a knife can be inserted easily. Drain and keep warm.

To Serve

1. Once the tripe is finished, add the potatoes. Transfer to a heated serving dish and finish with chopped parsley. Grind pepper onto the dish prior to serving.

Bande de tarte pomme
—Long apple tart

Bavarois rubané
—Tri-colored Bavarian cream

Beignets aux pommes, sauce abricot
—Apple fritters with apricot sauce

Brioches
—Brioche bread

Charlotte aux poires, coulis de framboises
—Pear charlotte

Charlotte aux pommes
—Traditional bread apple charlotte

Chaussons aux pommes
—Apple turnovers

Crèmes Glacées
—Ice creams

Crème renversée au caramel
—Baked caramel custard

Crêpes au sucre
—Sugar crêpes

Crêpes Soufflées
—Soufflé-filled crêpes

Pains aux Croissants, Raisins, Pains au Chocolat
—Croissants, raisin buns, and chocolate croissants

Éclairs au café et au chocolat
—Chocolate and coffee eclairs

Gâteau Basque
—Cream-filled butter cake

Galette des Rois
—Epiphany cake

Gâteau Forêt Noire
—Black Forest cake

Moka
—Coffee butter cream sponge cake

Succès
—Hazelnut buttercream meringue cake

Génoise Confiture
—Jam-filled sponge cake

Gratin de fruits rouges
—Red fruit Sabayon

Île flottante
—Floating island

Religieuses au café et chocolat
—Chocolate and coffee cream puffs

Madeleines
—Shell sponge cakes

Millefeuille
—Napoleon

Miroir cassis
—Black currant mousse cake

Mousse au chocolat
—Chocolate mousse

Mousse au citron
—Lemon mousse

Palmiers
—Palm leaf pastry

Petits fours secs
—Assorted tea biscuits

Pithiviers
—Puff pastry filled with almond cream

Poires pochées au vin rouge
—Pears poached in red wine

Profiteroles au chocolat
—Profiteroles with chocolate sauce

Riz Condé
—Molded rice pudding

Riz à l'Impératrice
—Bavarian cream rice pudding

St. Honoré
—Cream puff cake with caramel and chiboust cream

Salambos
—Caramel coated cream puffs filled with kirsch cream

Savarin aux fruits et à la crème
—Savarin cake with fruit

Les sorbets
—Sorbets

Tarte au citron meringuée
—Lemon meringue tart

Tarte au sucre
— Sugar tart

Tarte aux fraises
—Strawberry tart

Tarte aux pommes
—Apple tart

BANDE DE TARTE POMME

Learning Outcomes

Feuilletage
Tarte en bande
Crème d'amandes
Abricoter

Equipment

Knives:
Chef knife (***couteau chef***),
paring knife (***office***),
vegetable peeler (***économe***)

Tools:
Rolling pin, corne,
docker (***pique pâte***),
pastry brush,
wooden spatula,
apple corer

Pans:
1 medium saucepan,
baking sheet

Serving

4 persons

BANDE DE TARTE POMME

Long Apple Tart

Method

1. Preheat the oven to 385°F (195°C).

Pâte Feuilletée

Note: *To obtain the correct amount of pâte feuilletée use the ingredients list in this recipe following the method in Les Bases on page 377 in Cuisine Foundations.*

1. Roll the puff pastry into a rectangle that is ⅛ in. (3 mm) thick and slightly longer than the width of a baking sheet. Cut a strip off this rectangle that is as wide as a plastic scraper [4 ½ in. (12 cm)].
   Note: *This strip will be referred to as the "band"* (***bande***). Lay the band across the width of an ungreased baking sheet, allowing the ends to hang off the edges. Next, without pressing down, fold the remaining pastry in half, then in half again and cut two strips that are 1 in. (2.5 cm) wide. Brush a 1 in. (2.5 cm) swath of egg wash along the edges of the band. Unfold the narrower strips directly on top of the egg wash to form two raised borders (line up the edges neatly). Using a fork or a docker (***pique pâte***), pierce holes in the central channel of the band and use the back of a paring knife to make indentations along the outer edge of the raised borders (***chiqueter***). Using a chef knife, slice off the pastry hanging over the sides of the baking sheet. Brush the raised borders with egg wash, being careful not to get any on the sides (over brushing may prevent even rising). Reserve the pastry on the baking sheet in the refrigerator until needed.

Garniture

1. Peel the apples. With each apple standing upright, cut out the core with a straight downward cut using an apple corer. Cut the cored apples in half. Lay the halves cut side down on the cutting board and trim off the top and bottom. The halves will now resemble dome-shaped rectangles. Reserve the halves in cold water and lemon juice until needed. Cut the leftover flesh off the cores and cut it into small dice to use in making the compote.

Crème d'Amandes

1. Cream (***crémer***) the butter and sugar together until light and fluffy.
2. Beat in the egg until well combined.
3. Using a small knife, split the vanilla bean in half lengthwise and scrape out the seeds. Whisk into the mixture.
4. Add the rum and finish by mixing in the almond powder.
5. Reserve the crème d'amande in a covered bowl in the refrigerator until ready to use.

Note: *Vanilla bean can be replaced by 1–2 tsp. of vanilla extract.*

Quantity		*Ingredient*
U.S.	*Metric*	*Pâte Feuilletée*
1 lb	500 g	Flour
7 fl oz	225 mL	Water
7 oz	200 g	Butter
1 ¾ tsp	10 g	Salt
7 oz	200 g	Butter
1 pc	1 pc	Egg for egg wash
		Garniture
4 pcs	4 pcs	Apples
		Crème d'Amandes
2 oz	60 g	Butter (***pommade***)
2 oz	60 g	Sugar, granulated
2 oz	60 g	Almond powder
1 pc	1 pc	Egg
½ fl oz	10 mL	Rum
1 pc	1 pc	Vanilla bean
		Compote de Pommes
3 pcs	3 pcs	Golden Delicious apples
½ pc	½ pc	Lemon, juice of
2 oz	60 g	Sugar, granulated
1 ¼ oz	40 g	Butter
		To Serve
3 oz	90 g	Apricot jelly
		Powdered sugar

Method

Compote de Pommes

1. Peel and core the apples. Cut the flesh into small dice and mix it with the dice left over from cutting the garnish. Toss the dice in lemon juice.
2. Cook the sugar over medium-high heat in a medium saucepan until it caramelizes. Remove the pan from the stove and deglaze with the butter, stirring with a wooden spatula until the caramel is soft. Add the diced apples and return the pan to the heat. Cook the apples gently, stirring them occasionally until they are cooked but retain some bite. Drain out any excess liquid and let the apples cool to room temperature.

Montage

1. Spread a thin, even layer of crème d'amandes along the central channel of the band. On top of the crème d'amandes spread a thin, even layer of apple compote. Drain the apple pieces and pat them dry. Thinly slice the apple halves from top to bottom. In the process of slicing the apples do not separate the slices but rather keep the shape of the apple halves intact.
2. Place the sliced apple halves side by side along the center of the band so that the slices run parallel to the raised borders of the tart.
3. Lightly push the apple halves to fan them out across the tart.

Finishing

1. Transfer the baking sheet to the oven and bake the tart until golden (*30 minutes*). Turn the tray around after 15 minutes of baking.
2. Remove the baking sheet from the oven and let the tart cool to room temperature.

To Serve

1. Heat the apricot ***nappage*** in a small pan over low heat until it is liquid. Using a pastry brush, apply the ***nappage*** to the apple slices and dust the raised border of the tart with powdered sugar. Transfer the tart to a serving dish.

BAVAROIS RUBANÉ

Learning Outcomes

Making a crème Anglaise
Making a bavarois
Layering mousses
Making a crème Chantilly
Blanchir
à la nappe
Soft peak

Equipment

Knives:
Paring knife (*office*)

Tools:
Bowls, balloon whisk, rubber spatula, wooden spoon,
chinois, charlotte mold, ice bath, pastry bag, large star tip

Pans:
Medium saucepan

Serving

6–8 persons

Quantity		Ingredient
U.S.	Metric	Bavarois
1 pt	500 mL	Milk
½ pc	½ pc	Vanilla pod
4 pcs	4 pcs	Egg yolks
4 oz	120 g	Sugar, granulated
5 to 6 pcs	5 to 6 pcs	Gelatin sheets
1 oz	30 g	Dark chocolate, chopped
		Coffee essence
		Vanilla essence
14 fl oz	400 mL	Whipping cream
		Crème Chantilly
3 fl oz	100 mL	Whipping cream
½ oz	30 g	Powdered sugar
		To Serve
		Chocolate couverture

BAVAROIS RUBANÉ

Tri-colored Bavarian Cream

Method

1. Place a charlotte mold in the freezer to chill.

Bavarois

1. Place the milk in a medium saucepan and bring to a low boil over medium-high heat. Using a small knife, split the vanilla bean lengthwise. Scrape the seeds from both sides and add to the milk along with the pod. Whisk well.
2. Place the egg yolks in a mixing bowl, add the sugar, and immediately begin to whisk it into the yolks. Continue whisking until the sugar is completely dissolved and the mixture is pale in color (***blanchir***).
3. Once the milk is scalded whisk about one-third of the hot milk into the yolks to temper them. Whisk until the mixture is well-combined and evenly heated.
4. Stir the tempered egg yolks into the pan of remaining hot milk and stir with a wooden spatula. Place the pan over low heat and stir in a figure 8 motion. As you stir, the foam on the surface will disappear; at the same time, the liquid will begin to thicken and become oil-like in resistance. Continue cooking until the mixture is thick enough to coat the back of a wooden spatula and when your finger leaves a clean trail (***à la nappe***). DO NOT ALLOW IT TO COME TO A BOIL. Remove the pan from the heat.
5. Bloom the gelatin in a bowl of ice water. Once it is completely softened, squeeze out the excess water and add the gelatin to the hot crème Anglaise. Stir it in until completely dissolved, then strain the hot crème Anglaise through a fine mesh sieve (***chinois***).
6. Place the chocolate couverture in a small mixing bowl and the coffee extract in another. Pour one-third of the crème Anglaise onto the chocolate, let it rest 1 minute to melt the chocolate, and stir gently with a wooden spatula until the mixture is smooth and homogeneous. Pour another one-third of the crème Anglaise onto the coffee extract and stir gently with a wooden spatula until combined. Stir the vanilla essence into the remaining crème Anglaise. Scrape the sides of all 3 bowls clean and reserve at room temperature.
7. Whip the cream in a large bowl over an ice bath, using a large whisk, until it reaches the soft peak stage. Divide the whipped cream into 3 equal portions and reserve in the refrigerator in 3 separate bowls.

Montage

1. Place the bowl of vanilla crème Anglaise in an ice bath and stir it back and forth with a wooden spatula (***vanner***) until it is cool to the touch but still liquid. Remove the bowl from the ice bath and remove 1 bowl of whipped cream from the refrigerator. Add one-third to the vanilla mixture to lighten its texture and gently stir it in with a whisk until the mixture is homogeneous. Add the rest of the whipped cream and fold it in with a rubber spatula until combined. Remove the charlotte mold from the freezer and being careful not to get any drips down the sides, pour the mixture into the bottom of the mold. Gently tap the mold on the work surface to even out the layer of bavarois and remove any air bubbles and return it to the freezer to set before going on with the next layering.

Method

2. Meanwhile, prepare the coffee bavarois by cooling the crème Anglaise over an ice bath and adding the whipped cream to it following the same procedure as for the vanilla bavarois. Once the bottom layer has set, remove the mold from the freezer and being careful not to get any on the sides of the mold, pour in the coffee bavarois. Gently tap the mold on the work surface to even out the layer of bavarois and remove any air bubbles and return it to the freezer to set the second layer.
3. Prepare the chocolate bavarois by following the same method as for the vanilla and coffee bavarois. Remove the mold from the freezer and pour in the chocolate bavarois. Gently tap the mold on the work surface to even out the layer and remove any air bubbles and return it to the freezer to allow the bavarois to set. The bavarois can be kept in the refrigerator once the layering is done; the use of the freezer in the process is to accelerate the setting of the different layers.

Chocolate Shavings

1. With a clean vegetable peeler, scape a bar or piece of chocolate to make chocolate shavings. Scrape into a small bowl and set aside.

Finishing

1. *Prepare a crème Chantilly:* Whip the cream in a large mixing bowl in an ice bath using a large whisk. Whip until it reaches the soft peak stage then add the powdered sugar and continue to whip the cream until it is stiff. Reserve the crème Chantilly in the refrigerator until needed.

To Serve

1. Attach a medium star tip to a pastry bag and half fill it with the crème Chantilly. Half fill a large mixing bowl with hot water.
2. Remove the charlotte mold from the freezer and dip it in the hot water for 2 to 3 seconds to loosen the bavarois from the mold. Unmold the bavarois onto a chilled serving dish.
3. Once the bavarois is unmolded, pipe (***coucher***) crème Chantilly on the top and sprinkle with chocolate shavings.
4. Serve the bavarois immediately or return it to the freezer until ready to serve.

BEIGNETS AUX POMMES, SAUCE ABRICOT

Learning Outcomes

Making a pâte à beignets (*pâte à frire*)
Sweet marinades
Sweet sauces

Equipment

Knives:
Vegetable peeler (*économe*), apple corer, paring knife (*office*)

Tools:
Tamis, mixing bowls, rubber spatula, whisk, skimmer (*écumoire*), sugar dredger

Pans:
Small sauce pan, shallow pan

Serving

4 persons

FYI...

Since the Middle Ages, cooks have been coating both sweet and savory ingredients in batter and cooking them in hot oil. *Beignets*, also known as fritters in English, have evolved from region to region with some recipes calling for a choux paste and others, such as *beignets de pommes*, calling for beer-based dough. In terms of sweet fillings, apples are one of the oldest and best-loved filling ingredients for *beignets*.

BEIGNETS AUX POMMES, SAUCE ABRICOT

Apple Fritters with Apricot Sauce

Quantity		Ingredient
U.S.	*Metric*	
4 pcs	4 pcs	Golden Delicious apples
		Pâte à Beignets
8 oz	250 g	All-purpose flour
2 pcs	2 pcs	Eggs
½ oz	10 g	Sugar, granulated
¼ oz	5 g	Salt
¾ fl oz	25 g	Unsalted butter, melted
5 fl oz	150 mL	Beer, lukewarm
3 pcs	3 pcs	Egg whites
		Water (optional)
		Marinade
2 oz	60 g	Powdered sugar
1 fl oz	30 mL	Calvados or domestic apple brandy
1 pc	1 pc	Lemon, juice of
		To Finish
5 oz	150 g	Apricot jam
		Oil for deep-frying
		Powdered sugar for dusting

Method

1. Heat the oil in the deep fryer to 360°F (180°C).

Pâte à Beignets

1. Melt the butter in a small saucepan over low heat and set aside.
2. Sift the flour, place it in a large mixing bowl, and make a well (***fontaine***) in the center. Add the eggs, salt, and sugar into the well and stir them into the flour with a wooden spatula. ***Note***: *It is normal for the mixture to be lumpy at this stage.*
3. Pour in one-half the beer and stir it in with a whisk until the mixture resembles a smooth paste. Add the remaining beer and stir until incorporated. Pour a little of the batter into the melted butter and mix them together until smooth. Pour this back into the batter and stir it in with the whisk.

Note: *If the dough is difficult to stir with a spatula, add a bit of water.*

4. Scrape the sides of the bowl clean with a rubber spatula or plastic scraper (***corne***) and cover the bowl in plastic wrap. Let the batter rest in the refrigerator for 15 to 20 minutes (*minimum*). The dough can be reserved for longer if necessary.

Marinade

1. Peel and core the apples and slice them into 1 cm (½ in.) thick slices ***rouelles***. Arrange the rouelles in a single layer in a shallow pan. Dredge with powdered sugar then drizzle with Calvados. Reserve in the refrigerator for 15 minutes (*minimum*).

Sauce Abricot

1. Thin out the apricot jam with enough marinade to make the sauce just thick enough to coat the back of a spoon (***à la nappe***).

To Finish

1. Drain the apple ***rouelles*** on a paper towel to absorb excess moisture.
2. Whisk the egg whites to soft peaks, then fold into the beignet batter until just incorporated.
3. One by one, dip the ***rouelles*** in the ***pâte à beignets*** to cover them completely, then place them in the deep fryer. Turn them over halfway through the cooking process (*when they are just beginning to color*). Remove the ***beignets*** from the oil when they are golden and drain them on a paper towel.

To Serve

1. Arrange the ***beignets*** on a serving platter, dust them with powdered sugar, and serve the sauce on the side.

BRIOCHES

Learning Outcomes

Making brioche dough
Developing gluten
Kneading
Preparing various shapes of brioche

Equipment

Knives:
Chef knife (*couteau chef*)

Tools:
Mixing bowl, corne, whisk

Pans:
Small saucepan, baking sheet

Serving

6 pcs petites brioches à tête,
1 pc grande brioche à tête,
1 pc brioche tressée

FYI...

"Let them eat cake" is often quoted as an example of Marie Antoinette's arrogance at the plight of hungry Parisians during the French Revolution (1789–1799). In fact, it was not "cake" that she said but "brioche." This puts the last queen of France in a slightly better light; a simple brioche with no filling was then considered simply bread enriched with butter and eggs. This doesn't mean she wasn't being a little snarky when she said it, but it should nevertheless give historians pause to reflect.

BRIOCHES
Brioche Bread

Quantity		Ingredient
U.S.	*Metric*	
4 fl oz	125 mL	Milk
½ oz	15 g	Fresh yeast
4 pcs	4 pcs	Eggs
1 oz	30 g	Sugar, granulated
1 lb	500 g	Flour, sifted
¼ oz	10 g	Salt
½ lb	250 g	Butter, room temperature
1 pc	1 pc	Egg for egg wash

Method

1. Preheat the oven to 400°F (205°C).

Pâte à Brioche

1. Heat the milk in a small saucepan over low heat until it is warm to the touch. Remove it from the heat, mix in the yeast, and set aside.
2. Mix the eggs and the sugar together in a bowl using a whisk. Sift (***tamiser***) the flour onto a clean, dry work surface, add the salt, and make a large well (***fontaine***) in the center using a plastic scraper (***corne***). Into the well, pour the milk and yeast mixture as well as the egg mixture. Incorporate all the ingredients in the well with your fingertips, simultaneously using the ***corne*** to add small quantities of flour from the sides. Continue until the mixture in the center of the well resembles a thick paste. Gather all the ingredients together and work them with the heel of your palm until combined.
3. With the heel of your palm, push the dough away from yourself. ***Note***: *At this point, the dough should be wet enough to stick to the work surface. If not, work in a little more milk, a spoonful at a time, until the desired consistency is reached.*
4. Repeat the motion of stretching the dough out then folding it in on itself until the dough is elastic and no longer sticks.
5. Dot the dough with small pieces of softened butter and fold the dough over to enclose the butter. Knead it until the butter is incorporated. Throw the dough hard onto the work surface, fold it onto itself and repeat. Continue working the dough in this manner in order to develop the gluten.
6. Roll the dough into a ball, dust it lightly with flour, and place it in a clean bowl. Cover the bowl with a damp cloth and leave it to proof either in a warm area or in a proofer until doubled in size.
7. Transfer the risen dough to a lightly floured work surface and press out all the air bubbles. Roll the dough back into a ball, lightly dust it with flour, and return it to the bowl to rise a second time. Let it rise overnight in the refrigerator or, for faster results, in a warm place or proofer.
8. Butter the brioche molds and divide the risen dough into 3 equal parts (see pages 387-392 in *Cuisine Foundations*).

Petites Brioches à Tête

1. Lightly dust the work surface with flour (***fleurer***). Place one of the pieces of dough on the floured work surface and lightly press down with the flat of your hands, shaping the dough into a rectangle. Fold the upper one-third of the rectangle down over itself and apply pressure with the heel of your palm to seal the seam. Turn the dough around and

Method

repeat the procedure of folding down the top one-third and sealing the seam with the heel of your palm.

Roll the dough out into an even cylinder, about 2 in. (5 cm) thick, and bend it in half so that the two ends touch. Score the midpoint with a knife and straighten out the cylinder. Cut it in half at the scored midpoint and proceed to cut each section into equally sized pieces [1 ¾ oz (50 g) each].

2. One by one, dip the pieces in flour and roll them into balls; using the palm of your hand, roll the dough in a circular motion until the ball is smooth and tight. Once all the dough has been shaped into balls, roll them back and forth into oblong shapes. Dip the side of your hand in flour, and place it on the oblong, one-third of the way down. With a light sawing motion, use the floured side of your hand to roll the oblongs into bowling pin shapes. When doing so, guide the long end of the dough with your free hand to keep it from flopping. Continue this process until a defined "neck" is created. Repeat with all the oblongs.
3. Dip your thumb, index finger, and middle finger in flour and pick up a piece of dough by the "neck." Lower the dough, fat end first, into a buttered mold and press your fingers down through the dough to the bottom of the mold (*still holding the "neck"*). Let go and gently remove your fingers. The result should be a small ball of dough nestled in a larger one. Dip your index finger in flour and press it deeply into the seam, between the "head" and "body." Carefully remove your finger and repeat this motion, working your way around the head.
4. Repeat with the remaining brioches.
5. Set the brioches aside to rise in a warm place or in a proofer until they double in size.

Grande Brioche à Tête

1. Lightly dust the work surface with flour (***fleurer***). Place one of the pieces of dough on the floured work surface, cut off one-quarter, and set it aside.
2. Using your hands, flatten the remaining dough into a circle. Fold an edge of the circle into the center and seal it by pressing it with the heel of your palm. Repeat this motion, turning the dough a little every time, until you obtain a small, compact "round." Turn the dough over and cup your hands around it. Roll the dough around the work surface in a circular motion with your cupped hands until it forms a smooth, tight ball.
3. Create a deep depression in the center of the ball using your fingers or the end of a rolling pin and place the dough in a large brioche mold.
4. Roll the small piece of reserved dough into a ball. Cup your hand around the dough and roll it around in a circular motion until the ball is smooth and tight. Roll one end of the ball back and forth to create a pear shape. Dip your fingers in flour and pick up the pear (*holding it upside down*). Insert the upside-down pear into the hole in the center of the brioche, pressing down deeply until your fingers touch the bottom of the mold. Carefully remove your fingers and dip your index finger in flour. Press it deeply into the gap between the "head" and the "body," working your way around to seal the seam. Set the brioche aside to rise in a warm place or in a proofer until it doubles in size.

Brioche Tressée

1. Lightly dust the work surface with flour (***fleurer***). Place one of the pieces of dough on the floured work surface and divide it into 3 equal parts. Lightly press down one of the pieces of the dough with the flat of your hand to form a rectangle. Fold the upper one-third of the rectangle down over itself and apply pressure with the heel of your palm to seal the seam. Turn the dough around and repeat the procedure of folding down the top one-third and sealing the seam with the heel of your palm.
2. Roll the dough out into an even cylinder, about 1 in. (2.5 cm) thick. Repeat with the 2 remaining pieces of dough.
3. Place the 3 cylinders side by side and, if necessary, trim them to the same length. Attach them all at one end.
4. *Braid the brioche*: Lift the central strand of dough over the right hand one. Next, lift the strand that is now at the center over the left one. Again, lift the strand that is now at the center over the right one and continue this process, alternating right and left until you reach the end of the strands of brioche.
5. Tuck the ends under the braid and transfer it to a clean baking sheet. Set the brioche aside to rise in a warm place or in a proofer until it doubles in size (see pages 387–392 in *Cuisine Foundations*).

Finishing

1. Brush the brioches with egg wash, place them on the baking sheet, and transfer them to the preheated oven to bake. Once the brioches begin to turn golden, reduce the heat to 375°F (190°C) and continue to bake the brioches until they turn a deep shade of gold.
2. When the brioches are cooked, remove them from the oven and turn the brioches à tête upside down in their molds to cool.

CHARLOTTE AUX POIRES, COULIS DE FRAMBOISES

Learning Outcomes

Preparing a bavarois
Preparing biscuit cuillère
Montage of a charlotte
Monter les blancs
Coucher
Perler
Tamiser
Blanchir
Crème Anglaise
Vanner

Equipment

Knives:
Paring knife (*office*), chef knife (*couteau chef*)

Tools:
Scissors, charlotte mold, balloon whisk, rubber spatula, wooden spatula, chinois, pastry bag, plain 12-mm tip, medium star tip, mixing bowls, whisk, ice bath

Pans:
1 medium saucepan,
1 small saucepan,
2 baking sheets

Serving

6 persons

Quantity		Ingredient
U.S.	*Metric*	*Biscuit Cuillère*
4 oz	120 g	Flour, sifted (***tamiser***)
4 pcs	4 pcs	Egg whites
4 oz	120 g	Sugar, granulated
4 pcs	4 pcs	Egg yolks
		Powdered sugar, for dusting
		Bavarois
7 fl oz	200 mL	Milk
3 pcs	3 pcs	Egg yolks
1 ¾ oz	50 g	Sugar, granulated
4 pcs	4 pcs	Gelatin sheets
7 oz	200 g	Pear purée, room temperature
		Pear liqueur, room temperature (*optional*)
7 fl oz	200 mL	Cream
		Coulis
3 ½ oz	100 g	Raspberry purée
1 ¼ oz	35 g	Sugar, granulated
¼ fl oz	10 mL	Raspberry liqueur
		To Serve
3 ½ fl oz	100 mL	Cream
2 pcs	2 pcs	Pears, canned (*optional*)

CHARLOTTE AUX POIRES, COULIS DE FRAMBOISES

Pear Charlotte

Method

1. Preheat the oven to 380°F (195°C).
2. Place 2 large mixing bowls and whisks in the refrigerator to chill.
3. On the underside of a sheet of parchment paper, trace the contour of the open end of the charlotte mold.
4. Line a baking sheet with the parchment paper, upside down so that the pencil tracings are underneath.
5. Line a second baking sheet with a blank piece of parchment paper.
6. Cut a circle of parchment paper to line the bottom of the charlotte mold. Place it inside.

Biscuit Cuillère

1. *Prepare the biscuit cuillère batter:* Sift (***tamiser***) the flour onto a piece of parchment paper. Separate the eggs into 2 large mixing bowls. Beat the whites (***monter les blancs***) with a balloon whisk until they are white and frothy but still fluid. Set aside. Add the sugar to the egg yolks and whisk them together until the mixture is thick and pale in color (***blanchir***). Set aside.
2. Whip the egg whites (***monter les blancs***) with the balloon whisk until they reach the soft peak stage (***bec d'oiseau***). Gradually begin to add the sugar, beating it in with the whisk. Once all the sugar has been added, continue to beat the whites until they are thick and glossy and you can no longer feel the granules when the meringue is rubbed between your fingers. Stir the flour into the egg yolks until just incorporated. Add some of the meringue to lighten the yolk mixture, then fold in the remaining meringue. The end result should be a pale, homogeneous batter with a light airy texture that is ideal for piping.
3. Transfer the batter to a pastry bag fitted with a 12-mm plain tip.
4. *Pipe (**coucher**) the biscuit cuillère:* Create a flower shape (***marguerite***) by piping teardrop shapes from the exterior of one of the circles drawn onto the parchment paper to its center, rotating the tray with each "petal." Finish the flower by piping a "chou" into its center. Next, pipe a tight spiral, beginning at the center of a traced circle, going out to the edge of the circle. Pipe straight fingers, slightly longer than the depth of the charlotte mold, onto the second baking sheet until all the batter is used up. Dust all the piped shapes with powdered sugar until they are thickly coated (***perler***) and transfer the baking sheets to the oven to bake. After 5 minutes, rotate

Method

the baking sheets and turn the oven down to 365°F (185°C). Continue to bake the biscuits until they are lightly golden and dry to the touch. Remove the baking sheets from the oven and carefully slide the parchment paper onto a rack to cool. Dust the shapes with powdered sugar and leave them to cool.

Bavarois

1. Whip (***monter***) the cream to soft peaks in the chilled mixing bowl and whisk and reserve covered in the refrigerator.
2. Cut the pear into a small dice.

Prepare the Crème Anglaise

1. Place the milk in a medium saucepan and bring to a low boil over medium-high heat. Whisk well. Place the egg yolks in a mixing bowl, add the sugar, and immediately begin to whisk it into the yolks. Continue whisking until the sugar is dissolved and the mixture is pale in color (***blanchir***).
2. Once the milk is hot, whisk about one-third of the hot milk into the yolks to temper them. Whisk until the mixture is well combined and evenly heated.
3. Stir the tempered egg yolks into the pan of remaining hot milk and stir with a wooden spatula. Place the pan over low heat and stir in a figure 8 motion. As you stir, the foam on the surface will disappear; at the same time, the liquid will begin to thicken and become oil-like in resistance. Continue cooking until the mixture is thick enough to coat the back of a wooden spatula and your finger leaves a clean trail (***à la nappe***). Remove the pan from the heat.

Tip: *Crème Anglaise should be cooked to between 167°F and 185°F (75°C and 85°C).*

4. Bloom the gelatin in a bowl of ice water. Once it is completely softened, squeeze out the excess water and add the gelatin to the hot crème Anglaise. Stir it in until completely dissolved, then strain the crème Anglaise through a fine mesh sieve (***chinois***) into a large mixing bowl set in an ice bath. Stir it back and forth with a wooden spatula (***vanner***) until it is cool to the touch but still liquid. Remove the crème Anglaise from the ice bath and stir the pear purée and liqueur into it. Reserve at room temperature.

Montage

1. *Line the mold* (***chemiser***): Trim one end off all the ladyfingers, making sure they are all the same length as the depth of the charlotte mold. Line the sides of the mold, making sure that the biscuits fit snugly, cut side down.
2. Give the cream a final turn of the whisk and add one-half of it to the prepared crème Anglaise. Fold it in to lighten the texture of the custard and add the second half of the whipped cream. Delicately fold it in until the mixture is homogeneous. Pour it into the mold, filling it halfway up, and then add the diced pear. Fill the mold with the bavarois to just ½ in. (1 cm) from the top of the ladyfingers and gently tap the mold on the work surface to expel any air bubbles. Cut the spiral circle of biscuit to the right size to fit in the opening of the charlotte mold and gently place it in, so that it is level with the end of the ladyfingers.
3. Transfer the charlotte to the freezer to set for 30 minutes.

Coulis

1. Heat the raspberry purée and sugar together in a small saucepan over medium heat, stirring occasionally until it is thick enough to coat the back of a spoon (***à la nappe***). When the liquid has reached the desired consistency, remove the pan from the stove and stir the coulis back and forth with a wooden spatula (***vanner***) until it reaches room temperature. Add the liqueur and stir it in.

To Serve

1. Whip the cream to stiff peaks using the second chilled mixing bowl. Cover and reserve in the refrigerator.
2. When the charlotte is set, unmold it onto a chilled serving dish. Remove the parchment circle from the top.
3. Fill a pastry bag fitted with a medium star tip with the whipped cream and pipe (***coucher***) a line of teardrops into each depression between the ladyfingers. Pipe a small amount of cream onto the top of the charlotte and place the marguerite on top. Reserve the charlotte in the refrigerator until ready to serve.
4. Serve the charlotte with the coulis on the side.

Optional: Decorate the serving dish with sliced pears.

Learning Outcomes

Compoter
Chemiser a charlotte mold
Blanchir
Delayer
Cuisson à la nappe
Making a crème Anglaise
Vanner

Equipment

Knives:
Serrated knife, paring knife (*office*), vegetable peeler (*économe*)

Tools:
Wooden spatula, mixing bowl, charlotte mold, whisk, ice bath, chinois, pastry brush, apple corer

Pans:
Large saucepan, medium saucepan

Serving

6 persons

Originally an English dessert named after Queen Charlotte (wife of George III), Charlotte aux pommes was adapted for French tastes in the 18th century. Antoine Carême, enamored with the shape of the charlotte mold, dispensed with the English use of breadcrumbs, favoring instead ladyfingers or artfully trimmed white bread for the dessert's outer layer. While this recipe is the most traditional *charlotte* in that it is made with apple compote and is served hot, similar preparations may involve different fruit and can be served cold.

CHARLOTTE AUX POMMES

Traditional Bread Apple Charlotte

Quantity		Ingredient
U.S.	Metric	*Compote*
1 lb 12 oz	800 g	Golden Delicious apples
1 ½ oz	40 g	Sugar, granulated
		Chemisage
10 pcs	10 pcs	Slices of white sandwich bread
5 oz	150 g	Butter, clarified
1 ¾ oz	50 g	Apricot glaze
¼ fl oz	2 mL	Vanilla extract
		Crème Anglaise
7 fl oz	200 mL	Milk
2 pcs	2 pcs	Egg yolks
2 oz	60 g	Sugar, granulated
½ fl oz	15 mL	Calvados
¼ fl oz	2 mL	Vanilla extract

Method

1. Heat the oven to 375°F (190°C).

Compote

1. Peel and core the apples and cut them into a small dice. Place the apples and sugar together in a large saucepan over medium heat and cook them until they begin to soften. Cover the saucepan, turn the heat down to low, and cook the apples until they are completely soft and have turned a golden color. Remove the lid and cook the apples, stirring gently, until all the liquid has evaporated.
2. Transfer the compote to a clean bowl and set aside.

Montage

1. Cut the crusts off the bread slices. Cut 2 bread slices in half diagonally. Cut each of these triangles down the center into 2 equal halves (*giving a total of 4 equally shaped triangles per bread slice*). Round off one end of all the triangles to obtain teardrop shapes. Cut the rest of the bread into 1 in. (2 ½ cm) thick strips. Place the teardrop shapes in the bottom of the charlotte mold to form a flower (***marguerite***) and line the sides with the bread fingers, fitting them in snugly. If necessary, trim the shapes for a better fit.
2. Remove the bread shapes from the mold and, one by one, dip them in the clarified butter, then place them back into the mold to line it (***chemiser***).
3. Stir the apricot glaze and the vanilla extract into the compote and pour the mixture into the charlotte mold to fill it. Transfer the molds to the oven to bake for 40 to 45 minutes.

Crème Anglaise

1. Place the milk in a medium saucepan and bring to a low boil over medium-high heat. Whisk well. Place the egg yolks in a mixing bowl, add the sugar, and immediately begin to whisk it into the yolks. Continue whisking until the sugar is completely dissolved and the mixture is pale in color (***blanchir***). Once the milk is scalded, whisk about one-third of the hot milk into the yolks to temper them. Whisk until the mixture is well combined and evenly heated.
2. Stir the tempered egg yolks into the pan of remaining hot milk and stir with a wooden spatula. Place the pan over low heat and stir in a figure 8 motion. As you stir, the foam on the surface will disappear; at the same time, the liquid will begin to thicken and become oil-like in resistance. Continue cooking until the mixture is thick enough to coat the back of a wooden spatula and when your finger leaves a clean trail (***à la nappe***).

Tip: *Crème Anglaise should be cooked to between 167°F and 185°F (75°C and 85°C).*

3. Remove the pan from the heat and strain the sauce through a fine mesh sieve into a clean bowl set over an ice bath. Stir it back and forth with a wooden spatula (***vanner***) until it is cool to the touch. Add the calvados and vanilla and stir them in. Cover the bowl in plastic wrap and reserve it in the refrigerator.

To Serve

1. Once the charlotte is cooked and the bread is golden and crispy, trim any excess bread from the top of the mold. If the trimmings are not too dark or burned, flatten them on the surface of the charlotte. Let the charlotte rest for 5 minutes, then turn it out onto a serving dish.
2. Pour some crème Anglaise into the serving dish to coat the bottom and serve the rest on the side.

Optional: *Heat some apricot glaze in a small saucepan over medium heat until it is liquid. Apply it to the unmolded charlotte using a pastry brush to add shine.*

CHAUSSONS AUX POMMES

Learning Outcomes

Pâte feuilletée
Compote de pommes
Making a chausson
Decorating pâte feuilletée
Chiqueter
Videler
Tamiser
Détrempe
Abaisser
Fleurer

Equipment

Knives:
Vegetable peeler (*économe*), paring knife (*office*), chef knife (*couteau chef*)

Tools:
Rolling pin, corne, baker's brush, mixing bowls, wooden spoon, wooden spatula, 4-inch cookie cutter, pastry brush

Pans:
Medium saucepan, small saucepans, baking sheet

Serving

12 persons

FYI...

In 1580, a plague swept through the French town of Saint-Calais. Those too poor to flee were left in the town to starve. According to the legend, an aristocratic lady baked an enormous pastry filled with apples which she served to the remaining townspeople—presumably saving them all from starvation. To commemorate the benevolence of this unnamed lady, the people of Saint-Calais still conduct a yearly march through town on her behalf. On this day, the bakers and pastry chefs of the town distribute chaussons aux pommes (single-serving sizes, of course!) to the cheering crowd.

CHAUSSONS AUX POMMES
Apple Turnovers

Quantity		Ingredient
U.S.	*Metric*	*Pâte Feuilletée*
1 lb	500 g	Flour
7 fl oz	225 mL	Water
7 oz	200 g	Butter, room temperature
1 ¾ tsp	10 g	Salt
7 oz	200 g	Butter
1 pc	1 pc	Egg for egg wash
		Compote
3 pcs	3 pcs	Golden Delicious apples
½ pc	½ pc	Lemon, juice of
2 oz	60 g	Sugar, granulated
1 ½ oz	40 g	Butter
		Vanilla (*optional*)
		Syrup
1 ¾ oz	50 g	Sugar, granulated
1 ¾ fl oz	50 g	Water
		Powdered sugar for dusting

Method

1. Preheat the oven to 400°F (205°C).

Pâte Feuilletée

Note: *To obtain the correct amount of pâte feuilletée use the ingredients list in this recipe following the method in Les Bases on pages 377–379.*

Compote

1. Peel and core the apples. Cut the flesh into a small dice and toss with the lemon juice. Cook the sugar over medium-high heat in a medium saucepan until it caramelizes. Remove the pan from the stove and add the butter to deglaze. Stir with a wooden spatula until the caramel is soft. Add the diced apples and return the pan to low heat. Cook the apples gently, stirring them occasionally until they are cooked but retain some texture. Drain out any excess liquid and set the compote aside to cool to room temperature.

Montage

1. Lightly dust the work surface with flour (***fleurer***). Place the puff pastry on the work surface and roll it out (***abaisser***) with a rolling pin to a thickness of 3/16 in. (5 mm). Transfer the dough to a refrigerator to relax for 5 minutes. Remove the dough from the refrigerator and cut it into circles with a 4-inch (10 cm) cookie cutter. Roll each circle of dough into an oval with a couple of strokes of the rolling pin. Neatly arrange the ovals on a clean baking sheet, brush off any excess flour using a baker's brush, and leave the ovals to rest for 5 minutes in the refrigerator. Remove the baking sheet from the refrigerator and fold the ovals in half to create a guideline crease along the widest point of the oval. Brush the edge of one-half of each oval with egg wash. Place a tablespoon of compote in the center of the each egg-washed half and fold the other half over it, gently pressing out any air bubbles. Carefully press down the rim of each chausson to secure the seal. Score the edges with the back of a paring knife (***chiqueter***). The goal is to seal the seam while creating a decorative pattern.
2. Let the chaussons rest in the refrigerator for 10 minutes, then brush them in egg wash and lightly score the tops with a paring knife. Make a small hole (***cheminée***) in the top of each chausson to allow steam to escape during baking. Transfer the chaussons to the oven and bake until golden (*rotating the baking sheet as soon as the pastry begins to color*).

Method

Finishing

1. *Prepare a syrup:* Pour the water into a small saucepan and add the sugar. Bring the mixture to a boil over medium-high heat and continue to boil it until all the sugar has dissolved. Remove the pan from the stove and set aside to cool.
2. When the chaussons aux pommes are a deep golden color, remove them from the oven and transfer them to a wire rack to cool. While they are still warm, brush them with the syrup using a pastry brush.
3. Once the chaussons are cool, dust them with powdered sugar and transfer them to a serving dish.

CRÈMES GLACÉES

Learning Outcomes

Crème Anglaise
Vanner
Blanchir
à la nappe

Equipment

Knives:
Serrated knife, paring knife (*office*)

Tools:
Wooden spatula, mixing bowls, whisk, chinois, ice cream churner, ice bath

Pans:
Medium saucepans

Yield

Vanilla ice cream yield
.5 qt (0.5 L)
Coffee ice cream yield
1 qt (1 L)
Chocolate ice cream yield
1 qt (1 L)

FYI...

In 336 BC, Alexander the Great is reported to have enjoyed a combination of fruit, wine, and honey frozen in snow. Pliny the Elder reports in his work *Naturalis Historia* (AD 77) that Nero had his troops collect snow for the purposes of making sweet frozen preparations. Upon returning from his 1292 exploration of China, Marco Polo introduced vital Chinese ice-cream-making technology to Europe. This technology was a method for super cooling ice cream with saltpeter and salt. In Europe, ice-cream-like preparations have evolved dramatically since the 13th century: from frozen beverages brought to France by Catherine De'Medici, to a frozen molded dessert created by Francesco Procopio, to the recipes we have before us here.

CRÈMES GLACÉES
Ice Creams

Quantity		Ingredient
U.S.	*Metric*	*Vanilla Ice Cream*
8 fl oz	250 mL	Milk
4 fl oz	125 mL	Cream
3 pcs	3 pcs	Egg yolks
3 ½ oz	100 g	Sugar, granulated
1 pc	1 pc	Vanilla bean
		Coffee Ice Cream
8 fl oz	250 mL	Milk
8 fl oz	250 mL	Cream
¼ fl oz	5 mL	Vanilla extract
4 pcs	4 pcs	Egg yolks
4 oz	125 g	Sugar, granulated
½ fl oz	15 mL	Instant coffee
		Chocolate Ice Cream
1 pt	500 mL	Milk
5 oz	150 g	Dark chocolate
3 ½ oz	100 g	Sugar, granulated
6 pcs	6 pcs	Egg yolks

Method

Vanilla Ice Cream

1. *Prepare a crème Anglaise:*
2. Place the milk and heavy cream in a medium saucepan and bring to a low boil over medium-high heat. Using a small knife, split the vanilla bean lengthwise. Scrape the seeds from both sides and add to the milk along with the pod. Whisk well.
3. Place the egg yolks in a mixing bowl, add the sugar, and immediately begin to whisk it into the yolks. Continue whisking until the sugar is completely dissolved and the mixture is pale in color (***blanchir***).
4. Once the milk is hot whisk about one-third of the hot milk into the yolks to temper them. Whisk until the mixture is well combined and evenly heated.
5. Stir the tempered egg yolks into the pan of remaining hot milk and stir with a wooden spatula. Place the pan over low heat and stir in a figure 8 motion. As you stir, the foam on the surface will disappear; at the same time, the liquid will begin to thicken and become oil-like in resistance. Continue cooking until the mixture is thick enough to coat the back of a wooden spatula and when your finger leaves a clean trail (***à la nappe***). DO NOT ALLOW IT TO COME TO A BOIL.

Tip: *Crème Anglaise should be cooked to between 167°F and 185°F (75°C and 85°C).*

6. Remove the pan from the heat and strain the crème Anglaise through a fine mesh sieve (***chinois***) into a clean bowl set in a bowl of ice. Stir it back and forth (***vanner***) with the spatula until cooled.
7. Cover the bowl in plastic wrap and reserve in the refrigerator for 24 hours to develop the flavors.
8. Pour the cold custard into a running ice cream churner and let it run until the ice cream is set.
9. Once churned, reserve the ice cream in an airtight container in the freezer.

Coffee Ice Cream

1. *Prepare a crème Anglaise* as described above, with the addition of the instant coffee to the milk and replacing vanilla bean with extract.
2. Once the crème Anglaise is cooked, remove the pan from the heat and strain the crème Anglaise through a fine mesh sieve (***chinois***) into a clean bowl set in a bowl of ice. Stir it back and forth (***vanner***) with the spatula until cooled.

Method

3. Cover the bowl in plastic wrap and reserve in the refrigerator for 24 hours to develop the flavors.
4. Pour the cold custard into a running ice cream churner and let it run until the ice cream is set.
5. Once churned, reserve the ice cream in an airtight container in the freezer.

Chocolate Ice Cream

1. Finely chop the chocolate (***hacher***) and place it in a large mixing bowl.
2. *Prepare a crème Anglaise:*
3. Once the crème Anglaise is cooked, remove the pan from the heat and pour it into the mixing bowl containing the chocolate. Let it sit for 1 to 2 minutes to melt the chocolate, then stir gently with a wooden spatula until the chocolate has melted. Strain the mixture through a fine mesh sieve into a mixing bowl set on an ice bath. Stir the custard back and forth with a wooden spatula (***vanner***) until it is cold. Cover the bowl in plastic wrap and place the bowl in the refrigerator for 24 hours to develop the flavors. Pour the cold custard into a running ice cream churner and let it run until the ice cream is set.
4. Once churned, reserve the ice cream in an airtight container in the freezer.

CRÈME RENVERSÉE AU CARAMEL

Learning Outcomes

Making a caramel
Cooking a set cream
Démouler

Equipment

Knives:
Paring knife (*office*)

Tools:
Whisk, chinois, bowls, ladle (*louche*)

Pans:
Roasting pan, ramekins, small pan, small sauce pan, medium sauce pan

Serving

4 persons

FYI...

Crème renversée au caramel is a dessert preparation that is in the same family as crème brulée, the main difference being that crème brûlée contains yolks and cream whereas crème caramel is made using whole eggs and milk. Also, in this preparation caramel is poured into the bottom of the mold before adding the appareil. The caramel becomes liquid as the appareil sets during cooking; it then acts as a sauce when the dessert is turned out of its mold onto the plate.

CRÈME RENVERSÉE AU CARAMEL

Baked Caramel Custard

Quantity		Ingredient
U.S.	*Metric*	*Caramel*
2 ½ oz	75 g	Sugar, granulated
¾ fl oz	25 mL	Water
2 to 3 drops	2 to 3 drops	Lemon juice
		Crème Renversée
8 fl oz	250 mL	Milk
¼ fl oz	2 mL	Vanilla extract
2 pcs	2 pcs	Eggs
2 ¼ oz	65 g	Sugar, granulated

Method

1. Preheat the oven to 340°F (170°C).

Caramel

1. Place the sugar, water, and lemon juice together in a small saucepan and cook them over medium-high heat until they reach the caramel stage. When the caramel becomes a deep amber color, remove the pan from the heat, dip the bottom of the pan in cold water to stop the cooking, and pour a thin layer of caramel into the bottom of each ramekin. Set aside.

Crème Renversée

1. Bring the milk and vanilla extract to a boil in a medium saucepan over medium-high heat.
2. Meanwhile, break the eggs into a large mixing bowl, add the sugar, and beat with a whisk until combined. As soon as the milk begins to boil, remove it from the heat and pour it into the egg mixture in a thin stream while stirring with a whisk. Strain the mixture through a fine mesh sieve (***chinois***) into a clean bowl and skim (***skim***) the froth off the surface. Ladle the mixture into the ramekins filling them half an inch from the top.
3. Transfer the ramekins to a roasting pan and fill it with hot water to two-thirds up their sides (bain marie). Transfer the pan to the oven and cook the custards until they no longer ripple when lightly shaken (*30 minutes*) and the tip of a knife inserted in the middle comes out clean.
4. Remove the pan from the oven and let the custards cool a little before removing them and wiping the ramekins dry with a clean cloth. Let the custards cool to room temperature before covering them in plastic wrap and reserving them in the refrigerator until needed (*minimum 2 hours*).
5. Crème caramel is best prepared the day before to allow the caramel a chance to melt into a sauce.

To Serve

1. Cut around the edge of the custard with a paring knife to loosen it from the ramekin and flip it onto a plate. Remove the ramekin.

CRÊPES AU SUCRE

Learning Outcomes

Pâte à crêpes
Cooking crêpes

Equipment

Knives:
Chef knife (*couteau chef*), paring knife (*office*)

Tools:
Zester, whisk, mixing bowls, tamis

Pans:
Small saucepan, non stick 10-inch crêpe pan

Serving

4 persons

FYI...

The word *crêpe* comes from the Latin *crispus,* which means "wavy or curly" and describes the lacy pattern on the surface of the crêpe. Rich with eggs and butter, this preparation is equally rich in lore and symbolism: In feudal France, it was a sign of allegiance for a farmer to serve his landowner crêpes; to hold a coin while flipping a crêpe was to bring you financial luck; and on the morning after a wedding, a successful flip of the crêpe indicated the beginning of a happy marriage. To this day, serving crêpes on Shrove Tuesday is a way of celebrating family life. The Brittany region in France is known for its crêpes made from buckwheat flour.

CRÊPES AU SUCRE

Sugar Crêpes

Quantity		Ingredient
U.S.	***Metric***	***Pâte à Crêpes***
4 oz	125 g	Flour
¾ oz	20 g	Sugar, granulated
2 pcs	2 pcs	Eggs
2 oz	60 g	Butter
1 pc	1 pc	Vanilla bean
1 pc	1 pc	Lemon, zest of
1 pc	1 pc	Orange, zest of
8 fl oz	250 mL	Milk
		Cuisson
2 oz	60 g	Butter
		To Serve
3 ½ oz	100 g	Butter
1 ¾ oz	50 g	Sugar, granulated

Method

Pâte à Crêpe

1. Melt the butter in a small saucepan over medium heat. Set aside. Sift (***tamiser***) the flour into a large mixing bowl and make a well (***fontaine***) in the center. Add the sugar to the well and pour in the eggs. Whisk the eggs, gradually incorporating the flour from the sides. When half of the flour is incorporated, add the melted butter. Continue to whisk until all the flour is incorporated and the mixture is smooth. Cut the vanilla bean in half lengthwise and scrape out the seeds using a paring knife. Add the vanilla seeds, lemon zest, and orange zest to the batter and stir them in. Pour in the milk in a thin stream while stirring the batter with a whisk. Cover the mixing bowl with plastic film and let the batter rest for a minimum of 1 hour in the refrigerator.

Cuisson

1. Melt the butter in a small saucepan over medium heat and set aside. Preheat a non stick 10-inch crêpe pan over medium-high heat. Using a pastry brush, brush the hot pan with melted butter. Pour a ladleful [3 fl oz (100 mL)] of crêpe batter into the center of the pan while turning it so that the batter evenly covers the bottom of the pan. Cook the crêpe until the edges begin to brown and curl away from the sides of the pan. Slip a spatula beneath the crêpe and turn it over. Cook the second side for 30 seconds to a minute, then slip the crêpe onto a clean warm plate and cover it with a clean cloth. Repeat the cooking process, stacking the crêpes one on top of the other until the batter is used up.

To Serve

1. Melt some butter in an omelette or crêpe pan. Place the crêpe to warm and sprinkle with sugar. Fold into quarters and arrange on a buttered serving dish. Keep warm until ready to serve.

CRÊPES SOUFFLÉES

Learning Outcomes

Pâte à crêpes
Crème pâtissière
Appareil à soufflé
Tamiser
Fontaine

Equipment

Tools:
Grater, whisk, balloon whisk, wooden spoon, rubber spatula, mixing bowls

Pans:
Crêpe pan

Serving

4 persons

CRÊPES SOUFFLÉES

Soufflé-Filled Crêpes

Quantity		Ingredient
U.S.	**Metric**	***Pâte à Crêpes***
4 oz	125 g	All-purpose flour
¾ oz	20 g	Sugar, granulated
Pinch	Pinch	Salt
2 pcs	2 pcs	Eggs
2 oz	60 g	Unsalted butter, melted
1 pc	1 pc	Vanilla bean
1 pc	1 pc	Orange, zest of, grated
8 fl oz	250 mL	Milk
		Crème Pâtissière
8 fl oz	250 mL	Milk
¼ fl oz	5 mL	Vanilla extract
2 pcs	2 pcs	Egg yolks
2 oz	60 g	Sugar, granulated
1 oz	30 g	Flour (or cornstarch)
¾ fl oz	20 mL	Cointreau
		Finishing
2 pcs	2 pcs	Egg whites
		Unsalted butter, for cooking
1 pc	1 pc	Orange, zest of, cut into ***julienne***
		Powdered sugar, for dusting

Method

Pâte à Crêpe

1. Sift (***tamiser***) the flour into a large mixing bowl and make a well (***fontaine***) in the center. Add the sugar and salt to the well and pour in the eggs. Whisk the eggs, gradually incorporating the flour from the sides. When half the flour is incorporated, add the melted butter. Continue to whisk until all the flour is incorporated and the mixture is smooth. Cut the vanilla bean in half lengthwise and scrape the seeds into the batter using a paring knife. Add the orange zest and stir. Pour in the milk in a thin stream while stirring the batter with a whisk. Cover the mixing bowl and let the batter rest for a minimum of 1 hour in the refrigerator.

Crème Pâtissière

1. Prepare a shallow tray or baking sheet by lining with plastic wrap.
2. Pour the milk into a medium saucepan, add the vanilla extract, and bring to a boil over medium-high heat. Add about one-quarter of the sugar to the milk and stir to dissolve it.
3. Meanwhile, place the egg yolks in a small mixing bowl and add the remaining sugar. Whisk the sugar into the eggs until it completely dissolves and the yolks lighten in color (***blanchir***). Add the flour or cornstarch to the yolks and stir until well combined.
4. When the milk begins to come to a boil, remove it from the stove and pour one-third of it into the egg yolks. Stir well to temper the yolks, then whisk the tempered mixture into the remaining hot milk. Place back onto the heat and cook until the crème pâtissière begins to bubble. Continue whisking (being sure to press the whisk around the corners of the pan) and allow to cook for 1 minute in order to cook the starch. The crème pâtissière will become very thick. Once cooked, pour the crème patissiere onto the prepared tray and pat the surface with a piece of cold butter held on the end of a fork (***tamponner***) to create a protective film. Cover with plastic, pressing out any air bubbles and seal the sides. Let the crème pâtissière cool to room temperature.

Crêpes

1. Melt the butter in a small saucepan over medium heat. Set aside. Preheat a small non stick crêpe pan over medium-high heat. Using a pastry brush, brush the hot pan with melted butter. Pour a ladleful [3 fl oz (100 mL)] of crêpe batter into the center of the pan while turning it so that the batter evenly covers the bottom of the pan. Cook the crêpe until the edges begin to brown and curl away from the sides of the pan. Slip a spatula beneath the crêpe and turn it over. Cook the second side for 30 seconds to a minute, then slip the crêpe onto a clean warm plate and cover it with a clean cloth. Repeat the cooking process, stacking the crêpes one on top of the other until the batter is used up.

Appareil à Soufflé

1. Whisk the crème pâtissière until it is smooth and elastic, then mix in the Cointreau. Beat the egg whites (***monter les blancs***) in a large mixing bowl with a balloon whisk to soft peaks. Using a rubber spatula, carefully fold in one-half of the egg whites into the crème pâtissière until almost completely incorporated. Add the remaining egg whites and fold them in until just incorporated.

Finishing

1. Lay the crêpes out flat and spread a layer of soufflé mixture ¼ to ½ in. (0.5 to 1 cm) thick on half of each crêpe. Gently fold the crêpes in 4 and arrange them staggered on a buttered ovenproof serving dish. Sprinkle the ***julienne*** of orange zest on top and place them in the oven to bake for 8 minutes, being careful not to open the door until they are cooked.
2. Remove the crêpes from the oven and sprinkle them with powdered sugar. Serve immediately.
3. There are several variations on serving crêpes souffles. You can butter and then line a small gratin dish with crêpes, then fill and bake the souffle mixture. You can butter and line individual ramekins with trimmed crêpes, then fill and bake the souffle.

PAINS AUX CROISSANTS, RAISINS, PAINS AU CHOCOLAT

Learning Outcomes

Pâte à croissant
Viennoiseries

Equipment

Knives:
Paring knife (*office*), slicing knife (*éminceur*), serrated knife

Tools:
Corne, rolling pin, baker's brush, whisk, wooden spoon, rubber spatula, mixing bowls, pastry brush

Pans:
1 small saucepan, 1 medium saucepan, baking sheet

Serving

8 to 10 portions of each

FYI...

One account of the birth of the croissant is that a pastry chef in Vienna in the late 1600s saved his city from invasion by alerting officials that he had heard sounds coming from underground. According to the story, the Turks were attempting to invade through underground tunnels. Depending on the source, the pastry chef was either granted his wish to have exclusive rights to sell commemorative pastries or was one of many who were asked to create a crescent-shaped pastry in honor of the thwarted invasion. The details of this story, however, change dramatically from one source to another. Even if a crescent-shaped pastry was invented on this occasion, it was likely not made of the same flaky yeast-risen pastry as the modern French croissant. It was, more likely, a dense pastry made of almond and sugar. In any case, the origin of this staple French pastry remains obscure.

The crescent shape is normally only given to croissants made with margarine; the pure butter croissants are generally straight.

PAINS AUX CROISSANTS, RAISINS, PAINS AU CHOCOLAT

Croissants, Raisin Buns, and Chocolate Croissants

Quantity		*Ingredient*
U.S.	*Metric*	*Pâte à Croissants Détrempe*
1 lb	500 g	Flour
6 fl oz	180 g	Water
½ oz	15 g	Fresh yeast
2 oz	55 g	Sugar, granulated
¼ oz	10 g	Salt
5 fl oz	150 g	Milk
1 ¾ fl oz	50 g	Butter, melted, cooled
		Beurrage
8 oz	250 g	Butter
		Pains aux Raisins Crème Pâtissière
1 ½ pt	750 mL	Milk
6 pcs	6 pcs	Egg yolks
5 oz	150 g	Sugar, granulated
2 ½ oz	75 g	Flour (or cornstarch)
¼ fl oz	5 mL	Vanilla extract
7 oz	200 g	Raisins
		Pains au Chocolat
15 pcs	15 pcs	Chocolate sticks
		Finishing
1 pc	1 pc	Egg for egg wash

Method

Pâte à Croissant

1. *Détrempe:* In a small bowl, combine the warm water [*max 90°F (32°C)*] and the yeast.
2. Sift (***tamiser***) the flour onto a clean work surface and make a well (***fontaine***) in the center. Add the sugar and salt into the well, then stir in the melted butter, the dissolved yeast, and half of the milk. Stir the ingredients with your fingers, slowly incorporating the flour from the sides of the well. When the mixture becomes thick, add the rest of the milk and continue to mix.
3. Once all the liquid has been absorbed into the flour, gather the ingredients together and work them with a plastic scraper (***corne***) until combined into a dough. Lightly knead the dough and gather the dough into a ball. Score the top with a deep cross using a large knife and loosely wrap the finished ***détrempe*** in plastic before placing it in the refrigerator to rest for a minimum of 20 minutes (*preferably overnight*).

Note: Détrempe *refers to the dough before the layer of butter (**beurrage**) is added.*

4. Although the procedure for incorporating the butter is the same as feuilletage, croissant dough is made from yeast-leavened dough that is much more elastic than that of feuilletage.

Beurrage and Tourage

1. Place the cold butter between two sheets of parchment paper and pound it with a rolling pin until it is similar to the ***détrempe*** in consistency.
2. Using the ***corne***, shape the butter into a flat square about ¾ in. (2 cm) thick and about the same size as the ball of ***détrempe***. If the kitchen is hot, reserve the butter in the refrigerator, or set it aside on a cool surface.
3. Lightly dust a clean work surface with flour (***fleurer***). Unwrap the ***détrempe*** and place it on the floured surface.
4. Using the scored marks as a guide, roll out (***abaisser***) the corners of the ***détrempe*** into a cross shape. The center of the crisis should be slightly thicker than the rolled-out arms.
5. Place the square of butter in the center of the cross and fold the 2 side arms over it so that they overlap slightly in the center (*in the process, be careful not to trap any air bubbles*). Give the dough a quarter turn and fold the 2 remaining arms over the butter so that no butter is visible. Press the seams well to seal.
6. Lightly tap the dough with the length of the rolling pin to even out the distribution of the butter inside. Give the dough a quarter turn and repeat the process. This is called the ***enveloppe***.

Method

Tourage, 3 Turns (3 tours simples):

1. *First turn:* Roll out (***abaisser***) the dough in long even strokes forming a rectangle that is twice the original length of the ***enveloppe*** or ½ in. (1 cm) thick. Brush off any excess flour. Fold the bottom third of the dough up; then fold the top third down over the first fold. Make sure the edges are even. Give the dough a quarter turn to the right so the seam is on the left and make one finger impression in the top left corner of the dough.

Note: *These marks are a reminder of the number of turns that the dough has received; they also indicate the position for subsequent turns.*

2. Wrap the dough in plastic and transfer it to the refrigerator to rest for 15 to 20 minutes.
3. *Second turn:* Lightly dust the work surface with flour (***fleurer***). Remove the dough from the refrigerator and unwrap it onto the floured surface (*with the indent in the top left corner*). Repeat the folding process (*bottom third up, top third down over first fold*) and give the dough a quarter turn to the right. Make two finger impressions in the top left corner of the dough before wrapping it in plastic and returning it to the refrigerator to rest for 15 to 20 minutes.
4. *Third turn:* Lightly dust the work surface with flour (***fleurer***). Remove the dough from the refrigerator and unwrap it onto the floured surface (*with the 2 indents in the top left corner*). Proceed to give the dough a third turn (*rolling and folding in the same manner as the first and second turns*). Mark the dough with 3 imprints in the top left corner before wrapping it in plastic and returning it to the refrigerator to rest for 15 to 20 minutes. When the dough has finished resting, cut it into 3 equal parts, wrap each separately, and refrigerate until needed.

Croissants

1. Lightly dust the work surface with flour (***fleurer***). Remove one section of dough from the refrigerator and unwrap it onto the floured surface.
2. With a rolling pin, roll the dough out (***abaisser***) to a rectangle about 8 in. (20 cm) wide by 8 in. (20 cm) long (*the dough should be 5 mm thick*). Cut the dough into triangles with a 10-cm base using a large knife. Gently stretch the triangles lengthwise, and place on the work surface. Roll the triangles up from the large end to the point. If the croissant is rolled correctly, it should have 5 segments from one end to the other. Transfer the finished croissants to a parchment paper-lined baking sheet, making sure that the tip of the triangle is underneath the rolled croissant. Leave the croissants to proof in a warm area or a proofer until they double in size.

Pains aux Raisins

1. *Prepare a crème pâtissière:* See pages 412–413 in *Cuisine Foundations*.
2. *Montage:* Lightly dust the work surface with flour (***fleurer***). Remove a section of dough from the refrigerator and unwrap it onto the floured surface. Using a rolling pin, roll out the dough (***abaisser***) to a rectangle about 8 in. (20 cm) wide 18 in. [0.5 m (46 cm) long] (*the dough should be* 3⁄16 *in. (5 mm thick*). Trim the edges of the dough to obtain straight sides. Transfer it to a sheet of parchment paper and brush a 1 ½ in. (3 cm) wide line of egg wash onto the edge closest to you. Beat the crème pâtissière with a whisk until it becomes smooth. Avoiding the egg-washed section, spread the crème pâtissière thinly onto the rectangle of croissant dough using a small offset spatula. Sprinkle the raisins onto the crème pâtissière and roll the rectangle toward yourself using the parchment paper as a tool to ensure even shaping. Seal it closed using the eggwashed edge. Place the roll in the freezer for 30 minutes to harden the dough and ensure a neat cut. Cut the roll into 10 equal slices. Leave the *pains aux raisin* to proof in a warm area or a proofer until they double in size.

Pains au Chocolat

1. Lightly dust the work surface with flour (***fleurer***). Remove part of the dough from the refrigerator and unwrap it onto the floured surface.
2. Using a rolling pin, roll out the dough (***abaisser***) to a rectangle about 6 in. (15 cm) wide by 18 in. (0.5 m) long (*the dough should be* 3⁄16 *in. (5 mm) thick*). Square off the edges using a knife. Cut the dough into vertical strips that are 1 cm wider than the length of the chocolate sticks. Place a chocolate stick near the top of each strip. Loosely roll the strip over the chocolate stick. Place another chocolate stick under the seam and roll over again. Making sure that the seam is hidden underneath, place the *pains au chocolat* on a baking sheet lined with parchment paper.
3. Leave the *pains au chocolat* to proof in a warm area or a proofer until they double in size.

Cooking

1. When the pastries have doubled in size, brush them with egg wash and transfer them to the oven to bake until golden.
2. Transfer to a wire rack to cool.

ÉCLAIRS AU CAFÉ ET AU CHOCOLAT

Learning Outcomes

Pâte à choux
Piping
Crème pâtissière
Filling pâte à choux
Flavoring crème pâtissière and fondant
Glaçage with fondant

Equipment

Tools:
Wooden spoon, rubber spatula, pastry bag, 12-mm plain tip, small star tip, 6-mm plain tip, fork, whisk

Pans:
1 large russe, 1 medium russe, 2 small russes, baking sheet

Serving

12–16 pieces

FYI...

Conjecture surrounds the history of the éclair. In the *Oxford English Dictionary,* 1864 is cited as the earliest the word *éclair* was being used to describe this pastry. Some historians use this date to discredit claims that the inventor of the éclair was Antonin Carême; however, he still may have invented this filled puff pastry only not under the name of éclair. The pastry was also known in France as *bâton de Judas,* an appellation which may or may not predate the use of *éclair.* On a less quarrelsome note, it is said that the éclair (that is French for lightning) is named because it is so good it is eaten as fast as lightning!

ÉCLAIRS AU CAFÉ ET AU CHOCOLAT

Chocolate and Coffee Eclairs

Quantity		Ingredient
U.S.	*Metric*	*Pâte à choux*
4 fl oz	125 g	Water
4 fl oz	125 g	Milk
4 oz	125 g	Butter
½ oz	12 g	Sugar, granulated
½ tsp	5 g	Salt
5 oz	150 g	Flour
1 pc	1 pc	Egg for eggwash
		Crème Pâtissière
1 ½ pt	750 mL	Milk
6 pcs	6 pcs	Egg yolks
5 oz	150 g	Sugar, granulated
1 ½ oz	45 g	Flour
1 ½ oz	45 g	Cornstarch
To taste	To taste	Coffee extract
2 oz	60 g	Dark chocolate, finely chopped
2 oz	60 g	Unsweetened chocolate, melted
		Glaçage
7 oz	200 g	Coffee fondant*
7 oz	200 g	Chocolate fondant*
		*Fondant can be purchased at specialty stores.

Method

1. Preheat the oven to 425°F (220°C).

Pâte à Choux

1. *Prepare a panade*: Combine the water, milk, butter, sugar, and salt in a large pan and bring to a boil over medium-high heat. Once the butter has completely melted, remove the pan from the heat and add the flour. Stir with a wooden spatula until combined; then, over medium heat, stir until the mixture doesn't stick to the spatula and it makes a thin skin on the bottom of the pan (***dessêcher***). Transfer to a clean bowl and cool until warm. Beat the eggs into the panade one at a time. The dough should be stretchy and slightly sticky.
2. Transfer the mixture to a pastry bag fitted with a 12-mm plain tip.
3. On a lightly greased baking sheet, pipe out 4 ¾ in. (10 to 12 cm) long lines of pâte à choux. Brush the pâte à choux with egg wash, then run the prongs of a fork (dipped in egg wash) down the length of the éclairs to even out their shape.
4. Transfer the éclairs to the oven and immediately reduce the temperature to 400°F (205°C). Bake until golden (*20 to 25 minutes*), rotating the baking sheet as soon as they begin to color.
5. Remove the éclairs from the oven and transfer to a wire rack to cool.
6. Using a small star tip, pierce the bottoms of the éclairs with 3 holes (1 in the middle and 1 at each end).

Crème Pâtissière

1. Pour the milk into a medium saucepan and bring it to a boil over medium-high heat. Add about one-quarter of the sugar to the milk and stir to dissolve it.
2. Meanwhile, place the egg yolks in a small mixing bowl and add the remaining sugar. Whisk the sugar into the eggs until it completely dissolves and the yolks lighten in color (***blanchir***). Add the flour or cornstarch to the yolks and stir until well combined.
3. When the milk begins to come to a boil, remove it from the stove and pour one-third of it into the egg yolks. Stir well to temper the yolks, then whisk the tempered mixture into the remaining hot milk. Place back onto the heat and cook until the crème pâtissière begins to bubble. Continue whisking (being sure to press the whisk around the corners of the pan) and allow to cook for 1 minute in order to cook the starch. The crème pâtissière will become very thick.

Method

4. Separate the crème pâtissière into two equal batches in separate mixing bowls and mix coffee extract into one batch and the melted unsweetened chocolate into the other using whisks. Pat the surface of the crème pâtissière with a piece of cold butter held on the end of a fork (***tamponner***) to create a protective film. Let the crème pâtissière cool to room temperature and refrigerate it.

Montage

1. Beat the crème pâtissière until its texture loosens, then transfer it to a pastry bag fitted with a 6-mm plain tip. Fill half the éclairs with chocolate-flavored crème pâtissière and the other half with coffee-flavored crème pâtissière, piping it in through the holes underneath.
2. Gently heat the coffee fondant in a medium saucepan over the lowest heat setting, stirring it gently with a wooden spatula. If the fondant stays very thick even when heated, add a few drops of warm water and stir it in. Continue to gently stir the fondant until it coats the back of a spoon in a thick, shiny layer. The ideal temperature to work with fondant is approximately 98.6°F (37°C).
3. To ice the éclairs (using coffee fondant for the coffee éclairs and chocolate fondant for the chocolate éclairs):
   - Hold the éclair upside down over the pan of fondant and dip one end into it. Lower the rest of the éclair into the fondant to coat the top while simultaneously pulling the first end out. Once the éclair is covered, clean the edges of the fondant using your fingers.
4. Lay the éclairs out on a clean tray and transfer them to the refrigerator to set the fondant.
5. Reserve in the refrigerator until serving.

Learning Outcomes

Crèmer
Crème pâtissière
Filling a cake before baking
Scoring a surface

Equipment

Tools:
Sieve, whisk, corne, colander, pastry bag, medium round pastry tip, mixing bowls, pastry brush

Pans:
Medium saucepan, 8-inch cake pan

Serving

6 persons

TTT

Originating from the town of Cambo-les-Bains in the Aquitaine region of France, gâteau Basque was first sold under the Basque name Biskotxak. The anecdotal history of the gâteau is that a baker by the name of Marianne Hirigoyen had been handed down the recipe by her mother in the 1830s. With the growing fame of Cambo-les-Bains as a spa town, well-to-do people from all over France started coming into Marianne's shop to sample the unique delicacy of Biskotxak.

GÂTEAU BASQUE

Cream-Filled Butter Cake

Method

1. Preheat the oven to 370°F (185°C).
2. Butter the inside of an 8-inch round cake mold (***moule à manqué***) and reserve it in the refrigerator.
3. Place the cherries in a colander over a mixing bowl to drain.

Crème Pâtissière

1. Combine the milk and the vanilla extract in a medium saucepan and bring the mixture to a boil over medium-high heat. Add about one-quarter of the sugar to the milk and stir to dissolve it.
2. Place the egg yolks in a small mixing bowl and add the remaining sugar. Whisk the sugar into the eggs until it completely dissolves and the yolks lighten in color (***blanchir***). Add the flour or cornstarch to the yolks and stir until well combined.
3. When the milk begins to come to a boil, remove it from the stove and pour one-third of it into the egg yolks. Stir well to temper the yolks, then whisk the tempered mixture into the remaining hot milk. Place back onto the heat and cook until the crème pâtissière begins to bubble. Add the almond powder, and continue whisking (being sure to press the whisk around the corners of the pan) and allow to cook for 1 minute in order to cook the starch. The crème pâtissière will become very thick. Transfer the finished ***crème pâtissière*** to a clean bowl. Pat (***tamponner***) the surface with a piece of cold butter held on the end of a fork to create a protective film and let the ***crème pâtissière*** cool to room temperature.
4. Prepare a shallow tray or baking sheet by lining with plastic wrap. Once cooked, pour the crême patissière onto the prepared sheet. Pat (***tamponner***) the surface with a piece of cold butter to create a protective film then cover with plastic wrap, pressing out any air bubbles and seal the edges. Set aside to cool to room temperature.

Pâte à Gâteau

1. Sift the flour, baking powder, and salt together. Cream (***crémer***) the butter and sugar together until fluffy. Add the eggs one at a time, mixing well after each addition. Add the vanilla and lemon zest and stir well. Stir in the flour until incorporated and a thick batter is formed. Transfer to a pastry bag, fitted with a large round tip.

Montage

1. Remove the cake mold from the refrigerator and lightly dust with flour. Starting at the center of the prepared mold, pipe a tight spiral of the batter outward. When you reach the sides of the pan, continue piping around the circumference of the pan to the top so that the sides are evenly covered. Lightly tap the pan to fill in any gaps. Whisk the crème pâtissière until it is smooth and stretchy, then mix in the rum. Toss the drained cherries in the flour and sugar, then mix into the creme. The coating will help absorb any excess juice. Add the pastry cream to the center of the cake mold. Spread the crème pâtissière into an even layer. Pipe the remaining cake batter in a spiral over the creme. Smooth any gaps with a palette knife. Drag a finger around the inside of the rim of the cake mold to create a shallow channel. Smooth the surface of the cake using a palette knife dipped in hot water. Let the cake rest in the refrigerator for 30 minutes before baking.

Cuisson

1. Remove the cake from the refrigerator and brush the surface with egg wash. Using a fork, lightly score the surface with a decorative pattern. Place the cake in the oven to bake until golden, then turn the oven down to 340°F (170°C) and bake the cake for a further 25 minutes. Remove the cake from the oven and, leaving it in the mold, let it rest on a wire rack for 10 minutes. Turn the cake out of its mold and let it rest on the wire rack until cool.

Optional: *Decorate with a marzipan rose.*

Quantity		*Ingredient*
U.S.	*Metric*	*Crème Pâtissière*
8 fl oz	250 mL	Milk
¼ fl oz	5 mL	Vanilla extract
2 pcs	2 pcs	Egg yolks
2 oz	60 g	Sugar, granulated
1 oz	30 g	Flour (or cornstarch)
1 oz	30 g	Almond powder
¾ fl oz	20 mL	Rum or Izzara
½ oz	14 g	Butter
		Garniture
5 oz	150 g	Cherries in syrup, drained
1 oz	30 g	Flour
1 oz	30 g	Sugar, granulated
		Appareil
8 oz	230 g	Flour
¼ oz	6 g	Baking powder
¼ oz	4 g	Salt
5 oz	150 g	Butter, softened (***pommade***)
5 oz	150 g	Sugar, granulated
3 pcs	3 pcs	Eggs
¼ fl oz	5 mL	Vanilla extract
1 pc	1 pc	Lemon, zest of
1 pc	1 pc	Egg for egg wash

GALETTE DES ROIS

Learning Outcomes

Pâte feuilletée
Crème frangipane
Piping a tight spiral
Chevron design in egg wash
Chiqueter

Equipment

Knives:
Paring knife (*office*)

Tools:
Corne, baker's brush, pastry brush, bowls, rubber spatula, rolling pin, pastry bag, plain 10-mm tip

Pans:
2 baking sheets

Serving

6 persons

FYI...

Called king's cake or twelfth night cake in English, galette des rois is traditionally served on the 12th day of Christmas. As part of the festivities, a porcelain figurine is baked into the galette. Whoever gets the trinket is named either king or queen for the night—hence the name galette des rois (king).

GALETTE DES ROIS

Epiphany Cake

Quantity		Ingredient
U.S.	Metric	Pâte Feuilletée
1 lb	500 g	Flour
7 fl oz	225 mL	Water
7 oz	200 g	Butter
1 ¾ tsp	10 g	Salt
7 oz	200 g	Butter
1 pc	1 pc	Egg for egg wash
		Crème Frangipane
2 oz	60 g	Butter
2 oz	60 g	Sugar, granulated
2 oz	60 g	Almond powder
1 pc	1 pc	Egg
½ fl oz	10 mL	Rum
1 pc	1 pc	Vanilla bean
3 ½ oz	100 g	**Crème pâtissière**

Method

1. Preheat the oven to 400°F (205°C).

Pâte Feuilletée

1. To obtain the correct amount of pâte feuilletée use the ingredients list in this recipe following the method on pages 377–379 in *Cuisine Foundations*.

Crème Frangipane

1. *Prepare the almond cream* (***crème d'amandes***): Cream (***crémer***) the butter and sugar together until light and fluffy. Beat in the egg until well combined. Using a small knife, split the vanilla bean in half lengthwise, scrape out the seeds, and whisk into the mixture. Add the rum and finish by mixing in the almond powder. Reserve the crème d'amandes in a covered bowl in the refrigerator until ready to use.
2. In a mixing bowl, whisk the ***crème pâtissière*** (*prepared beforehand*) until it is smooth and stretchy. Add the almond cream and stir with the whisk until the two are combined. Transfer to a pastry bag fitted with a medium round tip.

Montage

1. Lightly dust a clean work surface with flour (***fleurer***). Unwrap the puff pastry onto the lightly floured work surface and roll it out (***abaisser***) (*using a rolling pin*) to a thickness of ⅛ in. (3 mm). Place the puff pastry on a parchment paper-lined baking sheet and return it to the refrigerator to rest for 10 minutes.
2. Remove the pastry from the refrigerator and lay it out flat on the work surface. Using a cake ring or ***vol-au-vent*** circle as a guide, cut two circles out of the puff pastry, one 8 in. (20 cm) and the other slightly larger (*refrigerate the trimmings for other uses*). Place the larger circle on the parchment paper-lined baking sheet and reserve it in the refrigerator.
3. Brush the surface of the second round with water and flip it onto a parchment paper-lined baking sheet (*wet side down*).

Note: *It is important that the round be placed wet side down as the water will cause the dough to stick to the baking sheet and prevent it from warping during cooking*. Brush a 1 in. (2 ½ cm) band of egg wash around the surface of the round.

4. Pipe (***coucher***) a tight spiral of *crème frangipane* onto the disc of pastry (*begin piping in the center and work your way outward, stopping just before the egg wash*). Transfer the baking sheet to the refrigerator for 10 minutes to let the *crème frangipane* harden.
5. Remove both baking sheets from the refrigerator. Lightly dust the larger round with flour and gently fold it in half. Place it on top of the disc with the frangipane spiral, taking care to line up the edges of the 2 discs. Unfold the top disc so that it completely covers the bottom one and no air is trapped inside.

Method

Gently press down the edges to secure the egg wash seal and, using the cake ring as a guide, cut off any irregularities from the edge of the galette. Transfer the galette to the freezer for 30 minutes before baking.

Cuisson

1. Remove the galette from the freezer, brush the surface with egg wash, and lightly score a chevron design on the top with the back of a small knife. Score the sides of the galette using the back of a paring knife (***chiqueter***).
2. Transfer the galette to the pre-heated oven to bake for 10 minutes, then reduce the heat to 340°F (170°C) and bake the galette for a further 20 minutes or until golden. Remove the galette from the oven and slide it onto a wire rack to cool.
3. Serve the galette des rois at room temperature on a serving dish.

GÂTEAU FORÊT NOIRE

Learning Outcomes

Chocolate génoise
Montage
Tamiser
Clarifying eggs
Crème Chantilly
Rosettes
Chocolate shavings
Puncher
Masking a cake

Equipment

Knives:
Vegetable peeler (*économe*), serrated knife, chef knife (*couteau chef*), paring knife (*office*)

Tools:
Pastry brush, balloon whisk, rubber spatula, tamis, corne, pastry bag, 10-mm round tip, 24-mm star tip, scissors, mixing bowls, tamis, ice bath

Pans:
Small saucepan, bain marie

Serving

8 persons

FYI...

The black forest cake (Schwarzwalder Kirschtorte) is named after the Schwarzwald (literally Black Forest) in Germany. Joseph Keller, of Café Agner in Bad Godesberg, is thought to be the creator of this cake, although this is not certain. Little is known until the mid-1930s, it is now one of the most famous of German cakes. In its original version, it uses sour cherries and large amounts of kirsch. However, in its modern American version, the sour cherries have been replaced by maraschino cherries and the kirsch is, more often than not, non existent.

GÂTEAU FORÊT NOIRE

Black Forest Cake

Quantity		Ingredient
U.S.	*Metric*	*Chemisage*
1 oz	30 g	Flour
1 oz	30 g	Butter
		Génoise
¾ oz	25 g	Butter
4 oz	120 g	Flour, sifted (***tamiser***)
1 oz	30 g	Cocoa powder, (***tamiser***)
4 pcs	4 pcs	Eggs
2 pcs	2 pcs	Egg yolks
5 oz	150 g	Sugar, granulated
		Crème Chantilly
1 pt 3 fl oz	600 mL	Cream, cold
2 oz	60 g	Powdered sugar
¼ fl oz	5 mL	Vanilla
		Imbibing Syrup
7 oz	200 g	Sugar, granulated
7 fl oz	200 mL	Water
1 oz	30 g	Kirsch
		Montage
10 oz	300 g	Bittersweet chocolate, tempered
4 ½ oz	130 g	***Griotte*** cherries

Method

1. Preheat the oven to 400°F (205°C).
2. Butter an 8-inch cake pan (***moule à manqué***) and place it in the freezer for 5 minutes to set the butter. Butter the mold a second time and coat it in flour (***chemiser***). Tap off any excess flour and reserve the mold in the refrigerator.

Génoise

1. Melt the butter in a small saucepan over low heat and set it aside.
2. Sift (***tamiser***) the flour and cocoa powder onto a sheet of parchment paper.
3. Fill a saucepan one-quarter full of water and bring it to a simmer over medium-high heat (***bain marie***).
4. Break the eggs and yolks into a large mixing bowl, add the sugar, and whisk them together with a balloon whisk until combined. Place the bowl on the simmering bain marie and continue whisking until the mixture lightens in color and feels hot to the touch. At this point the mixture should form a ribbon when the whisk is lifted from the bowl. Remove the bowl from the bain marie and continue to whisk the egg mixture until it cools to room temperature. Add the flour and gently fold it into the egg mixture with a rubber spatula (keep folding until the flour is just incorporated). Fold in the melted butter, then transfer the finished génoise into the cake mold and place it in the oven to bake. Reduce the oven temperature to 350°F (185°C) and bake the génoise for 15 to 18 minutes (test by inserting a knife into the center; if it comes out clean the cake is fully baked). Remove the génoise from the oven and let it cool in the mold for 2 to 3 minutes. Turn the cake out of the mold and let it cool upside down on a wire rack.

Crème Chantilly

1. Whip the cream in a large mixing bowl in an ice bath using a large whisk until it reaches the soft peak stage. Add the powdered sugar and vanilla, and continue to whip the cream until it is stiff. Reserve the crème Chantilly in the refrigerator until needed.

Imbibing Syrup

1. Bring the water and sugar to a boil together in a small saucepan over medium-high heat until the sugar has completely dissolved. Remove the pan from the heat and pour the syrup into a clean bowl. Let it cool to room temperature, then stir in the kirsch with a spoon. Cover and set aside.

Method

Montage

1. Turn the cake over so the domed side is on top. Trim it flat with a serrated knife. Turn the cake onto a cake board cut to the same dimensions as the cake and place on a rotating cake stand. Score the cake around the sides with a bread knife to create 3 even layers. Cut off the top layer, slicing around the cake, slowly cutting deeper until the layer comes off. Set aside. Repeat the cutting process with the middle layer and set it aside. Using a pastry brush, wet the entire surface of the bottom layer with syrup to imbibe it (***puncher***). Remove the crème Chantilly from the refrigerator and whisk it until it reaches the stiff peak stage (***serrer***). Transfer the crème Chantilly to a pastry bag fitted with a 10-mm plain tip. Starting from the center, pipe (***coucher***) a tight spiral all the way to the edges of the bottom layer of cake. Evenly scatter one-third of the cherries onto the crème Chantilly and pipe (***coucher***) a few lines of crème Chantilly over them to help the next layer hold in place. Place the middle layer on top and press it down lightly. Smooth any crème Chantilly that seeps out with a metal spatula. Repeat the imbibing (***puncher***), piping (***coucher***), and cherry scattering, followed by piping a few lines over the cherries. Place the last layer of cake on top and press down lightly. Smooth any creme Chantilly with a metal spatula. Lightly score the top of the cake with the tip of a small knife and imbibe it (***puncher***) with syrup.
2. Reserve the cake in the refrigerator. Empty the pastry bag into the bowl of crème Chantilly and reserve in the refrigerator.

Tip: *If you need to refill the pastry bag at any time, empty it completely into the bowl of crème Chantilly, re-whip it until stiff, and refill the pastry bag.*

Chocolate Shavings

1. Using a clean vegetable peeler or a sharp knife, scrape a bar or piece of chocolate for chocolate shavings.

Masquage

1. Remove the cake from the refrigerator. Brush the top with syrup and wait for it to soak in. Whip the crème Chantilly to stiffen it (***serrer***). Using a large metal spatula, cover the sides of the cake in crème Chantilly and smooth out, scraping off any excess. Place a large dollop of crème Chantilly on top and spread it out toward the edges in an even layer. Smooth the top of the cake by sweeping over it using the metal spatula. Slip the spatula under the cake board and lift it up onto one hand. Rotating the cake in one hand, smooth the edges and corners of the cake, removing excess crème Chantilly in a continuous downward scrape of the spatula. Scrape off any excess crème Chantilly.

Finishing

1. Transfer the leftover crème Chantilly to a pastry bag fitted with a 24-mm star tip. Pipe (***coucher***) 8 rosettes (***rosace***) on top of the cake. Lightly press chocolate shavings to the sides of the cake and scatter them on the top between the rosettes. Sprinkle the cake with powdered sugar and top each rosette with a cherry.
2. Transfer the cake either to a clean cake board or to a serving dish and refrigerate until ready to serve.

Learning Outcomes

Biscuit génoise
Crème au beurre française
Masking a cake
Puncher
Cornet
Rosettes
Tamiser
Pâte à bombe
Coucher

Equipment

Knives:
Serrated knife

Tools:
Pastry brush, balloon whisk, offset spatula, mixing bowls, corne, cornet, pastry bag, 10-mm plain tip, 24-mm star tip, pâtisserie comb, large offset spatula

Pans:
Bain marie, 8-inch round cake mold, medium saucepan, small saucepan

Serving

8 persons

FYI...

Mocha coffee beans were first discovered by Europeans in the 17th century in the port city of Mocha (in Yemen). The popularity of the mocha bean, with its undertones of chocolate and its medium- to full-bodied taste, was fully entrenched by the mid-18th century. According to Pierre Lacam in *Le Mémorial Historique De La Pâtisserie,* gâteau mocha was invented by Guignard of the Carrefour de l'Odéon in Paris in 1857.

MOKA
Coffee Butter Cream Sponge Cake

Quantity		Ingredient
U.S.	Metric	*Chemisage*
1 oz	30 g	Flour
1 oz	30 g	Butter
		Biscuit Génoise
¾ oz	25 g	Butter
5 oz	150 g	Flour (***tamiser***)
4 pcs	4 pcs	Eggs
2 pcs	2 pcs	Egg yolks
5 oz	150 g	Sugar, granulated
		Crème au Beurre Française
6 oz	180 g	Sugar, granulated
2 fl oz	60 mL	Water
6 pcs	6 pcs	Egg yolks
12 oz	360 g	Butter
¼ fl oz	2 mL	Coffee extract
		Imbibing Syrup
7 fl oz	200 g	Water
7 oz	200 g	Sugar, granulated
¼ fl oz	2 mL	Coffee extract
		Finishing
1 ¾ oz	50 g	Chocolate couverture, melted
8 pcs	8 pcs	Chocolate-coated coffee beans
		Toasted almonds, chopped

Method

1. Preheat the oven to 400°F (205°C).
2. Butter an 8-inch cake pan (***moule à manqué***) and place it in the freezer for 5 minutes to set the butter. Butter the mold a second time and coat it in flour (***chemiser***). Tap off any excess flour and reserve the mold in the refrigerator.

Génoise

1. Melt the butter in a small saucepan over low heat and set it aside.
2. Sift (***tamiser***) the flour onto a sheet of parchment paper.
3. Fill a saucepan one-quarter full of water and bring it to a simmer over medium-high heat (***bain marie***).
4. Break the eggs and yolks into a large mixing bowl, add the sugar, and whisk them together with a balloon whisk until combined. Place the bowl on the simmering bain marie and continue whisking until the mixture lightens in color and feels hot to the touch. At this point the mixture should form a ribbon when the whisk is lifted from the bowl. Remove the bowl from the bain marie and continue to whisk the egg mixture until it cools to room temperature. Add the flour and gently fold it into the egg mixture with a rubber spatula (keep folding until the flour is just incorporated). Fold in the melted butter then transfer the finished génoise into the cake mold and transfer it to the oven to bake. Reduce the oven temperature to 350°F (185°C) and bake the génoise for 15 to 18 minutes (test by inserting a knife into the center; if it comes out clean the cake is fully baked). Remove the génoise from the oven and let it cool in the mold for 2 to 3 minutes. Turn the cake out of the mold and let it cool upside down on a wire rack.

Imbibing Syrup

1. Bring the water and sugar to a boil in a small saucepan over medium-high heat until the sugar has completely dissolved. Remove the pan from the heat and pour the syrup into a clean bowl. Let it cool to room temperature, then stir in the coffee extract with a spoon. Cover and set aside.

Crème au Beurre Française

1. Cook the sugar and 2 fl oz (60 mL) of water in a medium saucepan over medium-high heat until they reach the soft-ball stage (***petit boule***) [250°F (121°C)]. Meanwhile, place the egg yolks in a mixing bowl. When the sugar syrup reaches

Method

the soft-ball stage (***petit boulé***), pour it onto the egg yolks in a steady stream while whisking continuously. Continue whisking until the mixture is thick and pale and creates ribbons when the whisk is lifted from the bowl. This mixture is a pâte à bombe.

2. Whisk the pâte à bombe until the mixing bowl is just warm to the touch then add all the butter. Mix vigorously with the whisk until the mixture is homogeneous and thick enough to hold its own shape. Whisk in the coffee extract and reserve at room temperature.

Tip: *To tell how well done the sugar is, dip 2 fingertips and your thumb in cold water, then rapidly scoop up a little drop of the boiling syrup before plunging your fingers straight back into the cold water. If the syrup forms a malleable ball, it is at the soft-ball stage (**petit boule**) [250°F (121°C)]. If the sugar is too soft to form into a ball, it needs more cooking. If the sugar forms a hard ball, then it is overcooked and you need to start again.*

Montage

1. Turn the cake over so the domed side is on top. Trim it flat with a serrated knife. Then turn the cake over onto a cake board so that the flat bottom is on top. Cut the board to the same dimensions as the cake and place on a rotating cake stand. Cut two even layers with a bread knife. Remove the top half and set aside. Using a pastry brush, wet the cut surface of the bottom layer with syrup to imbibe it (***puncher***). Transfer the crème au beurre to a pastry bag fitted with a 10-mm (¼ in.) plain tip. Starting at the center, pipe (***coucher***) a tight spiral all the way to the edges of the bottom layer of cake. Place the second layer on top and press down lightly. Smooth any crème au beurre from the sides with a metal spatula. Lightly score the top of the cake with the tip of a small knife and imbibe it (***puncher***) with syrup.
2. Reserve the cake in the refrigerator. Empty the pastry bag into the bowl of crème au beurre and reserve at room temperature.

Masquage

1. Remove the cake from the refrigerator. Brush the top with syrup using a pastry brush and wait for it to soak in. Using a large metal spatula, cover the sides of the cake in crème au beurre and smooth them, scraping off any excess. Place a large dollop of crème au beurre on top and spread it out toward the edges in an even layer.
2. Smooth the top of the cake by sweeping over it using the metal spatula. Slip the spatula under the cake board and lift it up onto one hand.
3. Rotating the cake in one hand, smooth the edges and corners of the cake, removing excess crème au beurre in a continuous downward scrape of the spatula. Scrape off any excess crème au beurre.

Finishing

1. Drag a pastry comb across the surface of the cake to create a textured top. Holding the cake in one hand, use the other to apply handfuls of the chopped almonds to the sides of the cake; create a scalloped or domed decoration by cupping your hand as you press in the nuts. Repeat this pattern all the way around the cake. Fill a paper cone (***cornet***) with crème au beurre and snip off the end. Use it to pipe (***coucher***) the word *Moka* onto the top of the cake.
2. Fill a second cornet with melted chocolate and snip off the end. Write over the crème au beurre with the chocolate.
3. Transfer the leftover crème au beurre to a pastry bag fitted with a 24-mm star tip. Pipe (***coucher***) 8 rosettes (***rosace***) on top of the cake and top each rosette with a coffee bean.
4. Transfer the cake either to a clean cake board or serving dish and refrigerate until ready to serve.

Learning Outcomes

Biscuit dacquoise
Crème au beurre
Chemiser
Piping a disc of biscuit
Soft-ball stage (***petit boule***)
Pâte à bombe
Marzipan
Writing in chocolate

Equipment

Knives:
Paring knife (*office*)

Tools:
Whisk, rubber spatula, corne, tamis, pastry bag, 10-mm plain tip, balloon whisk, rolling pin, cornet, mixing bowls

Pans:
Baking sheet, medium saucepan

Serving

8 persons

Quantity		Ingredient
U.S.	Metric	*Biscuit Dacquoise*
5 oz	150 g	Almond powder
7 oz	225 g	Powdered sugar
¼ fl oz	10 g	Vanilla
2 ½ oz	75 g	Flour
6 pcs	6 pcs	Egg whites
2 ½ oz	75 g	Sugar, granulated
		Crème au Beurre
3 ½ oz	100 g	Sugar, granulated
1 fl oz	30 mL	Water
3 pcs	3 pcs	Egg yolks
7 oz	200 g	Butter
1 ¾ oz	50 g	Hazelnut praline
		Garniture
3 ½ oz	100 g	Sliced almonds
3 ½ oz	100 g	Almond paste
		Green food dye
		Whole hazelnuts
		Chocolate, melted
		Cocoa powder, for dusting

SUCCÈS

Hazelnut Buttercream Meringue Cake

Method

1. Preheat the oven to 380°F (195°C).
2. Lightly grease a baking sheet with cold butter and place it in the refrigerator to chill for 5 minutes. Remove the baking sheet from the refrigerator, dust it with flour (***chemiser***), and shake off the excess. Mark the baking sheet with two circles using a 10 in. (25 cm) cake ring and return it to the refrigerator until needed.

Biscuit Dacquoise

1. Sift together the almond powder, sugar, flour, and vanilla. Set aside.
2. *Make a meringue:* Beat the egg whites to soft peaks. Gradually incorporate the sugar until the meringue is firm and glossy and the sugar granules cannot be felt when the meringue is rubbed between two fingers. Fold in the sifted ingredients until just combined.
3. Transfer the mixture to a pastry bag fitted with a 10-mm plain tip. Remove the baking sheet from the refrigerator and pipe (***coucher***) two tight 10-inch spirals of dacquoise onto it using the marked circles as guides. Tap the baking sheet lightly to fill in any gaps. Dust the discs of dacquoise with powdered sugar and transfer them to the oven to bake until lightly golden (*8 to 10 minutes*).
4. Remove the baking sheet from the oven and transfer it to a wire rack to cool. Reduce the oven temperature to 350°F (175°C).

Crème au Beurre Française

1. Cook the sugar and water in a medium saucepan over medium-high heat until they reach the soft-ball stage (***petit boulé***) [250°F (121°C)]. Meanwhile, place the egg yolks in a mixing bowl and stir. When the sugar syrup reaches the soft stage, pour it into the egg yolks in a steady stream while whisking continously. Continue whisking until the mixture is thick and pale and creates ribbons when the whisk is lifted from the bowl. This mixture is a pâte à bombe.
2. Whisk the pâte à bombe until the mixing bowl is just warm to the touch and add all the butter. Mix vigorously with the whisk until the mixture is homogeneous and thick enough to hold its own shape. Whisk in the praline and reserve at room temperature.

Tip: *To tell how well done the sugar is, dip 2 fingertips and your thumb in the cold water, then rapidly scoop up a little drop of the boiling syrup before plunging them straight back into the cold water. If the syrup forms a malleable ball, it is at the soft-ball stage (**petit boulé**) [250°F (121°C)]. If the sugar is too soft to*

Method

form into a ball, it needs more cooking. If the sugar forms a hard ball, then it is overcooked and you need to start again.

Montage

1. Using the cake ring as a guide, trim any excess off the edges of the dacquoise discs.
2. Transfer the crème au beurre to a pastry bag fitted with a 10-mm plain tip. Cut a cake board to a 10 in. (25 cm) circle and place a disc of dacquoise on top of it upside down. Starting from the center, pipe (***coucher***) a tight spiral of crème au beurre onto the dacquoise to completely cover it. Place the second disc on top, right way up and gently press on it to secure it in place. Scrape away any excess crème au beurre from the edges and place the cake in the refrigerator.

Decoration

1. Spread the sliced almonds out onto a baking sheet and place them in the oven to toast until lightly golden (*5 to 10 minutes*). Set aside.
2. Lightly dust the work surface with powdered sugar. Using a rolling pin, roll out the marzipan to a thickness of ⅛ in. (2 mm) on the sugared surface. Cut a rough rectangle out of the marzipan and set aside. Gather the rest of the marzipan into a ball and add a drop of green food coloring. Knead the marzipan until it is homogeneously colored. Roll out the marzipan again and cut out leaf shapes. Cut some small strips and wrap them around peeled hazelnuts to imitate the husks of fresh hazelnuts. Using a small dry paintbrush, dust the finished nuts, leaves, and rectangle with cocoa powder to add detail and definition. Pour some melted chocolate into a paper cone (***cornet***) and snip off the tip. Pipe the word *Succès* onto the marzipan rectangle. Set aside.

Finishing

1. Remove the cake from the refrigerator. Holding the cake in one hand, apply crème au beurre around the outside using a metal spatula. Smooth the crème au beurre and remove any excess, being careful not to apply any to the top of the cake. Place the cake on the work surface and dust the top with powdered sugar until it is completely white. Press the toasted almonds to the side of the cake to completely cover the crème au beurre and arrange the marzipan decorations on the top. Transfer the cake either to a clean cake board or serving dish and refrigerate until ready to serve.

Learning Outcomes

Biscuit génoise
Chemiser
Tamiser
Puncher
Decor en glace royale

Equipment

Knives:
Serrated knife

Tools:
Balloon whisk, rubber spatula, tamis, whisk, mixing bowls, wire rack

Pans:
Bain marie, medium saucepan, 8 inch round cake mold, baking sheet

Serving

6 persons

GÉNOISE CONFITURE

Jam-Filled Sponge Cake

Quantity		*Ingredient*
U.S.	*Metric*	*Chemisage*
1 oz	30 g	Flour
1 oz	30 g	Butter
		Biscuit Génoise
¾ oz	25 g	Butter
5 oz	150 g	Flour, sifted (***tamiser***)
4 pcs	4 pcs	Eggs
2 pcs	2 pcs	Egg yolks
5 oz	150 g	Sugar, granulated
		Glace Royale
7 oz	200 g	Powdered sugar, sifted (***tamiser***)
1 pc	1 pc	Egg white
1 pc	1 pc	Lemon, juice of
		Montage
		Almonds, blanched, sliced, toasted
		Raspberry jam, with seeds
		Simple Syrup
5 oz	150 g	Sugar
5 fl oz	150 mL	Water

Method

1. Preheat the oven to 400°F (205°C).
2. Butter an 8 in. (20 cm) cake pan/mold (***moule à manqué***) and place it in the freezer for 5 minutes to set the butter. Butter the mold a second time and coat it in flour (***chemiser***). Tap off any excess flour and reserve the mold in the refrigerator.

Génoise

1. To obtain the correct amount of biscuit génoise use the ingredients list in this recipe following the method on pages 403–404 in *Cuisine Foundations*.

Simple Syrup

1. Combine the sugar and water in a small pan. Bring to a boil and cook until the sugar is dissolved. Remove from the heat and set aside to cool.

Glace Royale

1. Sift (***tamiser***) the powdered sugar into a large mixing bowl and make a well in the center. Whisk the egg white and lemon juice in a small mixing bowl until frothy. Pour the mixture into the well in the powdered sugar. Whisk all the ingredients until they form a smooth consistency. Cover the bowl with a moist cloth and reserve until needed.

Montage

1. Spread the sliced almonds out onto a baking sheet and bake them until they are lightly golden (*5 to 10 minutes*). Set aside.
2. Cut a cake board into an 8-inch circle and spread a small spoonful of raspberry jam in the center to hold the génoise in place.
3. Turn the cake over so the domed side is on top and trim it flat with a serrated knife. Turn the cake over onto the cake board. Score the cake around the side with a bread knife, marking a guide line that indicates 2 even layers. Following the guidelines, slice around the cake, slowly cutting deeper, until there are 2 even layers. Lift the top half and set aside. Using a pastry brush, wet the entire surface of the bottom layer with syrup to imbibe it (***puncher***). Place a large spoonful of raspberry jam on the bottom layer of cake and spread it out to the edges using a metal spatula. Place the second layer on top and press it down lightly. Smooth away any overflows of jam with a metal spatula. Score the top of the cake with the tip of a small knife and imbibe it (***puncher***) with syrup.
4. Place the cake on a wire rack set over a baking sheet.
5. Heat the jam in a medium saucepan over medium heat until it has the consistency of a thick syrup. Meanwhile, spoon some glace royale into a paper cone (***cornet***) and snip off the end. Pour the jam on the cake to coat it and immediately pipe a spiral of glace royale from the center to the edge of the cake. Drag a toothpick from the center outward every 45° around the cake (*forming 4 evenly spaced spokes*). Drag the icing the opposite direction between each spoke to form a web pattern.
6. Press the toasted almond slices to the sides of the cake to cover them.
7. Let the cake rest for 10 minutes in the refrigerator to set the raspberry jam.
8. Transfer the ***génoise confiture*** either to a clean cake board or serving dish and refrigerate until ready to serve.

GRATIN DE FRUITS ROUGES

Learning Outcomes

Sabayon
Blanchir
Glacer under a salamander

Equipment

Knives:
Paring knife (*office*)

Tools:
Balloon whisk, large mixing bowl, small mixing bowl, spoon

Pans:
Small plats à gratin, bain marie

Serving

2 persons

GRATIN DE FRUITS ROUGES

Red Fruit Sabayon

Quantity		Ingredient
U.S.	*Metric*	*Sabayon*
3 pcs	3 pcs	Egg yolks
2 ½ oz	75 g	Sugar, granulated
2 ½ fl oz	75 mL	White wine
		To Serve
8 oz	250 g	Berries

Method

1. Clean and cut the berries.
2. Arrange in ***gratin dishes*** and reserve in the refrigerator until needed.

Sabayon

1. With the aid of a balloon whisk, mix the egg yolks and sugar in a large mixing bowl (***blanchir***), then incorporate the white wine. Place the mixing bowl in a gently simmering bain marie and beat the mixture until it becomes airy, makes ribbons when the whisk is lifted (***au ruban***), and turns light yellow in color.

Note: *To incorporate maximum air into the mix and get volume, it is important to whisk the sabayon correctly from the beginning. Close attention to this technique will also prevent the egg yolks from cooking too quickly.*

To Serve

1. Spoon the ***sabayon*** onto the berries and place under a hot salamander or broiler until the sabayon takes on a golden color (***glacer***).
2. Serve immediately.

ÎLE FLOTTANTE

Learning Outcomes

Meringue française
Crème Anglaise
Caramel
Pocher

Equipment

Tools:
Balloon whisk, mixing bowls, whisk, pastry bag, large plain tip, wooden spoon, ice water bath

Pans:
1 medium russe,
1 small russe,
1 large gratin dish, roasting pan, baking sheet

Serving

4 persons

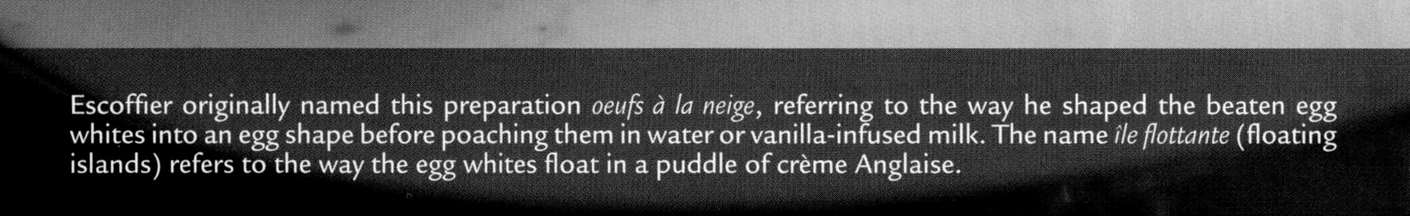

FYI...

Escoffier originally named this preparation *oeufs à la neige*, referring to the way he shaped the beaten egg whites into an egg shape before poaching them in water or vanilla-infused milk. The name *île flottante* (floating islands) refers to the way the egg whites float in a puddle of crème Anglaise.

ÎLE FLOTTANTE
Floating Island

Quantity		*Ingredient*
U.S.	*Metric*	*Meringue Française*
3 ½ oz	100 g	Powdered sugar
3 ½ oz	100 g	Egg whites
3 ½ oz	100 g	Sugar, granulated
		Crème Anglaise
½ pt	250 mL	Milk
1 pc	1 pc	Vanilla bean
3 pcs	3 pcs	Egg yolks
2 oz	60 g	Sugar, granulated
		Caramel
3 ½ oz	100 g	Sugar, granulated
1 ¾ fl oz	50 g	Water
		To Serve
¾ oz	20 g	Almonds, blanched and sliced

Method

1. Preheat the oven to 300°F (150°C).

Meringue Française

2. Sift (***tamiser***) the powdered sugar into a bowl. In a separate bowl, whip the egg whites using a balloon whisk (***monter les blancs***) until soft peaks form. Whisking continuously, add the sugar a little at a time, and continue to whisk the meringue until it is thick and glossy. Add the powdered sugar and, using a rubber spatula, fold it in until incorporated.
3. Place the silicone molds in a roasting pan. Transfer the meringue to a pastry bag fitted with a large plain tip and pipe the meringue into the molds. Fill the molds to the top and scrape the meringue flat using a plastic scraper (***corne***) or a metal spatula. Fill the roasting pan with enough water to ensure that the molds are two-thirds immersed. Cover the molds with foil, transfer the pan to the oven, and cook the meringues until set (*30 minutes*). Remove the pan from the oven and remove the foil. Let the meringues cool, then remove the molds from the roasting pan. Wrap the molds in plastic and reserve them in the refrigerator. Turn the oven up to 350°F (175°C).

Garniture

1. Spread the sliced almonds out onto a baking sheet and place them in the oven to toast until lightly golden (*5 to 10 minutes*). Set aside.

Crème Anglaise

1. Place the milk in a medium saucepan and bring to a low boil over medium-high heat. Using a small knife, split the vanilla bean lengthwise. Scrape the seeds from both sides and add to the milk along with the pod. Whisk well.
2. Place the egg yolks in a mixing bowl, add the sugar, and immediately begin to whisk it into the yolks. Continue whisking until the sugar is completely dissolved and the mixture is pale in color (***blanchir***).
3. Once the milk is scalded, whisk about one-third of the hot milk into the yolks to temper them. Whisk until the mixture is well combined and evenly heated.
4. Stir the tempered egg yolks into the pan of remaining hot milk and stir with a wooden spatula. Place the pan over low heat and stir in a figure 8 motion. As you stir, the foam on the surface will disappear; at the same time, the liquid will begin to thicken and become oil-like in resistance. Continue cooking until the mixture is thick enough to coat the back of a wooden spatula and when your finger leaves a clean trail (***à la nappe***).

Tip: *Crème Anglaise should be cooked to between 167°F and 185°F (75°C and 85°C).*

5. Remove the pan from the heat and strain the crème Anglaise through a fine mesh sieve (***chinois***) into a clean bowl set in a bowl of ice. Stir it back and forth (***vanner***) with the spatula until cooled.
6. Cover the bowl in plastic wrap and reserve the bowl in the refrigerator until needed.

Finishing

1. Unmold the meringue domes onto individual plates and fill the bottom of each plate with crème Anglaise. Bring the sugar and water to a boil in a small saucepan over medium-high heat and cook this mixture until it turns a light caramel color. Immediately pour the caramel onto the meringue domes and sprinkle them with sliced almonds.
2. Serve immediately or refrigerate until needed.

RELIGIEUSES AU CAFÉ ET CHOCOLAT

Learning Outcomes

Pâte à choux
Crème pâtissière
Filling pâte à choux
(*coucher des choux*)
Flavoring pastry cream and fondant
Glaçage with fondant
Rosettes with butter cream

Equipment

Tools:
Wooden spoon, rubber spatula, pastry bag, 12-mm plain tip, small star tip, 6-mm plain tip, fork, whisk, small offset spatula

Pans:
1 large russe, 1 medium russe, 2 small russes, baking sheet

Serving

8 persons

FYI...

According to Pierre Lacam in "Le Mémoriale Historique," la religieuse was created at Frascati, a 19th-century Paris restaurant that was shut down for gambling. Such immoral beginnings for a pastry so piously named! La religieuse translates as "the nun" and the pastry itself is modeled after the habit worn by nuns.

RELIGIEUSES AU CAFÉ ET CHOCOLAT

Chocolate and Coffee Cream Puffs

Quantity		Ingredient
U.S.	**Metric**	***Pâte à Choux***
4 fl oz	125 g	Water
4 fl oz	125 g	Milk
4 oz	125 g	Butter
½ tsp	5 g	Salt
½ oz	12 g	Sugar, granulated
5 oz	150 g	Flour
4–5 pcs	4–5 pcs	Eggs
1 pc	1 pc	Egg, slightly beaten, for eggwash
		Crème Pâtissière
1 ½ pt	750 mL	Milk
1 pc	1 pc	Vanilla bean
6 pcs	6 pcs	Egg yolks
5 oz	150 g	Sugar, granulated
1 oz	45 g	Flour
1 ½ oz	45 g	Cornstarch
		Coffee extract
		Dark chocolate, melted
		Montage
3 ½ oz	100 g	Coffee fondant*
3 ½ oz	100 g	Chocolate fondant*
3 ½ oz	100 g	Crème au beurre
		Cocoa powder
		*Available at specialty stores.

Method

Pâte à Choux

1. *Prepare a panade*: Combine the water, milk, butter, sugar, and salt in a large pan and bring to a boil over medium-high heat. Once the butter has completely melted, remove the pan from the heat and add all the flour at once. Stir with a wooden spatula until combined; then, over medium heat, stir until the mixture forms a smooth ball and comes cleanly away from the sides of the pan (***dessêcher***). Transfer to a clean bowl and allow to cool slightly. Beat the eggs into the panade one at a time. The dough should be stretchy and slightly sticky. Transfer the mixture to a pastry bag fitted with a 12-mm plain tip. Pipe 2 rows of large choux [*approximately 2 in. (5 cm) across*] and 2 rows of small choux [*about 1 in. (2.5 cm) across*] onto a very lightly greased baking sheet. Brush the choux with egg wash, then dip a fork in egg wash and lightly press to even their shape.
2. Transfer the choux to the oven and immediately reduce the temperature to 400°F (205°C). Bake until golden (*20 to 25 minutes*), rotating the baking sheet as soon as they begin to color.
3. Remove the choux from the oven and transfer to a wire rack to cool.
4. Using a small star tip, pierce a hole in the bottoms of the choux.

Crème Pâtissière

1. Pour the milk into a medium saucepan. Split the vanilla bean in half lengthwise and scrape out the seeds. Stir the seeds into the milk and bring it to a boil over medium-high heat. Add about one-quarter of the sugar to the milk and stir to dissolve it.
2. Meanwhile, place the egg yolks in a small mixing bowl and add the remaining sugar. Whisk the sugar into the eggs until it completely dissolves and the yolks lighten in color (***blanchir***). Add the flour and cornstarch to the yolks and stir until well combined.
3. When the milk begins to come to a boil, remove it from the stove and pour one-third of it into the egg yolks. Stir well to temper the yolks, then whisk the tempered mixture into the remaining hot milk. Place back onto the heat and cook until the crème pâtissière begins to bubble. Continue whisking (being sure to press the whisk around the corners of the pan) and allow to cook for 1 minute in order to cook the starch. The crème pâtissière will become very thick. Separate the crème pâtissière into 2 equal batches in separate mixing bowls and mix coffee extract into one batch and melted chocolate into the other using whisks. Pat (***tamponner***) the surface of the pastry cream with a piece of cold butter held on the end of a fork to create a protective film. Let the pastry cream cool to room temperature and refrigerate.

Montage

1. Beat the pastry cream until its texture loosens, then transfer it to a pastry bag fitted with a 6-mm plain tip. Fill half the choux with chocolate-flavored crème pâtissière and the other half with coffee-flavored crème pâtissière, piping it in through the holes underneath. *Fondant:* Stirring gently with a wooden spatula, heat the coffee fondant in a medium saucepan over the lowest heat setting. If the fondant stays very thick even when heated, add a few drops of simple syrup and stir it in. Continue to stir the fondant until it coats the back of a spoon in a thick, shiny layer. The ideal temperature for use is 98.6°F (37°C).
2. Repeat all the same steps to prepare the chocolate fondant.
3. To ice the choux, dip the tops in the fondant corresponding to their filling; make sure to clean off any drips with your fingers. When all the choux have been iced, place the crème au beurre in a pastry bag fitted with a small star tip and pipe a small rosette on top of each large choux. For the chocolate religieuses, mix cocoa powder into the crème au beurre. Stick 1 small choux on each rosette. Pipe a collar (***collerette***) around each little choux, then finish them with a small rosette on top.

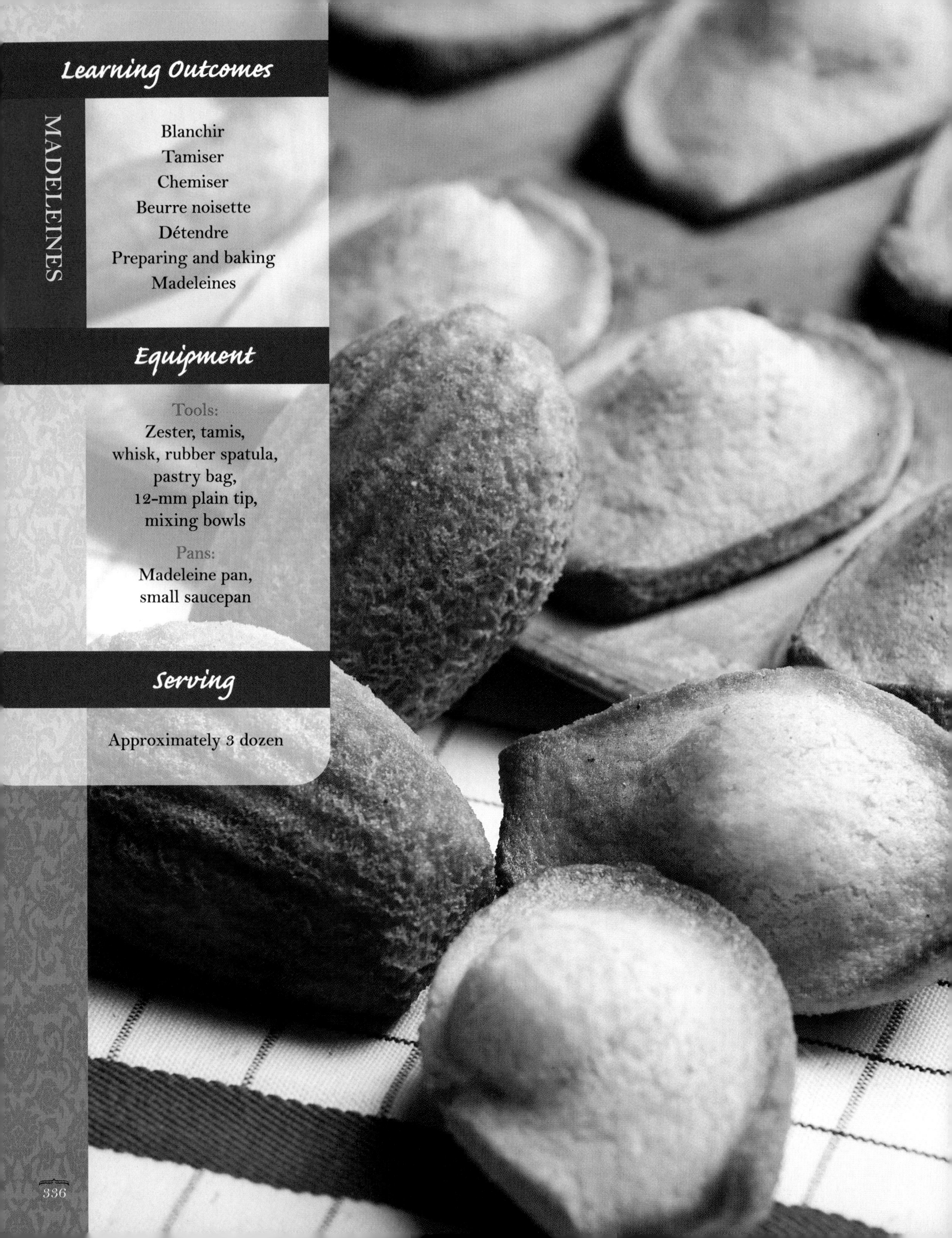

MADELEINES

Learning Outcomes

Blanchir
Tamiser
Chemiser
Beurre noisette
Détendre
Preparing and baking
Madeleines

Equipment

Tools:
Zester, tamis, whisk, rubber spatula, pastry bag, 12-mm plain tip, mixing bowls

Pans:
Madeleine pan, small saucepan

Serving

Approximately 3 dozen

MADELEINES

Shell Sponge Cakes

Quantity		Ingredient
U.S.	*Metric*	*Chemisage*
1 oz	30 g	Butter
1 oz	30 g	Flour
		Appareil à Madeleines
12 oz	360 g	Flour
5 ¾ oz	170 g	Butter
½ oz	10 g	Baking powder
		Lemon zest, grated
8 oz	240 g	Sugar, granulated
4 pcs	4 pcs	Eggs
2 fl oz	60 mL	Milk

Method

1. Preheat the oven to 465°F (240°C).
2. Melt the butter in a small saucepan over medium heat. Brush the Madeleine molds with butter and place in the refrigerator for 1 or 2 minutes to set. Apply a second layer of butter, then coat the molds with flour (***chemiser***) and set aside.

Appareil à Madeleines

1. Melt the butter in a small saucepan over medium heat and cook it until it begins to brown (***beurre noisette***). Set aside.
2. Sift (***tamiser***) the flour and baking powder together onto a piece of parchment paper. Sift a second time. Combine the lemon zest and sugar in a large mixing bowl and add the eggs. Using a whisk, beat the eggs and sugar together until the mixture is pale in color and creates ribbons when the whisk is lifted (***blanchir***). Pour in the milk and stir until combined. Add the flour and carefully fold it in using a rubber spatula. Add 2 tablespoons of the batter to the melted butter and stir until homogeneous (***détendre***). Fold this mixture back into the batter. Cover the bowl in plastic wrap and let the batter rest in the refrigerator for a minimum of 1 hour.

Cuisson

1. Transfer the batter to a pastry bag fitted with a large plain tip. Pipe teardrop shapes that half fill the Madeleine molds. Transfer the Madeleines to the oven to bake for 6 minutes, then turn the temperature down to 390°F (200°C) and bake them until golden (*3 to 4 minutes*). When the Madeleines are golden and have formed a bump on the top, remove them from the oven and unmold them onto a wire rack. Serve the Madeleines warm or let them cool to room temperature.

FYI...

The origin of this inconspicuous little shell-shaped cake has caused more than its fair share of controversy. It is attributed by some to Cordon Bleu chef Madeleine Paumier in 1843, and is also referred to by Carême as being a creation of the renowned pastry chef Jean Avice. There is also much earlier evidence that the cake was the innovation of a peasant girl named Madeleine who worked for the Duke of Commercy. Despite the controversy, the Madeleine holds a special place in the hearts of the French as a fondly remembered childhood treat. This fondness is epitomized by the venerable French writer Marcel Proust who describes a childhood memory in which eating a Madeleine is akin to a religious experience.

MILLEFEUILLE

Learning Outcomes

Pâte feuilletée
Crème pâtissière
Fleurer
Making a layered confection
Glaçage au fondant
Marbling
Blanchir
Masquer

Equipment

Knives:
Serrated knife, paring knife (*office*)

Tools:
Corne, rolling pin, baker's brush, docker, pastry brush, wire rack, whisk, mixing bowls, rubber spatula, pastry bag, 10-mm plain tip, large offset spatula, cornet

Pans:
Medium russe, 2 small russes, bain marie, baking sheet

Serving

8 persons

FYI...

Invented by French pastry chef Rouget, millefeuille was met with resounding approval by the *Jurys dégustatures* (a 19th-century Parisian jury that tested innovative preparations). Characterized by its layers of puff pastry floating on crème pâtissière, this pastry is generally made into single-serving portions but can also be prepared as larger gâteaux.

If prepared correctly, one should be able to cut a millefeuille with one firm stroke of the knife, where none of the cream will squeeze out.

MILLEFEUILLE
Napoleon

Quantity		*Ingredient*
U.S.	*Metric*	*Pâte Feuilletée*
1 lb	500 g	Flour
1 ¾ tsp	10 g	Salt
7 fl oz	225 mL	Water
7 oz	200 g	Butter, room temperature
7 oz	200 g	Butter
		Sugar for dusting
		Crème Pâtissière
½ pt	250 g	Milk
¼ fl oz	10 mL	Vanilla extract
1 ¾ oz	50 g	Sugar, granulated
2 pcs	2 pcs	Egg yolks
		Cornstarch
¾ oz	20 g	*Montage*
7 oz	200 g	Fondant*
¾ oz	20 g	Dark chocolate, melted
3 ½ oz	100 g	Blanched sliced almonds, toasted
		*Available at specialty food stores.

Method

1. Preheat the oven to 400°F (205°C).

Pâte Feuilletée

Détrempe:

1. Sift (***tamiser***) the flour onto a clean work surface and make a well in the center using a plastic scraper (***corne***). Add the salt and water to the well. Stir with your fingertips until the salt is dissolved.
2. Add the butter, cut into pieces, and begin to incorporate the flour using your fingertips. As the flour, butter, and water begin to combine, use the ***corne*** to cut the ingredients together, until it resembles a coarse dough. Sprinkle with additional water if the dough is too dry.
3. Once there are barely any traces of flour left, gather the dough into a ball and score the top of it with a deep cross using a large knife.
4. Loosely wrap the finished ***détrempe*** in plastic and transfer it to the refrigerator to rest for a minimum of 1 hour (preferably overnight).

Note: Détrempe *refers to the dough before the layer of butter (**beurrage**) is added.*

Beurrage and Tourage

1. Place the cold butter between two sheets of parchment paper and pound it with a rolling pin until it is similar to the ***détrempe*** in consistency.
2. Using the ***corne***, shape the butter into a flat square about ½ in. (1 cm) thick. Set the butter aside. If the kitchen is warm, place it in the refrigerator.
3. Lightly dust a clean work surface with flour (***fleurer***), then unwrap the ***détrempe*** and place it on the floured surface.
4. Using the scored marks as a guide, roll out (***abaisser***) the corners of the ***détrempe*** into a cross shape. Be careful to keep the center of the cross thicker than its outer arms (this will be important when rolling out the dough and the butter).
5. Place the square of butter in the center of the cross and fold the two side arms over it so that they overlap slightly in the center (in the process be careful not to trap any air bubbles). Give the dough a quarter turn and fold the two remaining arms over the butter so that the butter is completely enclosed. Press the seams well to seal.
6. Lightly tap the dough with the length of the rolling pin to even out the distribution of the butter inside. Give the dough a quarter turn and repeat the tapping process. This is called the ***enveloppe***.

Tourage, 6 Turns (6 tours simples)

1. *Turns 1 and 2:* Roll out (***abaisser***) the dough in long even strokes to form a rectangle that is three times the original length of the ***enveloppe*** or ½ in. (1 cm) thick. Brush off any excess flour.
2. Fold the bottom third of the dough up; then fold the top third down over the first fold. Make sure the edges are even. Give the dough a quarter turn to the right and repeat the same rolling process. Make sure to always brush away any excess flour.

Method

3. Repeat the folding process (top third up, top third down over first fold) and give the dough a quarter turn to the right. Make two finger impressions in the top left corner of the dough. ***Note***: *These marks are a reminder of the number of turns that the dough has received; they also indicate the position for subsequent turns.* Wrap the dough in plastic and transfer it to the refrigerator to rest for a minimum of 20 minutes. With two turns, the dough is now referred to as the ***pâton***.
4. *Turns 3 and 4:* Lightly dust the work surface with flour (***fleurer***).
5. Remove the dough from the refrigerator and unwrap it onto the floured surface (with the 2 indents in the top left corner). Proceed to give the dough a third and fourth turn (rolling and folding in the same manner as the first and second turns). Mark the dough with 4 imprints in the top-left corner before wrapping it in plastic and returning it to the refrigerator to rest for a minimum of 20 minutes.
6. *Turns 5 and 6:* Lightly dust the work surface with flour (***fleurer***).
7. Remove the dough from the refrigerator and unwrap it onto the floured surface (with the 4 indents in the top left corner). Proceed to give the dough its final 2 turns, folding and rolling as in previous turns. Wrap it in plastic and return it to the refrigerator to rest for a minimum of 20 minutes before rolling it out (the longer the dough rests the better it will perform).

Tip: *Because the détrempe and the butter are at the same consistency, it is necessary to complete the turns as explained above. If you allow the dough to over chill between turns, the butter may become too hard and crack when rolled out. Make sure you have allotted the necessary time to complete the turns.*

8. Lightly dust a clean work surface with flour (***fleurer***). Unwrap the puff pastry onto the lightly floured surface and, using a rolling pin, roll it out to a thickness of ⅛ in. (3 mm) and the same dimensions as a baking sheet.
9. Transfer it to a clean baking sheet and prick it thoroughly using a docker (***pique vite***). Brush the pastry with cold water and sprinkle it with sugar.
10. Transfer it to the oven to bake until lightly golden (*20 minutes*). Reduce the oven temperature to 370°F (190°C) and bake the puff pastry 5 to 10 minutes more to dry it out. Remove the baking sheet from the oven and transfer the pastry to a wire rack to cool to room temperature. Lower the oven temperature to 350°F (175°C).

Crème Pâtissière

1. Line a shallow tray or baking sheet with plastic wrap. Set aside.
2. Pour the milk into a medium saucepan, add the vanilla extract to the milk and bring the mixture to a boil over medium-high heat. Add about one-quarter of the sugar to the milk and stir to dissolve it.
3. Meanwhile, place the egg yolks in a small mixing bowl and add the remaining sugar. Whisk the sugar into the eggs until it completely dissolves and the yolks lighten in color (***blanchir***). Add the flour or cornstarch to the yolks and stir until well combined.
4. When the milk begins to come to a boil, remove it from the stove and pour one-third of it into the egg yolks. Stir well to temper the yolks, then whisk the tempered mixture into the remaining hot milk. Place back onto the heat and cook until the crème pâtissière begins to bubble. Continue whisking (being sure to press the whisk around the corners of the pan) and allow to cook for 1 minute in order to cook the starch. The crème pâtissière will become very thick. Once cooked, pour the crème pâtissière onto the prepared tray or baking sheet. Pat (***tamponner***) the surface with a piece of cold butter held on the end of a fork to create a protective film then cover with plastic, smoothing out any air bubbles and sealing at the edges. Let the crème pâtissière cool to room temperature then refrigerate until needed.

Toasted Almonds

1. Spread the sliced almonds out onto a baking sheet and place them in the oven to toast until lightly golden (*5 to 10 minutes*). Set aside.

Montage

1. Transfer the puff pastry to a clean work surface and trim the edges with a bread knife. Cut the pastry widthwise into 3 equally sized pieces. Lay 1 piece upside down on the work surface and set the others aside. Whisk the cooled crème pâtissière until it is smooth and elastic, then transfer it to a pastry bag fitted with a 10-mm plain tip. Pipe parallel rows of crème pâtissière side by side lengthwise onto the puff pastry rectangle to completely cover it. Place a second piece of puff pastry upside down on top and press it down lightly using a wire rack to secure it in place. Cover it in crème pâtissière using the same technique as before and place the last layer of puff pastry on top, upside down so that the smooth side is facing upward. Press it down lightly with a wire rack to secure it in place. Apply a thin layer of crème pâtissière (***masquer***) to the sides of the millefeuille using a metal spatula.

Finishing

1. Fill a paper cone (***cornet***) with melted chocolate and snip off the end. Gently heat the fondant in a medium saucepan over the lowest heat setting (stirring it gently with a wooden spatula). If the fondant stays very thick (even when heated) add a few drops of simple syrup or warm water and stir it in. Continue to gently stir the fondant until it coats the back of a spoon in a thick, shiny layer. The ideal temperature to work with fondant is approximately 98.6°F (37°C). Once the fondant has reached the correct temperature and consistency, pour the fondant onto the top of the millefeuille and spread it out to the edges using an offset metal spatula. Using the cornet, pipe thin lines of chocolate lengthwise onto the fondant, then drag a toothpick back and forth through them widthwise to create a marbled effect. Press the toasted almonds to the sides of the millefeuille. Transfer the millefeuille either to a clean cake board or serving dish and refrigerate until ready to serve.

MIROIR CASSIS

Learning Outcomes

Biscuit cuillère
Mousse
Tamiser
Blanchir
Meringue Italienne
Pocher
Perler
Gros boulée
Glaçage
Crème fouettée
Chemiser
Demouler

Equipment

Knives:
Serrated knife, paring knife (*office*)

Tools:
Scissors, balloon whisk, rubber spatula, mixing bowls, pastry bag, corne, 5-mm plain tip, whisk, pastry brush, metal spatula

Pans:
2 small pans, bain marie, baking sheet

Serving

6–8 persons

FYI...

This cake derives its name from its glossy top surface that shines like a mirror and should be as flat as one.

Quantity		Ingredient
U.S.	*Metric*	*Biscuit Cuillère*
7 oz	200 g	Flour, sifted (***tamiser***)
8 pcs	8 pcs	Egg whites
5 ½ oz	160 g	Sugar, granulated
8 pcs	8 pcs	Egg yolks
2 ¾ oz	80 g	Sugar, granulated
		Mousse Base
7 oz	200 g	Black currant purée
½ oz	12 g	Sugar, granulated
½ fl oz	12 g	Water
4 pcs	4 pcs	Gelatin leaves
		Ice water for gelatin
		Meringue Italienne
3 pcs	3 pcs	Egg whites
3 ½ oz	100 g	Sugar, granulated
1 fl oz	30 g	Water
		Finition Mousse
7 fl oz	200 mL	Cream
		Decoration
7 oz	200 g	Glaçage miroir (***neutre***)
5 oz	150 g	White chocolate, melted, in a cornet
2 ¾ oz	80 g	Fresh black currants

MIROIR CASSIS

Black Currant Mousse Cake

Method

1. Preheat the oven to 380°F (195°C).
2. Cut a cake board into a disc just small enough to fit through a cake ring.
3. Using this disc, draw a circle onto the side of a piece of parchment paper. Turn the paper over onto a baking sheet.

Biscuit Cuillère

1. *Prepare the biscuit cuillère batter:* Sift (***tamiser***) the flour onto a piece of parchment paper. Separate the eggs into 2 large mixing bowls. Beat the whites (***monter les blancs***) with a balloon whisk until they are white and frothy but still fluid and set aside. Add the sugar to the egg yolks and whisk them together until the mixture is thick and pale in color (***blanchir***). Set aside.
2. Whip the egg whites (***monter les blancs***) with the balloon whisk until they reach the soft peak stage. Gradually add the sugar, beating it in with the whisk. Once all the sugar has been added, continue to beat the whites until they are thick and glossy. Add the sifted flour to the egg yolks previously blanched with the sugar and fold it in (***incorporer***) using a rubber spatula until it is almost completely incorporated. Add one-third of the egg whites to the mixture and fold it in (***incorporer***) delicately until the mixture is streaked with thin lines of white. Fold the remaining egg whites until just incorporated. The end result should be a pale, homogeneous batter with a light airy texture that is ideal for piping. Transfer the batter to a pastry bag fitted with a 5-mm plain tip.
3. Using the drawn circle as a size guide, pipe a tight spiral onto the parchment-lined baking sheet. On the other side of the baking sheet, pipe a row of 4 in. (10 cm) lines at a 45-degree angle along the length of the baking sheet. Fill in both corners.
4. Dust the piped batter with powdered sugar to completely coat it. Let it rest 30 seconds and dust it again (***perler***), then transfer it to the oven to bake until lightly colored (*12 to 15 minutes*). Once cooked, remove the baking sheets from the oven and slide the parchment paper off them onto a wire rack to cool. Immediately dust the biscuit with powdered sugar.

Mousse Base

1. Soak the gelatin sheets in cold water until completely softened. Meanwhile, heat the sugar, water, and black currant purée together in a small

Method

saucepan over medium heat, stirring occasionally until the sugar has melted and the liquid is hot to the touch. Remove the pan from the heat. Remove the gelatin from the cold water and squeeze out any excess liquid. Add the softened gelatin to the pan and stir it in gently until completely dissolved. Pour this mixture into the black currant purée while stirring. Continue to stir until completely combined. Reserve at room temperature.

Tip: *If the marble is refrigerated, keep the bowl on a dish cloth to insulate the bottom and stop the gelatin from setting.*

Meringue Italienne

1. Whisk the egg whites in a large mixing bowl using a balloon whisk, until frothy. Set aside. Cook the sugar and water together in a medium saucepan over medium-high heat until they reach the soft-ball stage (***petit boulé***) [250°F (121°C)]. Meanwhile, whip the egg whites until soft peaks form. As soon as the syrup reaches the soft-ball (***petit boulé***) stage, pour it into the egg whites in a thin stream, whisking constantly. Continue to whisk until the meringue is firm and cooled to room temperature.

Finition Mousse

1. Whip the cream to soft peaks in a large mixing bowl over an ice bath. Add the jellified fruit purée to the Italian meringue and fold it in carefully using a rubber spatula. Add half the whipped cream to the bowl and fold it in until incorporated. Add the other half and fold it in gently until the mixture is homogeneous.

Montage

1. Turn the biscuit over onto a clean piece of parchment paper and peel off the parchment paper it was baked on. Turn the biscuit over.
2. Place a cake ring on a cake board. Line the inside of the ring with a strip of acetate. Fit the cut cake board inside it. Cut 2 strips, the same thickness as a metal spatula, lengthwise from the biscuit using a serrated knife. Line the ring, with the sugared side facing outwards so that the biscuit fits snugly. Fit the biscuit disc, cooked side up into the bottom of the circle, trimming it if needed.
3. Fill the prepared cake ring with half the mousse. Using a large metal spatula, spread the mousse up the sides of the ring to completely coat them (***chemiser***). Sprinkle with some drained black currants, then cover with the remaining mousse. If desired, sprinkle the surface with additional black currants, then smooth the top of the cake flat with a metal spatula. Transfer the cake to the freezer to chill for a minimum of 30 minutes.

Finishing

1. Gently stir the ***glaçage miroir*** with a rubber spatula until smooth, being careful not to incorporate any air bubbles. Once the mousse is thoroughly chilled, remove it from the freezer and pour the glaçage miroir on top. Immediately spread it out using a metal spatula. Pour some melted white chocolate into a paper cone (***cornet***) and snip off the tip. Pipe a design onto the top of the cake. Once the glacage is set, carefully remove the ring. Place the cake on a serving dish and finish the decoration with a few black currants. Brush them with glaze using a pastry brush and place the miroir in the refrigerator to gently thaw for a minimum of 10 minutes. Before serving remove the strip of acetate.
2. The miroir pictured here is a more modern presentation. Traditionally, it is left ungarnished so one may admire the perfect mirror finish of the cake.

MOUSSE AU CHOCOLAT

Learning Outcomes

Making a mousse
Melting chocolate
Monter des blancs
Monter une crème
Folding

Equipment

Tools:
Bain marie, rubber spatula, mixing bowls, whisk, balloon whisk

Serving

6 persons

FYI...

The word *mousse* translates into English as both *foam* and as the plant *moss*. This little detail causes food historians quite a lot of grief: Is the preparation named after the plant? After all, moss is light, airy, and grows in clumps that almost appear to have been formed in molds. Or is it named for its foamy or frothy texture? While the latter seems to make more sense, there is evidence to the contrary. Until the mystery is properly solved, food historians will simply have to live with this torturous uncertainty.

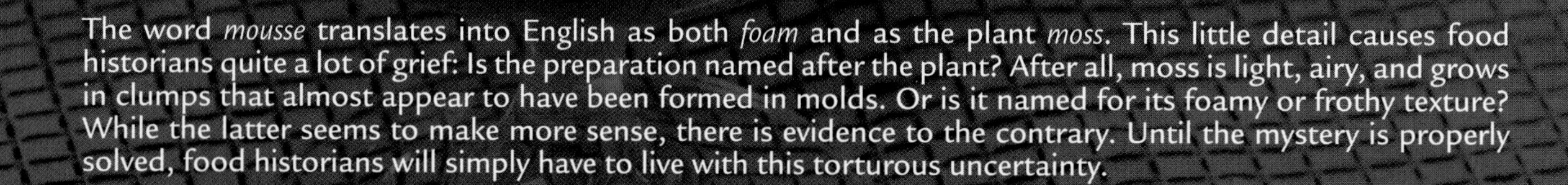

MOUSSE AU CHOCOLAT

Chocolate Mousse

Quantity		Ingredient
U.S.	*Metric*	*Base*
8 oz	250 g	Bittersweet chocolate
3 ½ oz	100 g	Butter
4 pcs	4 pcs	Egg yolks
6 pcs	6 pcs	Egg whites
1 ½ oz	45 g	Sugar, granulated
		Finishing
3 ½ fl oz	100 mL	Whipping cream
½ oz	15 g	Powdered sugar
		To Serve
		Cocoa powder
		Chocolate decorations

Method

Base

1. Melt the chocolate over a simmering bain marie. Once it has completely melted, remove the chocolate from the heat and incorporate the butter with a rubber spatula until melted. Add the egg yolks and stir until the mixture is homogeneous.
2. Beat the egg whites to soft peaks and, whisking continually, gradually add the sugar. Continue to whisk until firm and glossy.
3. Once the chocolate mixture has cooled to just above body temperature, add one-half the meringue and fold it in using a rubber spatula. Add the second half and fold it in until combined.

Finishing

1. Whip the cream and powdered sugar to soft peaks, then fold into the mousse.
2. Transfer the mousse to molds and place in the refrigerator to set for not less than 1 hour.

To Serve

1. Dust the mousse with cocoa powder and if desired, serve with chocolate decorations.

MIROIR AU CITRON

Learning Outcomes

Biscuit cuillère
Tamiser
Blanchir
Perler
Making a fruit mousse
Canneler
Meringue italienne
Cooking sugar
Glazing an entremet

Equipment

Knives:
Chef knife (*couteau chef*), channeling knife

Tools:
Whisk, bowls, spatula, drum sieve, tamis, ice bath, silicone mat, pastry bag, plain pastry large tip, 8-inch flan ring, flan ring, pastry brush

Pans:
3 saucepans, baking sheet

Serving

8 persons

FYI...

This mousse is a bavarois. There are two different types of bavarois: One has crème Anglaise for a base, the other has fruit purée. Both types work along the same principle: Add gelatin to the liquid (crème Anglaise or fruit purée) and fold in whipped cream. They can be served chilled or frozen and are the basis of many French mousse cakes otherwise known as entremets.

MOUSSE AU CITRON

Lemon Mousse

Method

1. Preheat the oven to 380°F (195°C).

Biscuit Cuillère

1. To obtain the correct amount of biscuit cuillère use the ingredients list in this recipe following the method on pages 400–402 in *Cuisine Foundation.*
2. Transfer the batter to a pastry bag fitted with a 5-mm plain tip. On the parchment paper-covered baking sheet, pipe the biscuit into a tight spiral 7 ½ in. (19 cm) in diameter. Dust the piped batter with powdered sugar to completely coat it. Let it rest 30 seconds and dust it again (***perler***), then transfer it to the oven to bake until lightly colored (*12 to 15 minutes*). Once cooked, remove the baking sheet from the oven and slide the parchment paper onto a wire rack to cool. Immediately dust the biscuit with powdered sugar.

Garniture

1. Channel (***canneler***) the lemons and slice them thinly (*discarding the end slices*). Remove and discard the seeds.
2. *Blanch the lemon slices:* Place the slices in a medium saucepan of cold water and bring to a low boil over high heat. Refresh quickly in an ice bath and drain. Be careful not to overboil the slices as they need to keep their shape for presentation.
3. Lay the slices out on paper towels to absorb the excess moisture.
4. Place an 8-inch flan ring on a baking sheet and line it with plastic wrap. Line the bottom and sides with lemon slices. Reserve in the freezer until needed.

Meringue Italienne

1. Cook the sugar and water in a medium saucepan over medium-high heat until they reach the soft stage (***petit boulé***) [250°F (121°C)]. Meanwhile, whip the egg whites until soft peaks form. As soon as the syrup reaches the soft-ball stage (***petit boulé***), pour it into the egg whites in a thin stream, whisking constantly. Continue to whisk until the meringue is firm and cooled to room temperature. Set aside.

Mousse au Citron

1. Whisk the cream in a large mixing bowl over an ice bath until soft peaks form.
2. Place the lemon pulp in a small pan over medium heat until just beginning to bubble. Remove from the heat. Meanwhile, soak the gelatin sheets in cold water.

Quantity		*Ingredient*
U.S.	*Metric*	*Biscuit Cuillère*
8 pcs	8 pcs	Egg yolks
5 ½ oz	160 g	Sugar
7 oz	200 g	Flour, sifted (***tamiser***)
8 pcs	8 pcs	Egg whites
		Powdered sugar, for dusting
		Meringue Italienne
6 oz	180 g	Sugar, granulated
2 oz	60 g	Water
3 pcs	3 pcs	Egg whites
		Mousse au Citron
10 oz	300 g	Lemon pulp
13 ½ oz	400 g	Cream
5 pcs	5 pcs	Gelatin sheets
		Finishing
6 pcs	6 pcs	Glaçage miroir
		Lemons, chaneled, ***rondelle***
		Strawberries
		Chocolate

Method

3. When the lemon pulp is hot, squeeze out the gelatin sheets and gently mix into the pulp until incorporated.
4. Place the bowl over an ice bath and stir back and forth with a spoon until the lemon mixture is oil-like in resistance and movement.
5. Slowly pour the lemon mixture into the meringue Italienne, folding it in with a spatula. When all the lemon mixture is incorporated, fold in the cream.

Montage

1. Pour the mousse into the lined flan ring, fill it to 3⁄16 in. (5 mm) from the top, and place it in the freezer to set for 10 minutes.
2. When the mousse is set, place the biscuit disc upside down on top and press down lightly enough to press it into the mousse. The goal is to stick the biscuit to the mousse, but not immerse it within the mousse.

Finishing

1. Warm the ***glaçage miroir*** in a small saucepan over low heat until just warm to the touch.
2. Remove the mousse from the freezer, turn it over onto a wire rack and remove the ring and plastic wrap. Place the rack over a baking sheet or pan and pour the glaçage over the top of the mousse to completely coat it. Tap the rack on the work surface a few times to smooth the surface.
3. Transfer the mousse to a cake dish and let it rest for 30 minutes minimum in the refrigerator; this will ensure that the center is not frozen. It might need more time depending on how long it was kept in the freezer.
4. Decorate with fanned strawberries, chocolate, and lemon slices.

Tip: *Cover any fruits used for decoration in glaçage to add shine (**lustrer**). Additionally, before the slices are used for the chemisage they can be soaked (**confire**) in simple syrup.*

Learning Outcomes

Pâte feuilletée
Rolling out a pâte feuilletée (*abaisser*)

Equipment

Knives:
Slicing knife (***éminceur***)

Tools:
Rolling pin, corne, baker's brush, pastry brush

Pans:
Baking sheet

Serving

Approximately 2 dozen

FYI...

Historical information on the palmier is as minimal as the ingredients in this classic pastry. Apart from the fact that it seems to have appeared in the late 1900s and is thought to have been invented as a way of using up puff pastry pairings, there appears to be no one inventor of this preparation. Palmier is essentially sugared, rolled, and cut puff pastry that takes its name from its resemblance to a palm leaf.

Quantity		Ingredient
U.S.	Metric	Pâte Feuilletée
1 lb	500 g	Flour
1 ¾ tsp	10 g	Salt
7 fl oz	225 mL	Water
7 oz	200 g	Butter, room temperature
7 oz	200 g	Butter
5 oz	150 g	Powdered sugar for dusting
1 pc	1 pc	Egg for egg wash
		Sugar, granulated

PALMIERS
Palm Leaf Pastry

Method

1. Preheat the oven to 400°F (205°C).

Pâte Feuilletée

Détrempe

1. Sift (***tamiser***) the flour onto a clean work surface and make a well in the center using a plastic scraper (***corne***). Add the salt and water to the well. Stir with your fingertips until the salt is dissolved.
2. Add the butter (cut into pieces) and begin to incorporate the flour using your fingertips. As the flour, butter, and water begin to combine, use the ***corne*** to cut the ingredients together, until it resembles a coarse dough. Sprinkle with additional water if the dough is too dry.
3. Once there are barely any traces of flour left, gather the dough into a ball and score the top of it with a deep cross using a large knife.
4. Loosely wrap the finished ***détrempe*** in plastic and transfer it to the refrigerator to rest for a minimum of 1 hour (preferably overnight).

Note: Détrempe *refers to the dough before the layer of butter (**beurrage**) is added.*

Beurrage and Tourage

1. Place the cold butter between two sheets of parchment paper and pound it with a rolling pin until it is similar to the ***détrempe*** in consistency.
2. Using the ***corne***, shape the butter into a flat square about ½ in. (1 cm) thick. Set the butter aside. If the kitchen is warm, place it in the refrigerator.
3. Lightly dust a clean work surface with flour (***fleurer***), then unwrap the ***détrempe*** and place it on the floured surface.
4. Using the scored marks as a guide, roll out (***abaisser***) the corners of the ***détrempe*** into a cross shape. Be careful to keep the center of the cross thicker than its outer arms (this will be important when rolling out the dough and the butter).
5. Place the square of butter in the center of the cross and fold the two side arms over it so that they overlap slightly in the center (in the process be careful not to trap any air bubbles). Give the dough a quarter turn and fold the two remaining arms over the butter so that the butter is completely enclosed. Press the seams well to seal.
6. Lightly tap the dough with the length of the rolling pin to even out the distribution of the butter inside. Give the dough a quarter turn and repeat the tapping process. This is called the ***enveloppe***.

Method

Tourage, 6 Turns (6 tours simples)

1. *Turns 1 and 2:* Roll out (***abaisser***) the dough in long even strokes to form a rectangle that is three times the original length of the ***enveloppe*** or ½ in. (1 cm) thick. Brush off any excess flour.
2. Fold the bottom third of the dough up; then fold the top third down over the first fold. Make sure the edges are even. Give the dough a quarter turn to the right and repeat the same rolling process. Make sure to always brush away any excess flour.
3. Repeat the folding process (top third up, top third down over the first fold) and give the dough a quarter turn to the right. Make two finger impressions in the top left corner of the dough.

***Note**: These marks are a reminder of the number of turns that the dough has received; they also indicate the position for subsequent turns.*

4. Wrap the dough in plastic and transfer it to the refrigerator to rest for a minimum of 20 minutes. With two turns, the dough is now referred to as the ***pâton***.
5. *Turns 3 and 4:* Lightly dust the work surface with flour (***fleurer***).
6. Remove the dough from the refrigerator and unwrap it onto the floured surface (with the 2 indents in the top left corner). Proceed to give the dough a third and fourth turn (rolling and folding in the same manner as the first and second turns). Mark the dough with 4 imprints in the top-left corner before wrapping it in plastic and returning it to the refrigerator to rest for a minimum of 20 minutes.
7. *Turns 5 and 6:* Lightly dust the work surface with flour (***fleurer***).
8. Remove the dough from the refrigerator and unwrap it onto the floured surface (with the 4 indents in the top left corner). Proceed to give the dough its final 2 turns, folding and rolling as in previous turns. Wrap it in plastic and return it to the refrigerator to rest for a minimum of 20 minutes before rolling it out (the longer the dough rests the better it will perform). Using powdered sugar to dust the surface instead of flour, roll out the puff pastry (***abaisser***) in a rectangle shape until ³⁄₁₆ in. (5 mm) thick.

***Tip:** Because the détrempe and the butter are at the same consistency, it is necessary to complete the turns as explained above. If you allow the dough to over chill between turns, the butter may become too hard and crack when rolled out. Make sure you have allotted the necessary time to complete the turns.*

Palmiers

1. Fold the pastry in half lengthwise to create a reference crease down the middle. Unfold the pastry, brush it with egg wash, and sprinkle it with sugar.
2. Fold each long edge in toward the center crease, leaving a small gap at the center. Brush with egg wash and sprinkle with more sugar then fold in two on the crease. Press firmly. Let rest in the freezer until hard (*approximately 30 minutes*).
3. Slice the roll of dough ½ in. (1 cm) thick. Lay the slices on a lightly greased baking sheet.

Cuisson

1. Transfer the baking sheet to the oven. When the ***palmiers*** are just turning golden, turn them over and return to the oven until deeply colored on both sides.
2. Transfer to a rack to cool.

PETITS FOURS SECS

Learning Outcomes

Sabler
Fraiser
Fleurer
Abaisser une pâte
Dorer
Tant pour tant
Using a pastry bag
Crèmer
Glacer
Tuiles
Using a goutière

Equipment

Tools:
Tamis, corne, rolling pin, cookie cutters, whisks, pastry brush, fork, wire rack, mixing bowls, rubber spatula, pastry bag, 10-mm plain tip, sieve, spoon, goutière

Pans:
4 baking sheets

Serving

8 persons

FYI...

Before the advent of the gas oven, a baker had to be resourceful in order to make the best use of the wood-fired oven. Requiring a lower heat, *petits fours* were the perfect preparation to bake once the oven had cooled down after the larger gâteaux were done. *Petits fours* translates loosely as "little flame," simply referring to the low flame in the oven. The word *sec* in "petits fours secs" indicates a dry biscuit, perfect as an accompaniment with tea, custard desserts, ice cream, and sorbets.

PETITS FOURS SECS

Assorted Tea Biscuits

Method

1. Heat the oven to 370°F (188°C).
2. Lightly grease a baking sheet with cold butter and reserve it in the refrigerator.

Petits Sablés Nantais

1. Sift (***tamiser)*** the flour and baking powder together onto a clean work surface. Sift the mixture once more. Sprinkle the salt on top of the sifted ingredients and make a large well in the center [*using a plastic scraper* (***corne***)]. Place the cold, diced butter in the center of the well. Work the butter into the flour using your fingertips while simultaneously cutting through the mixture with the ***corne***. Continue cutting until the butter is crumbly and coated in flour.
2. Rub the mixture between your fingers until it resembles fine sand (***sabler***).
3. Gather the flour-butter mixture into a neat pile and make a well (***fontaine***) in the center using the ***corne***.
4. Cut the vanilla bean in half lengthwise, scrape the seeds out, and set aside. Mix the two egg yolks and powdered sugar together with a whisk (***blanchir***) until the mixture becomes lighter in color. Stir the vanilla seeds into the egg yolk mixture and pour into the well. Stir these ingredients together, using your fingertips while simultaneously using the ***corne*** to gradually incorporate the dry ingredients from the sides. Continue this process until the mixture in the center of the well resembles a paste.
5. Gather all the ingredients together and cut through them repeatedly with the ***corne***. Continue this process until the mixture forms an homogeneous dough.
6. Using the heel of your palm, firmly smear the dough away from yourself to ensure that no lumps are left (***fraiser***). Shape the dough into a ball, wrap it in plastic, and flatten it out as much as possible. Let it rest in the refrigerator for at least 30 minutes (*preferably overnight*).
7. Lightly dust a clean work surface with flour (***fleurer***). Unwrap the dough onto the lightly floured surface and roll it out to a thickness of 1 cm using a rolling pin, giving it a quarter of a turn with each passage of the rolling pin.
8. Cut (***détailler***) circles out of the pastry using a 1 in. (2.5-cm) round cookie cutter and place them on the baking sheet. Brush them with the coffee egg wash and place them in the refrigerator for 5 minutes.

Quantity		*Ingredient*
U.S.	*Metric*	*Petits Sablés Nantais*
7 oz	200 g	Flour
¼ oz	2 g	Baking powder
Pinch	Pinch	Salt
3 ½ oz	100 g	Butter, in small cubes
1 pc	1 pc	Vanilla bean
3 ½ oz	100 g	Powdered sugar
1 pc	1 pc	Egg
1 pc	1 pc	Egg yolk
¼ fl oz	2 mL	Coffee extract
		Bâtons de Maréchaux
4 ½ oz	130 g	Powdered sugar
4 ½ oz	130 g	Almond powder
5 oz	145 g	Egg whites
1 oz	30 g	Sugar, granulated
		Palet des Dames
4 oz	125 g	Powdered sugar
4 oz	125 g	Butter
Pinch	Pinch	Salt
2 pcs	2 pcs	Eggs
5 oz	150 g	Flour
2 ¾ oz	80 g	Raisins, macerated in rum
		Tuiles aux Amandes
1 ¼ oz	40 g	Flour
4 ¼ oz	125 g	Powdered sugar
1 pc	1 pc	Egg
2 pcs	2 pcs	Egg whites
5 oz	150 g	Sliced almonds
		Pâte à Cigarette
1 ¾ oz	50 g	Butter
3 ½ oz	100 g	Powdered sugar
2	2	Egg whites
1 ½ oz	45 g	Flour
		Crushed almonds

Method

9. Stir the egg yolk and coffee extract together in a small bowl. Remove the baking sheet from the refrigerator and brush the discs with a second layer of egg wash (***dorer***). Dip a fork in the egg wash and score the tops of sablés to create a decoration.
10. Bake the sablés until golden (*10 to 15 minutes*).

Bâtons de Maréchaux

1. Sift (***tamiser***) the powdered sugar and almond powder together (***tamiser***) into a bowl.

Note: *The combination of equal amounts of powdered sugar and almond powder is referred to as* ***tant pour tant.***

2. Whisk the egg whites in a large mixing bowl until soft peaks form. Gradually add the sugar and continue to whisk until the whites are stiff and glossy. Add the ***tant pour tant*** and fold it in until combined.
3. Transfer the mixture to a pastry bag fitted with a 10-mm plain tip. Pipe small dollops of the batter at each corner of the baking sheet and glue down a sheet of parchment paper. On the parchment, pipe short lines of batter, about 3 in. (8 cm) long in rows along the length of the baking sheet.
4. Generously sprinkle with crushed almonds. Tip the baking sheet and tap to shake off any excess almonds.
5. Bake the bâtons de maréchaux until lightly golden (*10 to 15 minutes*).
6. As soon as they are out of the oven, slip them straight off the baking sheet onto the marble.
7. When they are cold, dip the bottom in tempered chocolate and lay them flat on a parchment paper-covered baking sheet in the refrigerator to set.

Pâte à Cigarette

1. Cream the butter and sugar together. Stir in the egg whites until smooth. Add the flour. Allow to rest at least one hour before using. Line a baking sheet with a silicone mat. Spread a 4–6 small spoonfuls of the pâte à cigarette on the mat, leaving at least 3 in. (8 cm) in between. Place in the oven and cook about 6 minutes or until they spread out and the edges begin to brown. Remove from the oven, and working very quickly, remove each cookie with a metal spatula and roll around the handle of a wooden spoon. If the cookies become too stiff, return to the oven to soften. Adjust the number of cookies you can roll for the remaining dough.

Palets des Dames

1. Grease and flour a baking sheet. Reserve it in the refrigerator.
2. Sift (***tamiser***) the powdered sugar into a bowl.
3. With a whisk, cream (***crémer***) the butter in a large mixing bowl until light and fluffy. Add the powdered sugar and salt to the butter and cream them together. Add the eggs one at a time and beat until the mixture is light in color. Sift the flour and add it to the mixture. Fold it in using a rubber spatula. Let the batter rest in the refrigerator for 30 minutes.
4. Drain the rum out of the raisins.
5. Transfer the batter to a pastry bag fitted with a 10-mm plain tip. Onto the baking sheet, pipe rows of batter in the form of circles, ¾ in. (2 cm) in diameter. Make sure that the palets are at least 2 in. (5 cm) apart on the baking sheet as they will spread. Place 3 raisins on each circle, pressing them in a little, and transfer the baking sheet to the oven to bake until the palets are golden around the edges (*6 to 8 minutes*).
6. Mix some water into powdered sugar to create a paste and brush it onto the palets using a pastry brush as soon as the palets come out of the oven.
7. Transfer the palets to a wire rack to cool.

Tuiles aux Amandes

1. Preheat the oven to 350°F (175°C).
2. Toast the almonds in the oven until lightly golden on the edges.
3. Sift (***tamiser***) the flour and powdered sugar together.
4. Prepare baking sheets by lining with silicone pads.
5. When the almonds are cool, add them to the flour and sugar. Mix in the egg and egg whites. Form the batter into circles with a wet fork (make sure to leave enough room for the tuiles to spread during baking). Transfer the baking sheet to the oven and bake the tuiles until golden.
6. Immediately after the tuiles are removed from the oven lift them off the baking sheet with a metal spatula and place them in the tuile or gutter mold. (***goutière***).

Learning Outcomes

Pâte feuilletée
Crème d'amandes

Equipment

Knives:
Paring knife (*office*)

Tools:
Corne, baker's brush, rolling pin, rubber spatula, mixing bowls, 6-inch (15-cm) cake ring, 10-inch (25-cm) cake ring, offset spatula, pastry brush

Pans:
Baking sheet

Serving

8–10 persons

FYI...

This cake, made with puff pastry and filled with crème d'amandes, was named after the French town of Pithiviers, which is located just southwest of Paris. Similar preparations existed in various forms until Chef Antoine Carème formalized the recipe for Pithiviers in approximately 1805.

Quantity		Ingredient
U.S.	***Metric***	***Pâte Feuilletée***
1 lb	500 g	Flour
1 ¾ tsp	10 g	Salt
7 fl oz	225 mL	Water
7 oz	200 g	Butter, room temperature
7 oz	200 g	Butter
		Crème d'Amandes
2 oz	60 g	Butter (***pommade***)
2 oz	60 g	Sugar, granulated
1 pc	1 pc	Egg
1 pc	1 pc	Vanilla bean
½ fl oz	10 mL	Rum
2 oz	60 g	Almond powder
		Finishing
1 pc	1 pc	Egg for egg wash

PITHIVIERS

Puff Pastry Filled with Almond Cream

Method

1. Preheat the oven to 400°F (205°C).

Détrempe (for Pâte Feuilletée)

1. Sift (***tamiser***) the flour onto a clean work surface and make a well in the center using a plastic scraper (***corne***). Add the salt and water to the well. Stir with your fingertips until the salt is dissolved.
2. Add the butter (cut into pieces) and begin to incorporate the flour using your fingertips. As the flour, butter, and water begin to combine, use the ***corne*** to cut the ingredients together, until it resembles a coarse dough. Sprinkle with additional water if the dough is too dry.
3. Once there are barely any traces of flour left, gather the dough into a ball and score the top of it with a deep cross using a large knife.
4. Loosely wrap the finished ***détrempe*** in plastic and transfer it to the refrigerator to rest for a minimum of 1 hour (preferably overnight).

***Note: Détrempe** refers to the dough before the layer of butter (**beurrage**) is added.*

Beurrage and Tourage

1. Place the cold butter between two sheets of parchment paper and pound it with a rolling pin until it is similar to the ***détrempe*** in consistency.
2. Using the ***corne***, shape the butter into a flat square about ½ in. (1 cm) thick. Set the butter aside. If the kitchen is warm, place it in the refrigerator.
3. Lightly dust a clean work surface with flour (***fleurer***), then unwrap the ***détrempe*** and place it on the floured surface.
4. Using the scored marks as a guide, roll out (***abaisser***) the corners of the ***détrempe*** into a cross shape. Be careful to keep the center of the cross thicker than its outer arms (this will be important when rolling out the dough and the butter).
5. Place the square of butter in the center of the cross and fold the two side arms over it so that they overlap slightly in the center (in the process be careful not to trap any air bubbles). Give the dough a quarter turn and fold the two remaining arms over the butter so that the butter is completely enclosed. Press the seams well to seal.
6. Lightly tap the dough with the length of the rolling pin to even out the distribution of the butter inside. Give the dough a quarter turn and repeat the tapping process. This is called the ***enveloppe***.

Tourage, 6 Turns (6 tours simples)

1. *Turns 1 and 2:* Roll out (***abaisser***) the dough in long even strokes to form a rectangle that is three times

Method

the original length of the ***enveloppe*** or ⅜ in. (1 cm) thick. Brush off any excess flour.

2. Fold the bottom third of the dough up; then fold the top third down over the first fold. Make sure the edges are even. Give the dough a quarter turn to the right and repeat the same rolling process. Make sure to always brush away any excess flour.
3. Repeat the folding process (top third up, top third down over first fold) and give the dough a quarter turn to the right. Make two finger impressions in the top left corner of the dough. ***Note***: *These marks are a reminder of the number of turns that the dough has received; they also indicate the position for subsequent turns.* Wrap the dough in plastic and transfer it to the refrigerator to rest for a minimum of 20 minutes. With two turns, the dough is now referred to as the ***pâton***.
4. *Turns 3 and 4:* Lightly dust the work surface with flour (***fleurer***).
5. Remove the dough from the refrigerator and unwrap it onto the floured surface (with the 2 indents in the top left corner). Proceed to give the dough a third and fourth turn (rolling and folding in the same manner as the first and second turns). Mark the dough with 4 imprints in the top-left corner before wrapping it in plastic and returning it to the refrigerator to rest for a minimum of 20 minutes.
6. *Turns 5 and 6:* Lightly dust the work surface with flour (***fleurer***).
7. Remove the dough from the refrigerator and unwrap it onto the floured surface (with the 4 indents in the top left corner). Proceed to give the dough its final 2 turns, folding and rolling as in previous turns. Wrap it in plastic and return it to the refrigerator to rest for a minimum of 20 minutes before rolling it out (the longer the dough rests the better it will perform).

Tip: *Because the détrempe and the butter are at the same consistency, it is necessary to complete the turns as explained above. If you allow the dough to over-chill between turns, the butter may become too hard and crack when rolled out. Make sure you have allotted the necessary time to complete the turns.*

Crème d'Amandes

1. Cream the butter and sugar together until light and fluffy (***crémer***).
2. Beat in the egg until well combined. Using a small knife, split the vanilla bean in half lengthwise and scrape out the seeds. Whisk into the mixture. Add the rum and finish by mixing in the almond powder.
3. Reserve the crème d'amandes in a covered bowl in the refrigerator until ready to use.

Montage

4. Cut the pâte feuilletée in half and place one-half back in the refrigerator. Dust the marble with flour (***fleurer***) and place the dough on top. Roll out (***abaisser***) the dough, giving it a one-quarter turn with each stroke of the rolling pin. Continue rolling until the dough is 3 fingers wider than the widest of the two cake circles. Let it rest for 5 minutes in the refrigerator on a sheet of parchment paper; if it shrinks during this time then roll it out again.
5. After letting it rest, transfer the rolled out dough to a parchment paper–lined baking sheet. Mark it by pressing down on it lightly with the wider and smaller cake rings, one inside the other. Fill inside the smaller circle with a dome of crème d'amandes, neatly sculpting it with a scraper (***corne***) or offset spatula. Place the baking sheet in the refrigerator for 10 minutes to set the crème d'amandes.
6. Roll out the second piece of dough to the same size as the first and let it rest in the refrigerator for 5 minutes.
7. Bring both pieces of dough out of the refrigerator. Brush around the crème d'amandes dome with egg wash using a pastry brush. Brush all the flour off the other piece of dough and fold it in four. Unfold it over the crème d'amandes dome without trapping any air. Place the small cake ring on top and press down on it lightly. Remove the ring and use your fingers to press around the indentation to seal the pastry. Transfer the pithiviers to the freezer to rest for 5 to 10 minutes. When the dough is hard, bring the ***pithiviers*** out of the freezer. Using a very sharp paring knife, cut decorative scallop shapes around the edge of the pithiviers; give the cake quarter turns (following the fold lines) to ensure even shaping of the scallops. Brush the entire pithiviers with egg wash. Find the center and make a hole in the top (***cheminée***). Mark the dome with spiraling shapes from the center toward yourself using the tip of a paring knife. If right-handed, make them clockwise; if left-handed, make them counterclockwise.

Cuisson

1. Transfer the pithiviers to the oven to bake. After 10 minutes, open the oven and quickly rotate the baking tray. Close the door and reduce the temperature 370°F (185°C). Bake the pithiviers for a further 20 to 25 minutes, then transfer them to a wire rack to cool. Serve at room temperature.

POIRES POCHÉES AU VIN ROUGE

Learning Outcomes

Pocher
Making a dessert sauce

Equipment

Knives:
Paring knife (*office*), vegetable peeler (*économe*)

Tools:
Scissors, spoon

Pans:
1 medium pan, 1 small pan

Serving

4 persons

POIRES POCHÉES AU VIN ROUGE

Pears Poached in Red Wine

Quantity		Ingredient
U.S.	*Metric*	
4 pcs	4 pcs	Pears
8 pcs	8 pcs	Prunes, pitted
		Cuisson
1 pt	500 mL	Red wine
4 ¾ oz	140 g	Sugar, granulated
¼ fl oz	4 mL	Vanilla extract
2 pcs	2 pcs	Cinnamon sticks
		To Serve
2 oz	60 g	Red currant jelly
8 brs	8 brs	Fresh mint

Method

Cuisson

1. In a deep pan or stockpot (***marmite***), bring the red wine, sugar, vanilla and cinnamon to a boil. Cook until the sugar has completely melted. Peel the pears, keeping the stems on. Using the tip of a vegetable peeler, cut out the eye at the base of the fruits.
2. Place the pears in the wine and cover with a parchment paper lid (***cartouche***) over the contents of the pot. Add the prunes and bring the liquid to a gentle simmer over low to medium heat and cook until a knife can be easily inserted into the center of the pears.
3. Remove the pan from the stove, let the pears cool to room temperature, then refrigerate in their cooking syrup (preferably overnight).

To Serve

1. Mix 2 tablespoons of cooking syrup with the red currant jelly in a small saucepan and cook in a small pan over medium heat. Cook until the sauce is like syrup in consistence.
2. Drain the pears, and place in a chilled soup plate or shallow dish with the prunes. Trim the bottom if needed to maintain them upright. Drizzle the plate with the sauce and decorate with fresh mint.

PROFITEROLES AU CHOCOLAT

Learning Outcomes

Pâte à choux
Sauce chocolat
Crème Anglaise
Coucher des choux

Equipment

Knives:
Serrated knife, paring knife (*office*)

Tools:
Whisk, ice cream churner, mixing bowls, pastry bag, 10-mm plain tip, rubber spatula, wooden spoon, chinois

Pans:
1 medium pan,
1 large pan,
1 small pan

Serving

8 persons

FYI...

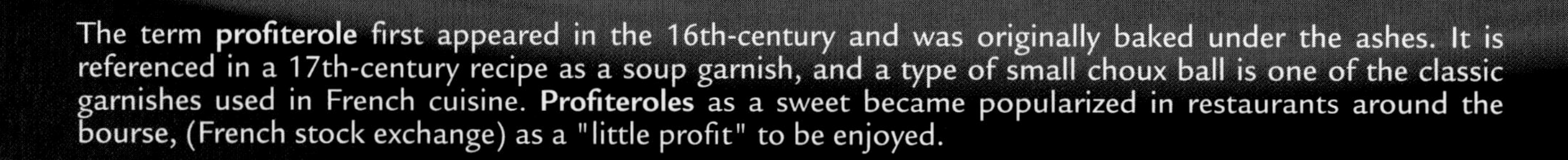

The term **profiterole** first appeared in the 16th-century and was originally baked under the ashes. It is referenced in a 17th-century recipe as a soup garnish, and a type of small choux ball is one of the classic garnishes used in French cuisine. **Profiteroles** as a sweet became popularized in restaurants around the bourse, (French stock exchange) as a "little profit" to be enjoyed.

PROFITEROLES AU CHOCOLAT

Profiteroles with Chocolate Sauce

Quantity		Ingredient
U.S.	Metric	Crème Glacée à la Vanille
1 pt	500 mL	Milk
8 fl oz	250 mL	Cream
2 pcs	2 pcs	Vanilla bean
6 pcs	6 pcs	Egg yolks
7 oz	200 g	Sugar, granulated
		Pâte à Choux
8 fl oz	250 mL	Water
4 oz	125 g	Butter
½ tsp	12 g	Sugar, granulated
½ tsp	5 g	Salt
5 oz	150 g	Flour
4–5 pcs	4–5 pcs	Eggs
1 pc	1 pc	Egg, slightly beaten, for egg wash
		Sauce au Chocolat
7 oz	200 g	Semisweet chocolate, cut into pieces
4 fl oz	125 mL	Milk
2 oz	55 g	Unsalted butter
		Decoration
1 ¾ oz	50 g	Powdered sugar
1 ¾ oz	50 g	Sliced almonds (roasted)

Method

1. Heat the oven to 400°F (205°C).

Crème Glacée à la Vanille

1. *Prepare a crème Anglaise:*
2. Place the milk and heavy cream in a medium saucepan and bring to a low boil over medium-high heat. Using a small knife, split the vanilla bean lengthwise. Scrape the seeds from both sides and add to the milk along with the pod. Whisk well.
3. Place the egg yolks in a mixing bowl, add the sugar, and immediately begin to whisk it into the yolks. Continue whisking until the sugar is completely dissolved and the mixture is pale in color (***blanchir***).
4. Once the milk is scalded, whisk about one-third of the hot milk into the yolks to temper them. Whisk until the mixture is well combined and evenly heated.
5. Stir the tempered egg yolks into the pan of remaining hot milk and stir with a wooden spatula. Place the pan over low heat and stir in a figure 8 motion. As you stir, the foam on the surface will disappear; at the same time, the liquid will begin to thicken and become oil-like in resistance. Continue cooking until the mixture is thick enough to coat the back of a wooden spatula and when your finger leaves a clean trail (***à la nappe***).

Tip: *Crème Anglaise should be cooked to between 167°F and 185°F (75°C and 85°C).*

6. Remove the pan from the heat and strain the crème Anglaise through a fine mesh sieve (***chinois***) into a clean bowl set in a bowl of ice. Stir it back and forth (***vanner***) with the spatula until cooled.
7. Cover the bowl in plastic wrap and reserve in the refrigerator for 24 hours. Once the flavors have developed, pour the crème Anglaise into a running ice cream churner and let it run until the ice cream is set.
8. Once churned, reserve the ice cream in an airtight container in the freezer.

Pâte à Choux

1. *Prepare a panade*: Combine the water, butter, sugar, and salt in a large pan and bring to a boil over medium-high heat. Once the butter has completely melted, remove the pan from the heat and add the flour. Stir with a wooden spatula until combined; then, over medium heat, stir until the mixture doesn't stick to the spatula and it makes a thin skin on the bottom of the pan (***dessêcher***). Transfer to a clean bowl and cool until warm. Beat the eggs into the panade one at a time. The dough should be stretchy and slightly sticky.
2. Transfer the mixture to a pastry bag fitted with a 15-mm plain tip.
3. On a lightly greased baking sheet, pipe out choux of about 1 in. (2.5 cm) in diameter. Brush the choux with egg wash, being careful not to get any on the tray. Using a fork dipped with egg wash, gently press down on the top of each choux to give them a more even shape. Bake the choux until golden. When cooked, remove the choux from the oven and let them cool on the baking sheet. When cold, remove them to a wire rack and let them dry overnight at room temperature.

Sauce au Chocolat

1. Place the chocolate in a mixing bowl. Heat the milk to just under the boiling point, then pour it over the chocolate. Let rest for 1 minute. Using a wooden spoon, stir gently until the chocolate has completely melted into the milk. Mix in the butter. Strain the sauce through a fine meshed sieve (***chinois***) and reserve it in a bain marie.

Finishing

1. Using an ice cream scoop, form small regular balls of vanilla ice cream and reserve them on a parchment paper–lined baking sheet in the freezer.
2. Cut the choux in half horizontally, place a scoop of vanilla ice cream on the bottom half and reserve (with the tops) in the freezer until needed.

To Serve

1. Place 4 choux on a plate and some of the chocolate sauce in a sauce boat. Pour a spoonful of chocolate sauce over the ice cream and fit the top section of the choux on top of the chocolate sauce. Dust the top of the choux with powdered sugar and finish with some lightly roasted sliced almonds.

RIZ CONDÉ

Learning Outcomes

Riz au lait
Blanchir
Sangler
Lier à l'œuf

Equipment

Knives:
Paring knife (*office*)

Tools:
Sieve, wooden spoon, fork, zester, ice bath, mixing bowls, savarin mold, pastry brush, blowtorch

Pans:
Large saucepan, small saucepan

Serving

8 persons

RIZ CONDÉ

Molded Rice Pudding

Quantity		*Ingredient*
U.S.	*Metric*	
7 oz	200 g	Rice
½ pc	½ pc	Vanilla pod
1 qt	1 L	Milk
Pinch	Pinch	Salt
4 pcs	4 pcs	Egg yolks
4 ¾ oz	140 g	Sugar, granulated
2 ¾ oz	80 g	Candied fruit (***fruits confits***)
¾ oz	20 g	Butter
		Sauce
7 fl oz	200 mL	Water
7 oz	200 g	Sugar, granulated
5 oz	150 g	Apricot ***nappage***
½ pc	½ pc	Lemon, juice of
½ pc	½ pc	Vanilla pod
		Decoration
8 pcs	8 pcs	Apricots, ***au sirop***
8 pcs	8 pcs	Green candied cherries (***bigarreaux verts confits***)
8 pcs	8 pcs	Red candied cherries (***bigarreaux rouges confits***)
1 ¾ oz	50 g	Apricot glaze, ***nappage***

Method

1. Preheat the oven to 360°F (180°C).
2. Wash the rice in cold water, drain it, then place it in a saucepan and cover it with cold water. Bring the water to a boil over high heat and blanch the rice for 2 minutes (***blanchir***). Rinse the rice again and drain it through a fine mesh sieve (***chinois***).
3. Slice the vanilla pod in half lengthwise and scrape out the seeds.
4. Bring the milk to a boil in a large saucepan over medium-high heat and add a pinch of salt, half of the vanilla pod, and all of the rice. Bring it back to a boil, cover the pot, and place it in the oven for 25 to 30 minutes.
5. When the rice is cooked, whisk the 4 egg yolks with the sugar (***blanchir***). Allow the rice to cool slightly before adding the egg mixture (*if the rice is too hot the yolks will cook too quickly*). Stirring gently with a fork, incorporate the egg mixture with the rice. Place the pan over low heat [*maximum temperature of between 176°F to 185°F (80°C and 85°C*)] and continue stirring until a custard-like texture is achieved. During this process, it is important to stir continuously to ensure an even temperature is maintained.
6. Remove the vanilla pod and incorporate the candied fruit (***fruits confits***) into the rice.
7. Transfer the mixture to a clean bowl set over an ice bath and chill it, stirring with a wooden spatula (***vanner***).
8. Meanwhile, grease a savarin mold with butter and place it in the freezer to set the butter (***sangler***). When the rice is cold press it into the buttered savarin mold, then cover it in plastic film and place it in the refrigerator to set overnight.

Sauce

1. Place the water and sugar in a medium saucepan over low heat and bring it to a boil to obtain a syrup. Add the apricot ***nappage***, the lemon juice, and the remaining half of the vanilla pod. Bring everything to a boil and reserve until needed.

To Serve

1. Unwrap the savarin mold and turn it onto a chilled plate. Lightly heat it with a blowtorch to soften the butter then carefully lift off remove the mold.
2. Heat 1 ¾ oz (50 g) of apricot glaze in a small saucepan over low heat until liquid.
3. Meanwhile, decorate the rice with the apricots, angelica, and cherries (***bigarreaux confits***). Once the glaze is liquid, coat the dessert with it using a pastry brush. Refrigerate before serving.
4. For presentation, pour some sauce in the hollow at the center of the dish and serve the remaining sauce on the side.

RIZ À L'IMPÉRATRICE

Learning Outcomes

Riz au lait
Crème Anglaise
Whipped cream
Pocher
Rosace
Making a bavarois

Equipment

Knives:
Paring knife (*office*)

Tools:
Wooden spoon, whisk, balloon whisk, rubber spatula, mixing bowls

Pans:
2 large pans, 1 small pan

Serving

8 persons

RIZ À L'IMPÉRATRICE

Bavarian Cream Rice Pudding

Quantity		*Ingredient*
U.S.	*Metric*	*Riz au Lait*
3 ½ oz	100 g	Rice
10 fl oz	300 mL	Milk
½ pc	½ pc	Vanilla pod
		Crème Anglaise Collée
7 fl oz	200 mL	Milk
4 pcs	4 pcs	Egg yolks
3 ½ oz	100 g	Sugar, granulated
½ pc	½ pc	Vanilla bean
5 pcs	5 pcs	Gelatin sheets
		Finishing
8 fl oz	250 mL	Whipped cream
4 oz	125 g	Candied fruit (***fruits confits***)
		Decoration
3 ½ fl oz	100 mL	Whipped cream
1 ¾ oz	50 g	Candied fruit (***fruits confits***)
8 fl oz	240 mL	Red currant jelly

Method

1. Preheat the oven to 350°F (175°C).
2. Place a jelly mold in the refrigerator to chill.

Riz au Lait

1. Wash the rice in cold water, then drain and place in a saucepan. Cover with cold water. Bring to a boil over high heat and blanch (***blanchir***) the rice for 2 minutes. Rinse the rice in cold water and drain it through a fine mesh sieve.
2. Bring the milk to a boil in a pan over medium-high heat and immediately remove the pan from the heat.
3. Slice the vanilla pod in half lengthwise and scrape the seeds directly into the milk. Add the rice to the milk, return to the heat, and cook for approximately 20 minutes.
4. When the rice is cooked, place it in a bowl over an ice bath and remove it once it cools to room temperature.

Note: *Do not let the rice cool to the point that it starts to solidify.*

Crème Anglaise Collée

1. To obtain the correct amount of crème Anglaise use the ingredients list in this recipe following the method on pages 405–406 in *Cuisine Foundations*.

Tip: *Crème Anglaise should be cooked to between 167°F and 185°F (75°C and 85°C).*

2. Once the crème Anglaise is cooked, add the gelatin and stir to make sure that it has melted completely. Remove the pan from the heat and strain the sauce through a fine mesh sieve into a clean bowl set over an ice bath. Stir it back and forth with a wooden spatula (***vanner***) until it is cool to the touch.

Finishing

1. Whip the cream to the consistency of soft peaks in a large mixing bowl over an ice bath.
2. Mix the candied fruit (***fruits confits***) with the cooked rice, then, using a rubber spatula, add the crème Anglaise collée, little by little. Finish by incorporating the whipped cream.

Note: *At this stage in the recipe, it is important to have all the elements at the right texture: If the gelatin has over set the crème Anglaise, place the bowl containing the crème over a bain marie to help return it to the desired consistency.*

Method

3. Fill the jelly mold with the rice mixture and gently tap the mold on the work surface to eliminate air bubbles. Smooth the top, cover in plastic wrap and let the rice mixture set in the refrigerator overnight.

To Serve

1. Using a large whisk, whip the cream in a large bowl over an ice bath until it reaches the soft peak stage. Transfer it to a pastry bag fitted with a large star tip.
2. Fill a large recipient with hot water and carefully dip the mold into the hot water to loosen its sides. Turn the jelly mold over onto a chilled serving dish and holding it securely to the plate, give it a good shake or two. Remove the mold. Decorate the riz à l'impératrice by piping out whipped-cream rosettes (***rosace***) and decorating with the candied fruits.
3. Melt the red currant jelly in a small saucepan over low heat and pour it into the bottom of the serving dish. Refrigerate before serving.

ST. HONORÉ

Learning Outcomes

Pâte brisée
Pâte à choux
Caramel
Crème chibouste
Using a St. Honoré tip
Making an Italian meringue
Coucher des choux

Equipment

Tools:
Corne, rolling pin, docker, cake ring, pastry bag, 10-mm plain tip, pastry brush, fork, whisk, wooden spoon, rubber spatula, large St. Honoré tip

Pans:
1 small saucepan, 1 large saucepan, 1 medium saucepan, 2 baking sheets

Serving

6–8 persons

FYI... This cake is named after the patron saint of bakers and is traditionally a cake made by bakers and not pastry chefs. The chibouste cream that is used in this cake was created by a pastry chef named Chibouste.

ST. HONORÉ

Cream Puff Cake with Caramel and Chiboust Cream

Quantity		*Ingredient*
U.S.	*Metric*	*Pâte à Choux*
4 fl oz	125 g	Water
2 oz	60 g	Butter
½ tsp	2 g	Salt
1 tsp	6 g	Sugar, granulated
4 oz	125 g	Flour
2 pcs	2 pcs	Eggs
		Pâte Brisée
7 oz	200 g	Flour
3 ½ oz	100 g	Butter, diced
Pinch	Pinch	Salt
¼ fl oz	10 mL	Water
1 pc	1 pc	Egg
1 pc	1 pc	Egg, for egg wash
		Chibouste
1 pt	500 mL	Milk
1 pc	1 pc	Vanilla bean
2 oz	60 g	Sugar, granulated
8 pcs	8 pcs	Egg yolks
1 ¼ oz	40 g	Cornstarch
6 pcs	6 pcs	Gelatin sheets
		Meringue Italienne
8 pcs	8 pcs	Egg whites
13 ½ oz	400 g	Sugar, granulated
4 fl oz	120 g	Water
		Caramel
3 ½ oz	100 g	Sugar, granulated
1 fl oz	30 mL	Water
½ pc	½ pc	Lemon

Method

1. Preheat the oven to 425°F (220°C).

Pâte à Choux

1. To obtain the correct amount of pâte à choux use the ingredients list in this recipe following the method on pages 373–374 in *Cuisine Foundations*.

Pâte Brisée

1. Sift (***tamiser***) the flour onto a clean work surface and make a large well (***fontaine***) in the center using a plastic scraper (***corne***).
2. Place the cold, diced butter in the center of the well. Work the butter into the flour using your fingertips while simultaneously cutting through the mixture with the plastic scraper. Continue cutting until the butter is crumbly and coated in flour.
3. Rub the mixture between the palms of your hands until it resembles fine sand (***sabler***).
4. Gather the flour-butter mixture into a neat pile and make a well in the center using the ***corne***. Add the salt, water, and the egg to the center of the well. Stir these ingredients together using your fingertips until combined; then gradually incorporate the dry ingredients from the sides until the mixture in the center of the well resembles a paste.
5. Using the ***corne***, incorporate the remaining dry ingredients using a cutting motion. Scoop up the mixture onto itself and continue cutting until a loose dough forms. Using the heel of your palm, firmly smear the dough away from yourself to ensure that no lumps of butter are left (***fraiser***). Scrape up the dough and repeat until the dough becomes uniform.
6. Shape the dough into a ball, wrap it in plastic, and flatten it into a disc. Let the dough rest in the refrigerator for at least 30 minutes (preferably overnight).

Base

1. Lightly dust the work surface with flour (***fleurer***). Unwrap the pâte brisée onto the floured surface and roll it out (***abaisser***) to a thickness of 3⁄16 in. (5 mm), giving it a quarter of a turn with each stroke of the rolling pin. Transfer the dough to a lightly greased baking sheet, prick it with the docker (***piquevite***), and, using a 8 in. (20 cm) cake ring, cut a circle out of it. Remove the trimmings and transfer the baking sheet to the refrigerator to let the round of dough rest. Meanwhile, transfer

Method

the pâte à choux to a pastry bag fitted with a 15-mm tip. On a lightly greased baking sheet, pipe out (***coucher***) about 10–12 balls (***choux***) 1 in. (2 ½ cm) in diameter and set aside. Remove the round of pâte brisee from the refrigerator and pipe a border around the endge of the round. Starting from the center, pipe a loose spiral of pâte a choux. Brush the pâte à choux with egg wash, being careful not to let any drip and lightly flatten the tops of the choux with a fork dipped in the egg wash.

2. Bake the pâte à choux and pâte brisée until they are golden (*about 12 to 15 minutes without opening the oven door*). When the pâte à choux are golden, rotate the baking sheets and continue to bake until slightly dried. When cooked, remove from the oven and transfer to a wire rack. Set aside to cool.

Chibouste

1. *Prepare a crème pâtissière:* Pour the milk into a medium saucepan. Split the vanilla bean in half lengthwise and scrape out the seeds. Stir the seeds into the milk and bring it to a boil over medium-high heat. Add about one-quarter of the sugar to the milk and stir to dissolve it.
2. Meanwhile, place the egg yolks in a small bowl and add the remaining sugar. Whisk the sugar into the eggs until it completely dissolves and the yolks lighten in color (***blanchir***). Add the cornstarch to the yolks and stir until well combined.
3. When the milk begins to come to a boil, remove it from the stove and pour one-third of it into the egg yolks. Stir well to temper the yolks, then whisk the tempered mixture into the remaining hot milk. Place back onto the heat and cook until the crème pâtissière begins to bubble. Continue whisking (being sure to press the whisk around the corners of the pan) and allow to cook for 1 minute in order to cook the starch. The crème pâtissière will become very thick. Transfer the finished crème pâtissière to a clean bowl.
4. Soak the gelatin sheets in cold water until completely softened. Squeeze out any excess water, then incorporate the sheets into the hot crème pâtissière using a whisk. Pat (***tamponner***) the surface of the crème pâtissière with a piece of cold butter held on the end of a fork to create a protective film. Set aside to cool to room temperature
5. *Prepare a meringue Italienne:* Whisk the egg whites in a large mixing bowl using a balloon whisk, until frothy. Set aside. Cook the sugar and water in a medium saucepan over medium-high heat until they reach the soft-ball stage (***gros boulé***) [250°F (121°C)]. Meanwhile, whip the egg whites to soft peaks. As soon as the syrup reaches the soft-ball (***gros boulé***) stage, pour it into the egg whites in a thin stream, whisking constantly. Continue to whisk until the meringue is firm and cooled to room temperature. Set aside.
6. Whisk the crème pâtissière until it is smooth and elastic. Add half the meringue Italienne and fold it in using a rubber spatula. Add the second half and fold it in until the mixture is homogeneous. Transfer half of the resulting chibouste to a pastry bag fitted with a 6-mm plain tip.
7. Using a small pastry tip, pierce the bases of the pâte à choux, then fill them with chibouste. Reserve them at room temperature.

Caramel

1. Bring the sugar and water to a boil in a small to medium saucepan over medium-high heat and cook this mixture until it turns a light caramel color. Stop the cooking process by briefly dipping the base of the saucepan in ice water. Dip the top of the pâte à choux in the caramel, holding them between 3 fingers and being careful not to touch the hot caramel. Place the caramel-covered side down on a silicone mat. Using the caramel as glue, stick the pâte à choux to the edge of the pastry disc.

Finishing

1. Transfer the leftover chibouste to a pastry bag fitted with a large St. Honoré tip and pipe (***coucher***) the chibouste onto the center of the cake and in between the pâte à choux. Refrigerate the St. Honoré until ready to serve.

SALAMBOS

Learning Outcomes

Pâte à choux
Piping salambos
Crème pâtissière
Filling choux
Caramel

Equipment

Tools:
Wooden spoon, rubber spatula,
pastry bag,
12-mm plain tip,
small star tip,
6-mm plain tip,
pastry brush, fork,
whisk, corne,
silicone mat

Pans:
1 large saucepan,
2 medium saucepans,
1 baking sheet

Serving

8 persons

FYI...

A small gâteau filled with rum-flavored crème pâtissière, the salambos was created in the late 1800s and was named after Reyer's Opera *Salammbô* which was itself named after Flaubert's novel of the same name.

SALAMBOS

Caramel Coated Cream Puffs Filled with Kirsch Cream

Quantity		Ingredient
U.S.	*Metric*	*Pâte à Choux*
4 fl oz	125 g	Water
4 fl oz	125 g	Milk
4 oz	125 g	Butter
½ oz	12 g	Sugar, granulated
Pinch	Pinch	Salt
5 oz	150 g	Flour
4–5 pcs	4–5 pcs	Eggs
1 pc	1 pc	Egg (for egg wash)
		Crème Pâtissière
1 ½ pt	750 mL	Milk
1 pc	1 pc	Vanilla bean
6 pcs	6 pcs	Egg yolks
5 oz	150 g	Sugar, granulated
1 ½ oz	45 g	Flour
1 ½ oz	45 g	Cornstarch
¼ fl oz	2 g	Rum or cognac
		Caramel
5 oz	150 g	Sugar, granulated
1 ¾ fl oz	50 mL	Water
		Sliced blanched almonds

Method

1. Preheat the oven to 425°F (220°C).

Pâte à Choux

1. *Prepare a panade:* Combine the water, milk, butter, sugar, and salt in a large pan and bring to a boil over medium-high heat. Once the butter has completely melted, remove the pan from the heat and add the flour. Stir with a wooden spatula until combined; then, over medium heat, stir until the mixture forms a smooth ball and comes cleanly away from the edge of the pan. Transfer to a clean bowl and cool until warm. Beat the eggs into the panade one at a time using a wooden spatula. The dough should be elastic and slightly sticky, and should create a V shape when the spatula is lifted out of the bowl. Transfer the dough to a pastry bag fitted with a 12-mm plain tip and pipe (***coucher***) rows of short, fat éclair shapes called salambos onto a lightly greased baking sheet.
2. Brush the salambos with egg wash using a pastry brush, and even them out with the prongs of a fork dipped in egg wash. Transfer the baking sheet to the oven and bake until golden, rotating the tray when the choux pastry begins to color. When cooked, transfer the salambos to a wire rack to cool. When cool, pierce the bottoms with two holes, using a small star tip.

Crème Pâtissière

1. Line a small serving tray or platter with plastic wrap. Pour the milk into a medium saucepan. Split the vanilla bean in half lengthwise and scrape out the seeds. Stir the seeds into the milk and bring it to a boil over medium-high heat. Add about one-quarter of the sugar to the milk and stir to dissolve it.
2. Meanwhile, place the egg yolks in a small mixing bowl and add the remaining sugar. Whisk the sugar into the eggs until it completely dissolves and the yolks lighten in color (***blanchir***). Add the flour and cornstarch to the yolks and stir until well combined.
3. When the milk begins to come to a boil, remove it from the stove and pour one-third of it into the egg yolks. Stir well to temper the yolks, then whisk the tempered mixture into the remaining hot milk. Place back onto the heat and cook until the crème pâtissière begins to bubble. Continue whisking (being sure to press the whisk around the corners of the pan) and allow to cook for 1 minute in order to cook the starch. The crème pâtissière will become very thick. Immediately transfer the finished crème pâtissière to the plastic-lined serving tray. Pat (***tamponner***) the surface with a piece of cold butter held on the end of a fork to create a protective film. Completely cover with a second piece of plastic wrap, pressing out any air bubbles. Let the crème pâtissière cool to room temperature before refrigerating.

Montage

1. Transfer the cooled crème pâtissière to a mixing bowl and whisk until smooth and elastic. Add the rum and stir well. Transfer the crème pâtissière to a pastry bag fitted with a 6-mm plain tip, and fill the salambos through the holes in the bottom of the pastry.
2. On a silicone mat, arrange little motifs of sliced almonds.

Caramel

1. Bring the sugar and water to a boil in a medium saucepan over medium-high heat and cook this mixture until it turns a light caramel color. Stop the cooking process by briefly dipping the base of the saucepan in ice water. Dip the salambos into the caramel and place them, caramel side down, on top of the almond motifs that have been arranged on the silicon mat. Be careful not to touch the hot caramel with your fingers. Let the caramel harden before removing the salambos from the mat. Serve the salambos caramel side up.

SAVARIN AUX FRUITS ET À LA CRÈME

Learning Outcomes

Pâte levée
Imbiber
Fontaine
Crème Chantilly
Napper
Pousser

Equipment

Knives:
Paring knife (*office*)

Tools:
Mixing bowls, corne, whisk, pastry brush, ladle, wire rack, pastry bag, large star tip

Pans:
Baking sheet, savarin mold

Serving

8 persons

SAVARIN AUX FRUITS ET À LA CRÈME

Savarin Cake with Fruit

Quantity		Ingredient
U.S.	Metric	Pâte à Savarin
1 lb	500 g	Flour
¼ oz	10 g	Salt
1 oz	30 g	Sugar, granulated
1 oz	30 g	Yeast
4 ¾ fl oz	140 mL	Water, warm
4 pcs	4 pcs	Eggs, lightly beaten, room temperature
5 fl oz	150 g	Butter, melted, room temperature
		Syrup for Imbibage
1 pt, 10 oz	800 mL	Water
13 ½ oz	400 g	Sugar, granulated
½ tsp	2 g	Vanilla extract
1 ¾ fl oz	50 mL	Rum
		Crème Chantilly
7 fl oz	200 mL	Cream
1 oz	30 g	Powdered sugar
¼ fl oz	5 mL	Vanilla
		Finishing
5 oz	150 g	Apricot glaze, warmed
8 oz	250 g	Fresh fruit

Method

1. Preheat the oven to 400°F (205°C).
2. Brush a savarin mold with melted butter and reserve it in a cool area.

Pâte à Savarin

1. Sift (***tamiser***) the flour onto a clean work surface and sprinkle it with salt and sugar. Mix the yeast into the warm water until it has completely dissolved. Make a well (***fontaine***) in the flour using a plastic scraper (***corne***) and pour the yeast mixture, eggs, and melted butter into it. Using your fingertips, gradually incorporate flour from the fountain into the center. As soon as all the flour is mixed in, work the dough by lifting it up off the marble and slapping it down onto the work surface, repeatedly. If the dough doesn't stick to the marble, it's too dry and needs more water. Continue working it until the gluten develops enough that the dough no longer sticks. Next, roll the dough into a ball, place it in a clean bowl, cover it with a humid cloth, and leave it in a warm place to proof until doubled in size. Once risen, punch down the dough and shape into a ring. Place it in the buttered savarin mold and leave it in a warm place to proof a second time until doubled in size.

Note: *While preparing the dough, never add the salt directly to the yeast/water mixture; the chemical reaction will kill the yeast. A pinch of sugar can be helpful.*

Sirop Pour Imbibage

1. Pour the water into a small saucepan and add the sugar. Bring the mixture to a boil over medium-high heat and continue to boil it until all the sugar has dissolved. Remove the pan from the stove and set aside.
2. Once the syrup has cooled to room temperature, stir in the vanilla extract and rum.

Cuisson

1. When the dough has risen a second time, transfer it to the oven to bake. After 20 minutes, reduce the temperature of the oven to 380°F (195°C). Bake for another 20 minutes or until golden. Once the savarin is cooked, remove it from the oven and let it rest in its mold for 5 minutes before unmolding it onto a wire rack set over a baking sheet. Immediately brush the savarin with syrup until it is drenched (***imbiber***) and all the syrup has been used up. Let the cake cool to room temperature.

Montage

1. Melt the apricot glaze in a small saucepan over medium heat until liquid. Using a pastry brush, apply it to the savarin to give it a shiny coating. Place the savarin in the refrigerator to chill.
2. Meanwhile, prepare a crème Chantilly: Using a large whisk, whip the cream in a large mixing bowl in an ice bath to soft peaks. Add the powdered sugar and the vanilla, and continue to whip the cream until stiff. Reserve the crème Chantilly in the refrigerator until needed.
3. Place the savarin on a chilled serving dish. Transfer the crème Chantilly to a pastry bag fitted with a medium star tip and fill the cavity in the center of the savarin. Arrange fresh fruit on top of the crème Chantilly and refrigerate the savarin before serving.

Learning Outcomes

Sorbet
Vanner

Equipment

Tools:
Wooden spatula, whisk, ice bath, ice cream churner, mixing bowls

Pans:
Small and medium saucepans

Yield

Sorbet à la fraise ¾ pt (400 g)
Sorbet au citron 1 pt: (500 g)
Sorbet à la pomme verte 1 qt (1 L)

LES SORBETS

Sorbets

Method

Sorbet à la Pomme Verte

1. Bring the water, sugar, and glucose to a boil in a small saucepan over medium-high heat and remove the pan from the stove once the sugar has completely dissolved. Pour the contents of the pan into a large mixing bowl set on an ice bath. Mix it back and forth using a wooden spatula (***vanner***), and once it is cold, add the lemon juice and apple purée. Stir the liquids together and continue stirring until the ingredients are cold.
2. Pour the mixture into a running ice cream churner and let it run until the sorbet is set.
3. Once churned, reserve the sorbet in an airtight container in the freezer.

Note: *The use of glucose in the recipe helps give the sorbet a consistency that is preferable to the granular texture that would result if regular sugar were used.*

Sorbet à la Fraise

1. Bring the water and sugar to a boil in a small saucepan over medium-high heat and remove the pan from the stove once the sugar has completely dissolved. Pour the contents of the pan into a large mixing bowl set on an ice bath. Mix it back and forth using a wooden spatula (***vanner***), and once it is cold, add the lemon juice and strawberry purée. Stir the liquids together and continue stirring until the mixture is cold.
2. Pour the mixture into a running ice cream churner and let it run until the sorbet is set.
3. Once churned, reserve the sorbet in an airtight container in the freezer.

Sorbet au Citron

1. Bring the water and sugar to a boil in a small saucepan over medium-high heat and remove the pan from the stove once the sugar has completely dissolved. Pour the contents of the pan into a large mixing bowl set on an ice bath. Mix it back and forth using a wooden spatula (***vanner***), and once it is cold, add the lemon juice and the extra water. Stir the liquids together and continue stirring until the mixture is cold.
2. Pour the mixture into a running ice cream churner and let it run until the sorbet is set. When the sorbet resembles a thick slush, remove 2 tablespoons and add mix it with the egg white in a small mixing bowl. Mix the two together with a whisk until combined and pour the mixture back into the churner. Once churned, reserve the sorbet in an airtight container in the freezer.

Note: *In this sorbet recipe, the egg whites serve as a natural stabilizer while at the same time adding texture to the preparation.*

3. *Generally, the basic syrup for sorbet has a density of 28 degrees (**Baumé**) and is characterized by equal amounts of water and sugar. The different densities of syrup can be measured using a hydrometer (**pèse sirop**).*

Quantity		*Ingredient*
U.S.	*Metric*	*Sorbet à la Pomme Verte*
5 fl oz	150 mL	Water
5 ½ oz	160 g	Sugar, granulated
1 ¾ oz	50 g	Glucose
1 lb 12 oz	800 g	Green apples, peeled, cored, and puréed
2 pcs	2 pcs	Lemons, juice of
		Sorbet à la Fraise
2 ½ fl oz	75 mL	Water
2 ½ oz	75 g	Sugar, granulated
½ pc	½ pc	Lemon juice
½ pt	250 mL	Strawberry purée
		Sorbet au Citron
7 ½ fl oz	225 mL	Water
7 ½ oz	225 g	Sugar, granulated
3 ½ fl oz	100 mL	Lemon juice
10 fl oz	300 mL	Water
1 pc	1 pc	Egg white

Learning Outcomes

Pâte sucrée
Crèmer
Crème d'amande
Fraiser
Monter au beurre
Fonçage
Meringue Italienne
Cooking sugar
Using a St. Honoré tip
Coloring meringue Italienne

Equipment

Knives:
Paring knife (*office*)

Tools:
Pastry brush, corne, mixing bowls, whisk, rubber spatula, zester, tamis, rolling pin, docker, balloon whisk, pastry bag, St Honoré tip

Pans:
Baking sheet, 8-inch tart mold, medium saucepan, small saucepan

Serving

6 persons

Cited in *Le Ménagier de Paris* (a 14th-century domestic guidebook), the general concept of the tart, or open pie in the United States, has undergone relatively few transformations through the centuries—aside from a multitude of fillings and the use of different types of pastry. While the technique for making meringue was still being developed up until the 17th century, versions of the lemon meringue pie go back to the Medieval period.

TARTE AU CITRON MERINGUÉE

Lemon Meringue Tart

Quantity		Ingredient
U.S.	*Metric*	*Pâte Sucrée*
7 oz	200 g	Flour
3 ½ oz	100 g	Powdered sugar
Pinch	Pinch	Salt
1 pc	1 pc	Lemon, zest of
3 ½ oz	100 g	Butter, diced
3 pcs	3 pcs	Egg yolks
		Crème D'Amande
2 oz	60 g	Butter (***pommade***)
2 oz	60 g	Sugar, granulated
1 pc	1 pc	Egg
1 pc	1 pc	Vanilla bean
½ fl oz	10 mL	Rum
2 oz	60 g	Almond powder
		Appareil Citron
3 pcs	3 pcs	Eggs
3 ½ oz	100 g	Sugar, granulated
2 pcs	2 pcs	Lemons, juice of
1 pc	1 pc	Lemon, zest of
3 ½ oz	100 g	Butter, diced
		Meringue Italienne
4 oz	120 g	Egg whites
8 oz	240 g	Sugar, granulated
2 ¾ fl oz	80 g	Water
		Sliced almonds

Method

1. Preheat the oven to 400°F (205°C).
2. Lightly grease an 8-inch tart mold with a removable bottom, with softened (***pommade***) butter using a pastry brush. Reserve the tart mold in the refrigerator.

Pâte Sucrée

1. Sift the flour (***tamiser***) onto a clean work surface and gather it into a neat pile. Sift the powdered sugar in a separate pile in front of the flour. Sprinkle the salt and lemon zest over the sugar and, using a plastic scraper (***corne***), make a large well in the center of these ingredients.

Note: *In the following steps, keep one hand clean and dry (for the plastic scraper) and use the other to stir in the wet ingredients.*

2. Add the butter to the center of the sugar and work it with your fingertips until it is soft. Using the ***corne,*** gradually add the powdered sugar from the edge of the well while simultaneously working it into the butter with your hands. Continue to mix the butter and sugar together until they are fully incorporated and creamy (***crèmer***).
3. Add the egg yolks to the butter and sugar and mix it in with your fingertips. The result will be slightly lumpy. With your clean hand, use the ***corne*** to gradually add some flour to the creamed ingredients while simultaneously mixing with your fingertips. Continue until the mixture resembles a thick paste. Using the ***corne***, cut in the remaining flour until a loose dough is formed.
4. Using the heel of your palm, firmly smear the dough away from yourself to ensure that no lumps of butter are left (***fraiser***). Scrape up the dough and repeat until a smooth dough forms. Form the dough into a smooth ball, wrap it in plastic, and flatten it into a disc. Let it rest in the refrigerator for a minimum of 30 minutes (preferably overnight).

Crème D'amandes

1. Cream the butter and sugar together until light and fluffy (***crèmer***).
2. Beat in the egg until well combined.
3. Using a small knife, split the vanilla bean in half lengthwise, scrape out the seeds, and whisk them into the mixture.
4. Add the rum and finish by mixing in the almond powder.
5. Reserve the crème d'amandes in a covered bowl in the refrigerator until ready to use.

Method

Appareil Citron

1. Whisk the eggs and sugar together in a round bottomed bowl until light in color (***blanchir***). Add the lemon juice and zest and stir them in. Place the bowl over a bain marie and whisk until the mixture begins to thicken. Continue cooking until very hot and thick. Remove from the heat and stir in the butter until completely melted (***monter au beurre***). Transfer to a clean bowl and cover it in plastic wrap. Let the mixture cool to room temperature, then reserve it in the refrigerator until needed.

Fonçage

1. Lightly dust the marble with flour (***fleurer***) and place the dough in the center. Roll out the pastry (***abaisser***), giving it quarter turns with each stroke of the rolling pin. Continue rolling and turning until the dough is ⅛ in. (3 mm) thick and 3 fingers wider than the tart mold. Prick it (***piquer***) with the docker (***pique pâte***) or a fork, roll the dough onto the rolling pin, and gently lay it on the tart mold. Lift the dough's edges and press it into the shape of the mold (***foncer***). To form an even border around the inside edge of the mold, simultaneously apply side and top pressure to the dough (*using a thumb and a finger*). Repeat this process all the way around the mold, then pass the rolling pin over the top of the tart to trim the excess dough. Gently pinch all around the top edge of the tart to create a decorative border (***chiqueter***).
2. Place the lined mold in the refrigerator to rest for at least 20 minutes.
3. Add the almond cream to the cavity of the tart shell and spread it out in an even layer. Place the tart mold on a baking sheet and transfer it to the oven. As soon as it is in the oven, reduce the temperature to 370°F (185°C). Bake the tart shell until the crust turns a light golden color and the almond cream doesn't move when the tart is given a light shake and the surface is dry to the touch. If the pastry cooks too fast, reduce the oven temperature or cover the tart in foil. Rotate the tart about 10 minutes into cooking. Once cooked, remove the tart from the oven and set it on a rack to cool. Turn the oven up to 500°F (260°C).

Meringue Italienne

1. In a large mixing bowl, whisk the egg whites until frothy using a balloon whisk. Set aside. Cook the sugar and water in a medium saucepan over medium-high heat to the soft-ball stage (***petit boule***) [250°F (121°C)]. Meanwhile, whip the egg whites to soft peaks. As soon as the syrup reaches the soft-ball (***petit boule***) stage, pour it into the egg whites in a thin stream, whisking constantly. Continue to whisk until the meringue is firm and has cooled to room temperature. Set aside.

Montage

1. Pour the chilled lemon mixture into the cooled tart shell and spread it out into an even layer. Transfer the Italian meringue to a pastry bag fitted with a large St. Honoré tip and pipe (***coucher***) a design onto the tart to completely cover the lemon cream. Dust the meringue with powdered sugar and sprinkle with sliced almonds. Transfer the tart to a baking sheet and place it in the oven just long enough for the almonds to lightly toast and the meringue to color on the edges. Remove the tart from the oven and set it on a wire rack to cool.
2. To unmold the tart, place the mold on an overturned saucepan (that is only slightly narrower than the mold) and slip the side of the mold down. Slide the tart off the metal disc onto a serving dish or cake board. Refrigerate before serving.

TARTE AU SUCRE

Learning Outcomes

Pâte levée
Tamiser
Fontaine
Pousser

Equipment

Tools:
Tamis, mixing bowls, whisk, rolling pin, corne, pastry brush

Pans:
Baking sheet, 8-inch deep tart pan, small saucepan

Serving

6–8 persons

FYI...

Using yeast-raised dough, this style of sugar pie appears to have originated in the northern part of France (Flanders in particular) where the production of sugar from sugar beets has been of economic importance for centuries.

Quantity		Ingredient
U.S.	Metric	*Pâte Levée*
8 oz	250 g	Flour
½ oz	10 g	Fresh yeast
1 oz	30 g	Sugar, granulated
2 ¾ fl oz	80 mL	Water, warm
1 pc	1 pc	Egg
¼ oz	5 g	Salt
3 ½ oz	100 g	Butter, softened (***pommade***)
1 pc	1 pc	Egg for egg wash
		Appareil
3 ½ oz	100 g	Sugar, granulated
1 pc	1 pc	Egg
1 ¾ fl oz	50 mL	Milk
1 ¾ fl oz	50 g	Butter, melted and cooled

TARTE AU SUCRE
Sugar Tart

Method

1. Preheat the oven to 350°F (175°C).
2. Lightly grease a high-sided 8 in. (20 cm) tart mold with softened butter and reserve it in the refrigerator.

Pâte Levée

1. Add the yeast and sugar to the warm water and stir them in until completely dissolved. Let the mixture rest until it is covered in a thin layer of foam. Meanwhile, sift (***tamiser***) the flour onto a clean dry work surface and make a well (***fontaine***) in the center. Pour the yeast mixture into the well and stir it using your fingertips while gradually adding flour from the sides using a plastic scraper (***corne***). When the center of the well resembles a thick paste, gather the flour on top and begin to work it into the yeast mixture with your hands. Once the ingredients are combined, make a well in the center. Whisk the egg and salt together in a small bowl and pour them into the well. Stir them with your fingertips, gradually adding the mixture from the sides of the well until all the egg is absorbed. Gather the mixture into a pile and knead it until it forms a dough. Continue to knead it until smooth and add the butter. Knead the dough until the butter is completely incorporated. Stretch and fold the dough repeatedly until it is elastic enough to form a thin skin (*diaphragm*) when stretched between the hands (*10 to 15 minutes*). Work the dough into a ball and place it in a lightly oiled mixing bowl. Cover the bowl with a damp cloth and leave it in a warm area to rise until doubled in size (***pousser***).

Appareil

1. Whisk the sugar and egg in a large mixing bowl until light in color (***blanchir***). Add the milk while stirring continuously. Pour 2 tablespoons of the sugar/egg/milk mixture into the melted butter and stir until combined. Recombine all the ingredients.

Cuisson

1. Lightly dust a clean work surface with flour (***fleurer***). Once the dough has doubled in size, turn it out onto the lightly floured surface and knead it for 10 to 15 minutes. Roll the dough out (***abaisser***) into a circle 2 in. (5 cm) wider than the tart mold. Place the dough in the tart mold and fold the edges in to create a raised border, pressing the seam

Method

down to secure it. Proof the dough a second time in the tart shell before cooking (*15 to 20 minutes*). Brush the surface of the dough with egg wash and transfer it to the oven to bake until lightly golden (*20 to 25 minutes*). Remove the tart base from the oven and press down the center using your knuckles or the back of a large spoon. Brush the edge with egg wash a second time and fill the depression with the custard mixture (***appareil***). Return the tart to the oven to bake for 10 to 15 minutes.

Finishing

1. Transfer the cooked tart to a wire rack to rest for 5 to 10 minutes. Unmold it and serve it warm or at room temperature.
2. Before serving, the tart can be brushed with simple syrup to give it shine and the center can be further decorated with a sprinkle of sugar.

TARTE AUX FRAISES

Learning Outcomes

Pâte sucrée
Crème Pâtissière
Tamiser
Crèmer
Fraiser
Crème mousseline
Blanchir
Abaisser
Foncer
Piquer
Chiqueter
Cuisson à blanc
Nappage

Equipment

Knives:
Paring knife (*office*)

Tools:
Pastry brush, corne, mixing bowls, tamis, whisk, wooden spatula, rolling pin, docker, baking beads, pastry bag, 10-mm plain tip, pastry crimper

Pans:
Baking sheet, 8-inch / 20-cm flan ring, small saucepan, medium saucepan

Serving

8 persons

FYI...

Unlike the citron meringuée, this tart involves pre baking the pâte sucrée. In order to avoid shrinkage and movement of the pastry as it bakes, ceramic beads are placed in the cavity of the shell to weigh it down. This process is called "baking blind."

TARTE AUX FRAISES
Strawberry Tart

Method

1. Preheat the oven to 400°F (205°C).
2. Using a pastry brush, lightly grease a 8 in. (20 cm) flan ring and a baking sheet with softened (***pommade***) butter and reserve them in the refrigerator.

Pâte Sucrée

1. Sift the flour (***tamiser***) onto a clean work surface and gather it into a neat pile. Sift the powdered sugar in a separate pile in front of the flour. Sprinkle the salt and lemon zest over the powdered sugar and, using a plastic scraper (***corne***), make a large well in the center of these ingredients.

Note: *In the following steps, keep one hand clean and dry (for the plastic scraper) and use the other to stir in the wet ingredients.*

2. Add the butter to the center of the sugar and work it with your fingertips until it is soft. Using the ***corne,*** gradually add the powdered sugar from the edge of the well while simultaneously working it into the butter with your hands. Continue to mix the butter and sugar together until they are fully incorporated and creamy (***crémer***).
3. Add the egg yolks to the butter and sugar and mix it in with your fingertips. The result will be slightly lumpy. With your clean hand, use the ***corne*** to gradually add some flour to the creamed ingredients while simultaneously mixing with your fingertips. Continue until the mixture resembles a thick paste. Using the ***corne***, cut in the remaining flour until a loose dough is formed.
4. Using the heel of your palm, firmly smear the dough away from yourself to ensure that no lumps are left (***fraiser***). Scrape up the dough and repeat until a smooth dough forms. Form the dough into a smooth ball, wrap it in plastic, and flatten it into a disc. Let it rest in the refrigerator for a minimum of 30 minutes (preferably overnight).

Crème Mousseline

1. *First make a crème pâtissière (the addition of butter at the final stages is what creates the crème mousseline):*
2. Line a small serving tray or platter with plastic wrap. Pour the milk into a medium saucepan. Split the vanilla bean in half lengthwise and scrape out the seeds. Stir the seeds into the milk and bring it to a boil over medium-high heat. Add about one-quarter of the sugar to the milk and stir to dissolve it.

Quantity		*Ingredient*
U.S.	*Metric*	*Pâte Sucrée*
7 oz	200 g	Flour
3 ½ oz	100 g	Powdered sugar
Pinch	Pinch	Salt
1 pc	1 pc	Lemon, zest of
3 ½ oz	100 g	Butter, diced
3 pcs	3 pcs	Egg yolks
		Crème Mousseline
8 fl oz	250 mL	Milk
1 pc	1 pc	Vanilla bean
2 oz	60 g	Sugar, granulated
2 pcs	2 pcs	Egg yolks
1 oz	30 g	Flour (or cornstarch)
½ fl oz	15 g	Kirsch
3 ½ oz	100 g	Butter, softened (***pommade***)
		Montage
10 oz	300 g	Strawberries, hulled, halved
1 oz	30 g	Apricot glaze (***nappage***)
¾ oz	20 g	Pistachio nuts, chopped (***concasser***)

Method

3. Meanwhile, place the egg yolks in a small mixing bowl and add the remaining sugar. Whisk the sugar into the eggs until it completely dissolves and the yolks lighten in color (***blanchir***). Add the cornstarch or flour to the yolks and stir until well combined.
4. When the milk begins to come to a boil, remove it from the heat and pour one-third of it into the egg yolks. Stir well to temper the yolks, then whisk the tempered mixture into the remaining hot milk. Place back onto the heat and cook until the crème pâtissière begins to bubble. Continue whisking (being sure to press the whisk around the corners of the pan) and allow to cook for 1 minute in order to cook the starch. The crème pâtissière will become very thick. Transfer the crème to a clean bowl and allow to cool until warm. While the crème is cooling, soften the butter. Beat in the softened butter to the warm crème.

Note: *The crème will now be referred to as crème mousseline.*

5. Transfer the crème mousseline to the plastic-lined serving tray. Pat (***tamponner***) the surface with a piece of cold butter held on the end of a fork to create a protective film. Completely cover with a second piece of plastic wrap, pressing out any air bubbles. Let the crème mousseline cool to room temperature before refrigerating it.

Fonçage

1. Place the flan ring on the baking sheet.
2. Lightly dust the marble with flour (***fleurer***) and place the dough in the center. Roll out the pastry (***abaisser***), giving it quarter turns with each stroke of the rolling pin. Continue rolling and turning until the dough is ⅛ in. (3 mm) thick and 3 fingers wider than the flan ring. Prick it (***piquer***) with the docker (***piquevite***) or a fork, roll the dough onto the rolling pin, and gently lay it on the flan ring. Lift the dough's edges and press it into the shape of the ring (***foncer***). To form an even border around the inside edge of the ring, simultaneously apply side and top pressure to the dough (*using a thumb and a finger*). Repeat this process all the way around the ring, then pass the rolling pin over the top of the tart to trim the excess dough. Gently pinch all around the top edge of the tart with a pastry crimper to create a decorative border (***chiqueter***).
3. Place the lined mold in the refrigerator to rest for at least 20 minutes.
4. Cut a circle of parchment paper larger than the tart shell. Place it in the cavity and fill it with baking beads. Transfer the tart to the oven. As soon as the door is closed, reduce the temperature to 370°F (185°C). Bake the tart shell until the crust turns a light golden color, rotating it 10 minutes into cooking. When the crust around the edge begins to color, remove the tart from the oven, take out the beads and paper, and return it to the oven to obtain an even color. When the entire shell is golden (***blond***), remove it from the oven and transfer it to a wire rack to cool completely. Once cooled, carefully remove the ring and place the baked tart shell on a cake board or serving plate.

Montage

1. Remove the crème mousseline from the refrigerator and whisk it until smooth and elastic. Whisk in the rest of the butter until completely incorporated, then transfer the crème mousseline to a pastry bag fitted with a 10-mm plain tip. Pipe (***coucher***) a tight spiral of crème mousseline into the tart shell to completely cover the bottom. Arrange the strawberry halves on top, pointing upward. Melt the apricot glaze over low heat in a small saucepan until it is completely liquid and apply it to the strawberries using a pastry brush. Sprinkle the tart with pistachio nuts and refrigerate it before serving.

TARTE AUX POMMES

Pâte sucrée
Tamiser
Crémer
Fraiser
Compoter
Foncer
Napper
Chiqueter
Abaisser

Equipment

Knives:
Paring knife (*office*), vegetable peeler (*économe*), chef knife (*couteau chef*)

Tools:
Apple corer, corne, pastry brush, rolling pin, mixing bowls, rubber spatula, pastry crimper, wooden spatula

Pans:
Tart mold, medium saucepan, small saucepan

Serving

8 persons

FYI...

Tarte aux pommes has provided warmth and comfort on cold winter nights from the Medieval period to the modern era. In the United States, apple pie (essentially a lidded tarte aux pommes) is so significant that it is seen as a national symbol, representing hardiness and strong moral fiber. There are many excellent recipe variations of tarte aux pommes. This particular recipe distinguishes itself, among other ways, with the technique of scraping whole vanilla-bean seeds directly from the pod into the compote.

Quantity		Ingredient
U.S.	*Metric*	*Pâte Sucrée*
7 oz	200 g	Flour
3 ½ oz	100 g	Powdered sugar
Pinch	Pinch	Salt
1 pc	1 pc	Lemon, zest of
3 ½ oz	100 g	Butter, diced
3 pcs	3 pcs	Egg yolks
		Compote de Pommes
3 pcs	3 pcs	Apples, peeled, cored and sliced
½ pc	½ pc	Lemon, juice of
1 pc	1 pc	Vanilla bean
2 oz	60 g	Sugar, granulated
1 ¼ oz	40 g	Butter
		Montage
3 pcs	3 pcs	Apples, peeled, cored, thinly sliced (***émincer***)
		Apricot glaze (***nappage***), heated

TARTE AUX POMMES

Apple Tart

Method

1. Preheat the oven to 400°F (205°C).
2. Lightly grease a 10 in. (25 cm) tart mold with a removable bottom, with softened butter using a pastry brush and reserve it in the refrigerator.

Pâte Sucrée

1. Sift the flour (***tamiser***) onto a clean work surface and gather it into a neat pile. Sift the powdered sugar in a separate pile in front of the flour. Sprinkle the salt and lemon zest over the sugar and, using a plastic scraper (***corne***), make a large well in the center of these ingredients.

Note: *In the following steps, keep one hand clean and dry (for the plastic scraper) and use the other to stir in the wet ingredients.*

2. Add the butter to the center of the sugar and work it with your fingertips until it is soft. Using the ***corne,*** gradually add the powdered sugar from the edge of the well while simultaneously working it into the butter with your hands. Continue to mix the butter and sugar together until they are fully incorporated and creamy (***crémer***).
3. Add the egg yolks to the butter and sugar and mix it in with your fingertips. The result will be slightly lumpy. With your clean hand, use the ***corne*** to gradually add some flour to the creamed ingredients while simultaneously mixing with your fingertips. Continue until the mixture resembles a thick paste. Using the ***corne***, cut in the remaining flour until a loose dough is formed.
4. With the heel of your palm, firmly smear the dough away from yourself to ensure that no lumps of butter are left (***fraiser***). Scrape up the dough and repeat until a smooth dough forms. Form the dough into a smooth ball, wrap it in plastic, and flatten it into a disc. Let it rest in the refrigerator for a minimum of 30 minutes (preferably overnight).

Compote de Pommes

1. Toss the apple slices in lemon juice as soon as it is cut to prevent discoloring. Slice the vanilla bean in half lengthwise and use the tip of a knife to scrape out the seeds. Cook the sugar in a large shallow pan over medium-high heat until it turns a light caramel color. Deglaze with the butter and add the apples and vanilla seeds and pod. Reduce the temperature and let the mixture stew (***compoter***)

Method

until the apples are cooked through but retain some texture. Transfer the compote to a clean bowl and let it cool to room temperature before covering it in plastic wrap and transferring it to the refrigerator.

Foncage (see pages 370–372 in Cuisine Foundations)

1. Lightly dust the marble with flour (***fleurer***) and place the dough in the center. Roll out the pastry (***abaisser***), giving it quarter turns as you roll. Continue rolling and turning until the dough is ⅛ in. (3 mm) thick and 3 fingers wider than the tart mold. Prick it (***piquer***) with the docker (***piquevite***) or a fork, roll the dough onto the rolling pin, and gently lay it on the tart mold. Lift the dough's edges and press it into the shape of the mold (***foncer***). To form an even border around the inside edge of the mold, simultaneously apply side and top pressure to the dough (*using a thumb and a finger*). Repeat this process all the way around the mold, then pass the rolling pin over the top of the tart to trim the excess dough. Gently pinch all around the top edge of the tart to create a decorative border (***chiqueter***).
2. Place the lined mold in the refrigerator to rest for at least 20 minutes.

Cuire à blanc (Blind bake)

1. Cut a circle of parchment paper larger than the tart shell, place it in the cavity and fill it with baking beads. Transfer the tart to the oven. As soon as the door is closed, reduce the temperature to 370°F (185°C). Bake the tart shell until the dough is cooked but not colored, rotating it 10 minutes into cooking. Remove the tart from the oven, take out the beads and paper, and transfer it to a wire rack to cool.

Montage

1. Remove the vanilla pod, then spread the compote into an even layer over the bottom of the tart shell. Arrange the sliced apples in concentric circles over the compote.
2. Return the tart to the oven and bake it until the apples begin to color (*25 minutes*). Remove the tart from the oven and place it on a wire rack to cool. Melt the apricot glaze (***nappage***) over low heat in a small saucepan until it is completely liquid and apply it to the top of the cooled apples using a pastry brush.
3. To unmold the tart, place the mold on an overturned saucepan (that is only slightly narrower than the mold) and slip the side of the mold down. Slide the tart off the metal disc onto a serving dish or cake board.

Note: *It is important to wait until the tart has cooled before brushing with nappage. If the apples are still hot, the heat will melt and absorb the nappage.*

CONVERSION CHART

A Note about Conversions

For cooking and baking, the metric system is probably the easiest to manage and an electronic scale can become your most valued tool in the kitchen! When making conversions, we took the liberty to sometimes round off the measurements as long as the proportions in the recipe were still respected.

Volume

U.S.	METRIC
1/4 fl oz	5 ml
1/2 fl oz	15 ml
3/4 fl oz	25 ml
1 fl oz	30 ml
2 fl oz	60 ml
3 fl oz	90 ml
4 fl oz	120 ml
5 fl oz	150 ml
6 fl oz	180 ml
7 fl oz	210 ml
8 fl oz	240 ml
9 fl oz	270 ml
10 fl oz	300 ml
11 fl oz	330 ml
12 fl oz	360 ml
13 fl oz	390 ml
14 fl oz	420 ml
15 fl oz	450 ml
1 pint (16 fl oz)	500 ml
1 quart (2 pints)	1L (1000 ml)
2 quarts	2 L (2000 ml)
3 quarts	3 L (3000 ml)
1 gallon (4 quarts)	4 L (4000 ml)

Weight

U.S.	METRIC
1/4 oz	5 g
1/2 oz	15 g
3/4 oz	20 g
1 oz	30 g
2 oz	60 g
3 oz	90 g
4 oz	120 g
5 oz	150 g
6 oz	180 g
7 oz	200 g
1/2 lb (8 oz)	250 g
9 oz	270 g
10 oz	300 g
11 oz	330 g
12 oz	360 g
13 oz	390 g
14 oz	420 g
15 oz	450 g
1 lb (16 oz)	500 g
1 1/2 lb	750 g
2 lb	1 kg

Common Household Equivalents

U.S.	METRIC
1/4 tsp	1 ml
1/2 tsp	3 ml
3/4 tsp	4 ml
1 tsp	5 ml
1 tbsp	15 ml
1/4 cup	60 ml
1/2 cup	120 ml
3/4 cup	180 ml
1 cup	250 ml
1/4 lb	120 g
1/2 lb	230 g
1 lb	450 g
1 pint	500 ml
1 quart	1 L
1 gallon	4 L

U.S. Measure Equivalents

3 tsp	1 tbsp	1/2 fl oz
2 tbsp	1/8 cup	1 fl oz
4 tbsp	1/4 cup	2 fl oz
5 tbsp + 1 tsp	1/3 cup	2 2/3 fl oz
8 tbsp	1/2 cup	4 fl oz
10 tbsp +2 tsp	2/3 cup	5 1/3 fl oz
12 tbsp	3/4 cup	6 fl oz
14 tbsp	7/8 cup	7 fl oz
16 tbsp	1 cup	8 fl oz
2 cups	1 pint	16 fl oz
2 pints	1 quart	32 fl oz
4 quarts	1 gallon	128 fl oz

INDEX